INTRODUCTION TO SUPERCONDUCTING CIRCUITS

INTRODUCTION TO SUPERCONDUCTING CIRCUITS

Alan M. Kadin

University of Rochester

A WILEY INTERSCIENCE PUBLICATION

JOHN WILEY & SONS, INC.

New York / Chichester / Weinheim / Brisbane / Singapore / Toronto

This book is printed on acid-free paper. ∞

Copyright © 1999 by John Wiley & Sons, Inc. All rights reserved.

Published simultaneously in Canada.

No part of this publication may be reproduced, stored in a retrieval system or transmitted in any form or by any means, electronic, mechanical, photocopying, recording, scanning or otherwise, except as permitted under Sections 107 or 108 of the 1976 United States Copyright Act, without either the prior written permission of the Publisher, or authorization through payment of the appropriate per-copy fee to the Copyright Clearance Center, 222 Rosewood Drive, Danvers, MA 01923, (978) 750-8400, fax (978) 750-4744. Requests to the Publisher for permission should be addressed to the Permissions Department, John Wiley & Sons, Inc., 605 Third Avenue, New York, NY 10158-0012, (212) 850-6011, fax (212) 850-6008, E-Mail: PERMREQ@WILEY.COM.

Library of Congress Cataloging-in-Publication Data:

Kadin, Alan M.
    Introduction to superconducting circuits/Alan M. Kadin.
      p. cm.
    Includes bibliographical references.
    ISBN 0-471-31432-3 (cloth: alk. paper)
    1. Superconductors. 2. Electronic circuits. I. Title.
TK7872.S8K26 1999
621.3815—dc21   98-38926

Printed in the United States of America.

10 9 8 7 6 5 4 3 2 1

To Sidney Shapiro,
for his pioneering steps
toward the development
of Josephson devices

CONTENTS

Preface xi

1. Preview 1

    1.1 Superconductor as Lossless Inductor / 1
    1.2 Low-Temperature and High-Temperature Superconductors / 3
    1.3 Superconductor as Flux Insulator / 5
    1.4 Superconductor as Macroscopic Atom / 8
    1.5 Josephson Junctions / 10
    1.6 Applications of Superconducting Devices / 13
    1.7 Overview of Book / 14
          Summary / 15
          References / 15
          Problems / 15

2. AC Properties and Superconducting Energy Gap 18

    2.1 Kinetic Inductance and Penetration Depth / 18
    2.2 Two-Fluid Model and RF Surface Resistance / 30
    2.3 RF Transmission Lines and Filters / 35
    2.4 Superconducting Energy Gap and Tunneling / 46
          Summary / 63
          References / 64
          Problems / 64

3. Magnetic Properties of Superconductors 67

    3.1 Flux Conservation and Diamagnetism / 67
    3.2 Critical Currents and Fields / 78
    3.3 Fluxoid Quantization and Vortices / 86
    3.4 Duality and Flux Flow / 94

3.5 Flux Pinning in Large Fields / 107
3.6 Magnet and Power Applications / 119
Summary / 136
References / 137
Problems / 137

4. Superconducting Materials and Thin-Film Technology — 141

4.1 Low-Temperature Metallic Superconductors / 141
4.2 High-Temperature Copper–Oxide Superconductors / 151
4.3 Other Superconducting Materials and Microstructures / 169
Summary / 175
References / 175
Problems / 176

5. Josephson Devices — 178

5.1 The Josephson Effect / 178
5.2 Shunted-Junction Models / 193
5.3 SQUIDs and Magnetic Detection / 207
5.4 Distributed Junctions and Arrays / 225
Summary / 244
References / 245
Problems / 245

6. Superconducting Digital Circuits — 249

6.1 Fast Switches and Memories / 249
6.2 Voltage-State Logic / 259
6.3 Single-Flux-Quantum Logic / 270
6.4 Applications to Digital Systems / 284
Summary / 295
References / 296
Problems / 297

7. Superconducting Radiation Detectors — 300

7.1 Modulation Detectors and Mixers / 301
7.2 Thermal and Quasi-Thermal Detectors / 312
7.3 Single-Photon and Particle Detectors / 324
Summary / 332
References / 332
Problems / 333

Epilogue: Future Prospects for Superconducting Circuits — 337

Bibliography	340
Appendix A. Transmission Lines and Electromagnetic Waves	342
Appendix B. Computer Simulations of Josephson Circuits	352
Appendix C. Cryogenic Technology	359
Appendix D. Electromagnetic Units and Fundamentals Constants	368
Appendix E. Symbols and Acronyms	371
Materials Index	375
Name Index	377
Subject Index	379

PREFACE

The subject of superconductivity has a reputation for being very difficult. And indeed, it can be presented in a way that requires advanced quantum formalism and statistical mechanics. On the other hand, I believe that a study of superconducting circuits and devices need be no more difficult than the study of semiconductor devices, which has become a standard part of the electrical engineering curriculum.

With the motivation of making the field more accessible to engineers and applied physicists, the goals of this book are twofold: first, to use circuit models to develop an understanding of the physics of superconductors and, second, to apply this understanding to superconducting circuits and systems. These build on a background with which most seniors in electrical engineering, for example, are already familiar. This includes linear circuits and systems in the time and frequency domains, the transmission line as a distributed circuit, electromagnetics through Maxwell's equations, and an introduction to semiconductor device physics. The only quantum mechanics used is at the level of a general physics course, such as the existence of a quantum wave function that oscillates at a frequency given by the Planck relation.

This is quite different from conventional approaches to the subject. Key features include the following:

- the central role of inductance and kinetic inductance;
- a transmission line model for rf and dc properties;
- dual-circuit transformations to follow motion of vortices and fluxons;
- a balanced emphasis on both low-temperature and high-temperature superconductors;
- discussion of both large-scale (power) and small-scale (electronic) applications;
- modern high-speed digital applications, including single-flux-quantum circuits;

- applications of superconducting devices to electromagnetic radiation detectors; and
- the use of SPICE to simulate Josephson junctions and circuits.

This book does not attempt to be encylopedic in scope; it is an introduction rather than a comprehensive review. In general, the simple examples and models that will illustrate a given phenomenon or application are chosen with liberal use of schematic diagrams to illustrate key concepts. Approximate treatments that provide for clear exposition are favored over mathematical rigor. In this way, the entire book is short enough to be covered in a single semester. It should be accessible both to upper level undergraduates and to graduate students in electrical engineering, applied physics, and related fields.

Why should students study superconducting circuits? Superconductivity has been an exciting research field in recent years, at the very forefront of modern technology, and all sorts of potential applications have been promoted. On the other hand, superconducting circuits have not yet had a significant impact on mainstream technologies, and most engineers will probably never have to deal with them. Partly for this reason, a course in superconducting devices has rarely been offered at most universities. The University of Rochester is unusual in this regard; we have a group of active researchers in the superconducting electronics field in our Electrical Engineering Department, and we have taught a course in superconducting electronics for some years. Furthermore, a number of our former students have gone on to key positions in the small (but growing) superconducting electronics industry.

This volume should also be useful to active engineers and scientists interested in determining whether superconductivity may have something to offer to their own specialities. All of the key superconducting applications are discussed, with an aim of informing the reader of both the advantages and the shortcomings of superconductors. This is particularly valuable for the newer high-temperature superconductors, which have often been promoted in the media with little regard for their limitations.

There are several other excellent textbooks on superconductivity and superconducting circuits. Two that we have used in our courses are *Foundations of Applied Superconductivity* by Orlando and Delin and *Principles of Superconductive Devices and Circuits* by Van Duzer and Turner. Together with Tinkham's *Introduction to Superconductivity*, these have provided a set of essential references in the preparation of the present volume. But of course any errors in the book are my own.

I have included a set of sample problems at the end of each chapter as well as some references for further reading. Given the voluminous literature in the field, there is no attempt to be comprehensive. The reader who wants the very latest results in the field should seek out the most recent Proceedings of the Biennial Applied Superconductivity Conference, published in the *IEEE Transactions on Applied Superconductivity*.

By combining a focus on basic principles with a description of current technology and future trends, *Introduction to Superconducting Circuits* aims to inspire the next generation of students to continue to explore the limits of the possible with superconducting technology. For that has been, and will continue to be, the allure of superconductivity.

ALAN M. KADIN

Rochester, New York
July, 1998

INTRODUCTION TO SUPERCONDUCTING CIRCUITS

1

PREVIEW

1.1 SUPERCONDUCTOR AS LOSSLESS INDUCTOR

In the course of this chapter, and throughout the book, we will be viewing a superconductor from the perspective of several fairly simple models. In the first model, a superconducting wire can be represented as an ideal inductor L, with the usual series resistance $R = 0$. Then, one can conduct a very large constant current I with zero voltage drop across the wire, which of course is the basis for the name "superconductor." On a microscopic level, one can set the electrical conductivity $\sigma \to \infty$ in the microscopic Ohm's law $J = \sigma E$, where J is the current density and E the electric field. Things are not quite as simple as this, but this single assertion yields several important consequences. Some of the details will be modified as we refine the model, but the general picture will continue to hold.

Consider, for example, a superconducting wire wound into a hollow solenoid, as shown in Fig. 1.1. The solenoid will have a self-inductance L so that there will be a voltage drop $V = L\,di/dt$ associated with initially turning on the current i. However, the current source is simply supplying energy that is being stored in the magnetic field: $\mathscr{E} = \frac{1}{2}Li^2$. When the desired current is reached, no additional voltage is needed from the supply. In fact, at this point, a superconducting short can be placed across the current supply, the current from the supply turned down to zero, and the supply disconnected. Since the series resistance in the superconductor is zero, the decay time L/R for the current in the solenoid is infinite; the current will circulate forever in this "persistent-current" mode. A superconducting inductor is therefore a true energy storage element like an ideal capacitor; it can hold its "charge" indefinitely.

Now consider what happens when one brings a magnet (say, an ordinary permanent magnet) near an isolated perfect conductor or conducting loop. By

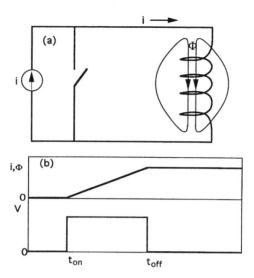

Figure 1.1. (*a*) Schematic circuit for supplying current to superconducting inductor. (*b*) Time dependence of voltage *V* across and current *i* (and flux Φ) through superconducting inductor.

Faraday's law $V = -d\Phi/dt$ (all electromagnetic formulas in this book are in SI units; see Appendix D), an induced current will be set up in the surface of the conductor that opposes the entry of magnetic flux Φ into the conductor. This, of course, is the basis for the electric generator. However, for a normal conductor, as soon as the magnet stops moving, the induced current will die out with a characteristic L/R time. In contrast, if $R = 0$, the induced surface currents never die out, and the applied flux can never enter. Furthermore, in order to prevent entry of magnetic flux, the induced current must be producing a magnetic field distribution that precisely opposes the applied magnetic field. This has a simple equivalent in the method of images. The applied magnet is looking at its reflection, a fictitious image magnet with the same polarity but pointing in the opposite direction. From this point of view, since like magnetic poles repel, it becomes clear that a perfect conductor will exert a repulsive force on any magnet brought near it. This can provide the basis for magnetic levitation of a magnet above a superconductor, as illustrated in Fig. 1.2. When the repulsive force is equal to the weight of the magnet, the magnet will levitate above the superconductor.

Furthermore, exclusion of electromagnetic fields from a perfect conductor is not limited to dc and low frequencies; it extends to high frequencies as well. We can see this from the standard formulas for the ac skin depth δ and surface resistance R_s of a normal metal at a frequency f:

$$\delta = \frac{1}{\sqrt{\pi f \mu_0 \sigma}} \qquad R_s = \frac{1}{\sigma \delta} = \sqrt{\frac{\pi f \mu_0}{\sigma}} \qquad (1.1)$$

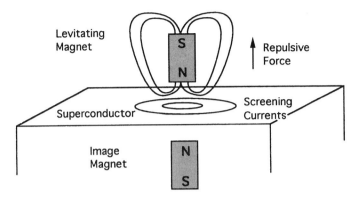

Figure 1.2. Configuration of screening current and magnetic flux lines for permanent magnet levitating above superconducting plate. The repulsive force can be obtained through consideration of an image magnet below the surface.

where $\mu_0 = 1.26$ µH/m is the permeability of free space. If we take $\sigma \to \infty$, then $\delta = 0$ and $R_s = 0$. In effect, a perfect conductor is a perfect reflector of electromagnetic waves, with no loss. This is brought about by ac surface screening currents, in the same way that a low-frequency magnetic field is screened out.

When we examine these screening currents more carefully, we will find that the surface screening currents have a small but finite depth (typically ~ 100 nm), which is known as the magnetic penetration depth λ_L. On this scale, one must model the superconductor not as a single lumped inductance but as a distributed inductance per unit length. Furthermore, this distributed inductance must include not only the usual magnetic inductance but also another contribution, known as the kinetic inductance, associated with the inertia of the superconducting electrons. The most familiar standard example of distributed impedances is the transmission line (see Appendix A), and we will make repeated use of a transmission line picture to help explain the behavior of currents and voltages in a superconductor. With this understanding, the model of a superconductor as a lossless inductor will continue to guide us.

1.2 LOW-TEMPERATURE AND HIGH-TEMPERATURE SUPERCONDUCTORS

Before going further, let us address some of the characteristics of superconductors in the real world. First of all, all superconductors suddenly revert to normal metallic behavior when heated above a critical temperature T_c. Unfortunately, the value of T_c for a superconductor is always well below room temperature (at least as of 1998) and is normally expressed in degrees kelvin above absolute zero. See Table 1.1 for some common examples. Historically, superconductivity is a phenomenon of the twentieth century; its discovery had to await the invention

Table 1.1. Maximum Superconducting Temperatures T_c for Selected Materials

Material	T_c (K)
Cu, Ag, Au	–
Al	1.2
Pb	7.2
Nb	9.2
$YBa_2Cu_3O_7$	92
$HgBa_2Ca_2Cu_3O_8$	130

of cryogenic technology (see Appendix C) that could cool samples near absolute zero. The key development, which was achieved in 1908 by the Dutch physicist and cryogenic pioneer Heike Kamerlingh Onnes, was the liquification of helium gas at 4 K. By immersing various materials in boiling liquid helium, their properties in this low-temperature range could be measured rather directly. He went on to discover superconductivity in a filament of mercury in 1911 at $T_c = 4$ K; several other simple elemental metals followed soon after that. At the time, it was expected that electrical resistance would decrease in a pure metal as it was cooled down; what was a surprise was that R should go sharply to zero below some temperature. Interestingly, the best metallic conductors (Cu, Ag, Au) never become superconducting, and a good conductor such as Al has a very low $T_c \approx 1$ K. The common elements with the highest values of T_c are Nb and Pb (with 9 and 7 K), which are not particularly good metallic conductors above T_c.

By 1986, seventy-five years later, many more superconducting alloys and compounds had been discovered, but none exhibited a value of T_c in excess of 23 K, and many scientists had come to believe that superconductivity was not possible at higher temperatures. Then in 1987, a tremendous breakthrough was achieved with the discovery of "high-temperature superconductivity" in the complex conducting oxide $YBa_2Cu_3O_7$, which has a critical temperature of 92 K, as shown in Fig. 1.3. This is safely above the 77 K boiling point of liquid nitrogen, a cheaper and more convenient cryogenic refrigerant than liquid helium. More recently, related compounds based on copper oxides with verified T_c up to 130 K (and up to 160 K at extremely high pressures) have been fabricated. Still higher temperatures may be possible, although this class of superconducting materials seems to have reached its limit.

At present, the term "low-temperature superconductor" (LTS) is applied to the older metallic materials with $T_c < \approx 20$ K and "high-temperature superconductor" (HTS) to the newer materials with $T_c > \approx 30$ K. Of course, even the "highest temperature" superconductor materials require cryogenic temperatures in order to operate; 130 K is still less than $-140°$C. It is important to appreciate that LTS materials did *not* suddenly become obsolete with the discovery of the

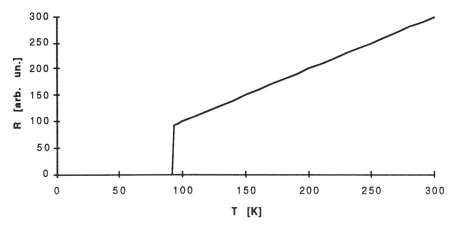

Figure 1.3. Temperature dependence of resistance for high-temperature superconductor $YBa_2Cu_3O_7$(YBCO), showing zero resistance below 93 K.

HTS. In fact, there is a substantial and growing technology of LTS devices and applications, designed to operate at liquid helium temperatures. The newer HTS materials are often difficult to fabricate into practical devices, so that despite the reduced cryogenic requirements of the HTS, it is likely that both LTS and HTS applications will continue to develop in the coming decades, and we will focus equally on both of them. There are some significant differences between the two classes of materials, which we will discuss in more detail in Chapter 4. However, in most respects both classes behave in a similar manner, and most of our analysis (unless otherwise noted) applies equally well to either type.

A real superconductor also has several other "critical parameters." There is a critical magnetic field H_c and a critical current I_c (or critical current density J_c) above which the material no longer superconducts. Typical values are $H_c \approx 1-10$ MA/m (10–100 kOe) and $J_c \sim 10^{10}$ A/m². (Actually, at these large values, most practical superconductors do not exclude magnetic flux, as discussed below.) Furthermore, there is also a critical frequency f_c (typically of order 10^{12} Hz) above which the ac electromagnetic behavior looks more like that of a normal metal. This is associated with the superconducting energy gap $\mathscr{E}_g = hf_c$, where h is Planck's constant.

1.3 SUPERCONDUCTOR AS FLUX INSULATOR

We pointed out earlier that one of the consequences of zero resistance is that magnetic flux cannot enter a superconductor. A complementary picture to the perfect conductor of electrical charge is one that regards the superconductor as an insulator with respect to the motion of magnetic flux. This is not merely an analogy; there is a formal duality between charge and flux in electrical circuits.

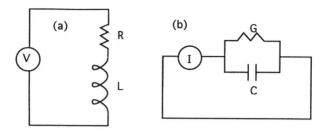

Figure 1.4. Simple example of dual circuits. (*a*) Schematic for voltage source driving series *R–L* circuit. (*b*) Dual circuit to that in (*a*) which represents the physical circuit for the flow of magnetic flux in (*a*).

Consider, for example, the simple circuit shown in Fig. 1.4*a*, in which a voltage supply is driving current through an ideal inductor and resistor in series. Since $V = d\Phi/dt$, the voltage source also actively pumps magnetic flux into the loop. This can be stored in the inductor, or alternatively it can leak out across the resistor. The dual circuit is shown in Fig. 1.4*b* and represents the path for the flow of magnetic flux. Topologically, the loop becomes a node, and elements in series transform to elements in parallel. The roles of voltage and current are switched, as are inductance and capacitance. A resistance R becomes a conductance G. The details of this dual mapping will be discussed in Chapter 3, but the main point is that for every electrical circuit, there is a unique dual circuit that describes the motion of the magnetic flux.

We can treat the case of the perfect conductor simply by setting $R = 0$ in Fig. 1.4*a*. Then, the flux that is pumped into the circuit cannot leave; it cannot cross a superconducting wire. So the current around the loop would continue to increase until either the current limit of the voltage source or the critical current of the superconducting wire is reached. This is analogous to the situation in the dual circuit in Fig. 1.4*b*, where a current source is charging up a capacitor (with leakage conductance $G = 0$). This will continue until either the voltage compliance of the current source or the breakdown voltage of the capacitor is reached.

So it would seem that an ideal superconductor can be regarded as a "flux insulator." However, even an electrical insulator can carry current if there are carriers (electrons or holes) in the appropriate band. The corresponding flux carrier in a superconductor is known as a fluxon (or vortex), with a flux $\Phi_0 = h/2e$ (a magnitude we will motivate below), where h is Planck's constant and e is the magnitude of the electronic charge. In the simplest case, consider a superconducting loop interrupted by a short, narrow, superconducting weak link, essentially what we call a Josephson junction (Fig. 1.5). If the current in the loop exceeds the maximum current of the weak link, then some of the fluxons will leak out of the loop by crossing the link until the circulating current becomes subcritical. But it is important to understand how they leak out: not at

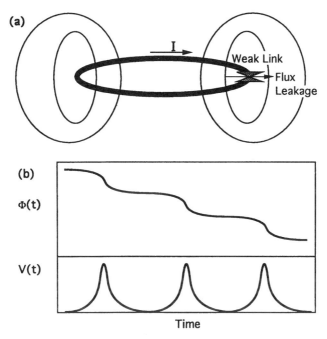

Figure 1.5. Superconducting loop containing circulating current and trapped magnetic flux, with weak link permitting magnetic flux to leak out, one fluxon at a time. (*a*) Configuration of loop, currents, and magnetic fields. (*b*) Time dependence of flux in loop and voltage across weak link.

random, but in a single-file, correlated sequence of fluxons. Each time a fluxon crosses the junction, there is a voltage pulse of integrated magnitude Φ_0. This follows from the equations for the Josephson junction, which we will describe in further detail below.

The other regime in which magnetic flux can cross a superconductor is in the presence of a large magnetic field. For most practical superconductors (known as type II superconductors), it is energetically unfavorable for the magnetic flux to be completely excluded. Instead, the flux penetrates via discrete fluxons, as indicated in Fig. 1.6*a*. (In this context, the fluxons are often called "vortices," since each one is surrounded by a screening current in the superconductor.) This is dual to the picture in Fig. 1.6*b*, which represents an insulator with the presence of a number of free carriers. In this regime, the superconductor is analogous to a doped semiconductor. But the question remains: Is the superconductor still superconducting, that is, does it still exhibit zero resistance? The analogy would suggest that since a doped semiconductor is not an insulator, a "magnetically doped superconductor" should permit the flow of fluxons, which in turn would lead to resistance. This analogy is true, provided that the fluxons are free to move. If they are all pinned to trap sites, then one has a flux insulator, that is, zero resistance, again. Strange as it seems, this does indeed provide the basis for the entire technology of high-field superconducting magnets.

8 PREVIEW

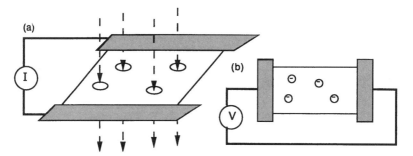

Figure 1.6. Duality between fluxons in superconductor and electrons in semiconductor. (*a*) Superconducting film or plate with perpendicular magnetic field penetrating via fluxons (vortices). (*b*) Dual circuit showing doped charge carriers in an otherwise insulating matrix. In either case, the material will exhibit resistance unless the carriers are immobilized.

1.4 SUPERCONDUCTOR AS MACROSCOPIC ATOM

This book will not focus on the microscopic theory of superconductivity, sometimes called the BCS theory (for Bardeen, Cooper, and Schrieffer). However, we can go a long way using just a few points from this theory. Key among them is the fact that supercurrent is carried not by single electrons but by bound pairs of electrons, called "Cooper pairs." We will discuss in Chapter 2 how this counterintuitive binding might take place. However, for the time being, let us assume that a superconductor can be represented by two "energy bands," as indicated in Fig. 1.7*a*. The binding energy, which is also the energy gap per pair, is given by $\mathscr{E}_g = 2\Delta$, where Δ is the gap parameter. The energy gap (typically of order 3–4 $k_B T_c$) depends on temperature and goes to zero at T_c, as shown in Fig. 1.7*b*. In the ground state of the superconductor at $T = 0$, all of the electrons are bound in pair states. At higher T, of course, there will be some thermally ionized (unpaired) electrons. These unpaired electrons act like normal, non-superconducting electrons and would exhibit resistance if they were not shorted out by the Cooper pairs. However, their effect is evident at higher frequencies. Furthermore, photoionization of the Cooper pairs is also possible if $hf > 2\Delta$. Above this characteristic frequency the superconductor is lossy and is no longer able to exclude electromagnetic fields.

Remarkably, all of the Cooper pairs are in the same quantum state, with not only the same energy but also the same quantum wave function. (This also seems counterintuitive and will be discussed further in Chapter 2.) This quantum wave function can be represented in the standard complex form $\Psi = \Psi_0 \exp(i\theta)$, where $n_s^* = \Psi_0^2$ is the number density of Cooper pairs and θ is their phase. Essentially, the wave functions for all of the Cooper pairs add up coherently, so that the phase information is preserved. The phase may vary continuously in position and time according to the usual quantum mechanical wave dependence $\theta = \mathbf{k} \cdot \mathbf{r} - \omega t$, where $\mathscr{E} = \hbar\omega$ is the energy and $\mathbf{p} = \hbar\mathbf{k}$ is the momentum of the

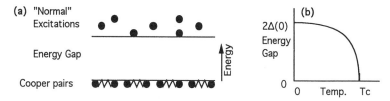

Figure 1.7. Energy gap in superconductor. (*a*) Schematic representation of energies of superconducting ground state and excited states, showing energy gap $\mathscr{E}_g = 2\Delta$. (*b*) Approximate temperature dependence of energy gap.

Cooper pair; these are essentially the de Broglie relations ($\hbar = h/2\pi$). The magnitude Ψ_0 is normally constant within a good superconductor, although it can be reduced near boundaries or in a weakened region. We can represent this quantum wave function by a phasor diagram, as shown in Fig. 1.8*a*. If we apply a voltage V to a superconductor, this changes the energy of each of the pairs by $\Delta\mathscr{E} = qV = 2eV$, and so causes the phasor to rotate with angular frequency $\omega = 2eV/\hbar$. If we apply a voltage pulse with integral $\int V\,dt = \Phi_0 = h/2e$ [a single-flux-quantum (SFQ) pulse], then the phase will change by $\Delta\theta = \int(2eV/\hbar)\,dt = 2e\Phi_0/\hbar = 2\pi$, rotating one complete revolution. Spatial variations in θ are associated with net motion of Cooper pairs, that is, supercurrent flow.

There is a close relationship between the quantum mechanical phase θ and the magnetic flux Φ in a superconductor. Consider a superconducting wire with no current and no nearby magnetic fields. The phase θ should be constant along the length of the wire. Now, connect a voltage source to the two ends and have it supply an SFQ pulse, corresponding to inserting one fluxon into the loop. This will have the effect of rotating the phase by 2π at one end relative to the other. Since only phase differences are relevant, it is immaterial whether the phase shift is applied clockwise at one end or counterclockwise at the other or an antisymmetric shift by π is applied at both ends. In all cases, one turn is inserted in the phase plot and distributes evenly around the loop, as shown in Fig. 1.8*b*. The ends of the wire can now be reconnected, since the phase matches at the two ends. Now we have one fluxon trapped in the loop. More generally, the phase difference around the loop changes by 2π for every fluxon entering the loop. Furthermore, the phase difference around the loop must be a multiple of 2π if the two ends are to be reconnected; if it is not, then transient effects will ensure that it quickly becomes so, since sharp "kinks" in the phase are energetically unfavored. This is the basis for the magnitude of the flux quantum Φ_0 in a superconducting loop. It follows simply from the fact that the phase must be single valued (modulo 2π) at all points along a closed superconducting loop. This is essentially the same as the basis for quantization of angular momemtum in the Bohr model of the hydrogen atom, and illustrates how in certain respects a superconductor can be regarded as a macroscopic atom.

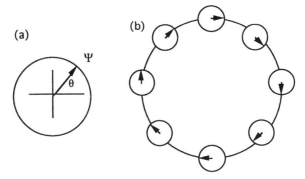

Figure 1.8. Behavior of quantum mechanical phase θ superconductor. (*a*) Phasor diagram for complex superconducting wave function $\Psi_0 \exp(i\theta)$. (*b*) Spatial dependence of phase θ around superconducting loop, showing phase shift of 2π corresponding to one flux quantum $\Phi_0 = h/2e$ inside the loop.

1.5 JOSEPHSON JUNCTIONS

We stated above that when two superconducting wires are brought together, their phases θ must match. A Josephson junction consists of two superconductors that are weakly coupled so that the Cooper pair wave functions may interact, but the phases of the two superconductors may differ by a phase $\phi = \theta_2 - \theta_1$. When this is the case, there is a supercurrent that passes between the two superconductors according to the two equations

$$I = I_c \sin\phi \qquad \frac{d\phi}{dt} = \frac{2eV}{\hbar} \qquad (1.2)$$

where I_c is a constant. A classic Josephson junction consists of an insulating layer on the nanometer scale between two superconductors, but other weak links such as those with a short resistive region or a small constriction in a superconductor behave in a similar fashion.

We can understand better the behavior of a Josephson junction in terms of the models that we introduced above. First, if there is zero voltage across the junction, then the phase difference ϕ is constant, and there may be a constant current. This should not be a surprise; even a weakly coupled superconductor can still carry a (small) current with zero resistance. Also, consider a superconducting loop with a single Josephson junction in it. If there are an integral number of flux quanta in the loop, then the phase difference across the junction is $\phi = 0$ (modulo 2π), and no current flows. But if the fluxon number is nonintegral, then there will be a circulating current, in the direction to try to cancel out the nonintegral part of the fluxon number, to reduce ϕ. So for $0 < \phi < \pi$, the circulating current goes in one direction, while for $-\pi < \phi < 0$, it goes in the other direction. However, a weak link cannot carry sufficient current to effectively screen the flux, so φ remains nonzero.

If, on the other hand, a nonzero dc voltage is applied across the junction, then we get an oscillating current $I_c \sin(\omega_J t)$, where $\omega_J = 2eV/\hbar$ is the Josephson (angular) frequency. For $V = 1$ mV, the junction current would oscillate at almost 500 GHz. This is sometimes known as the ac Josephson effect. This might initially seem odd, but it is consistent with our atomic picture (see Fig. 1.9). With a voltage between them, the ground-state energies of the two superconductors differ by $\Delta\mathscr{E} = 2eV$. In an atomic system, the only way for an electron to change its energy is by emission or absorption of a photon of energy $\mathscr{E} = \hbar\omega$; there is no dissipation permitted. The classical equivalent of photons in an electrical circuit requires having an oscillating current and an (in-phase) oscillating voltage at the same frequency, leading to ac power transfer. Here, the net transfer of a Cooper pair across the junction would have to couple the energy $2eV$ into a photon either emitted or absorbed. We have constrained the voltage to be constant, so that no ac power can be transferred; thus no photons can be coupled to the junction. Therefore, the current can only have an ac component; no dc current is permitted.

However, if we couple the Josephson junction to an antenna or transmission line, then ac power can be coupled in or out. This can sometimes be modeled as a resistive element shunting the junction, where the constraint of constant voltage can be relaxed. This does indeed permit a dc component of lossless supercurrent across the junction, with one photon of energy $2eV$ emitted for each net Cooper pair crossing the junction. This is analogous to spontaneous emission in an atomic system. In addition, if there is an externally applied ac voltage at the same frequency ω_J, then the junction oscillation will tend to synchronize to this external oscillation, and photons can be transferred in or out of the junction, together with net dc current in the appropriate direction. This is analogous to stimulated emission or absorption in the atomic system.

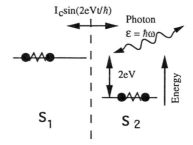

Figure 1.9. Atomic picture of a Josephson junction, with two energy levels for the Cooper pairs in two weakly connected superconductors S_1 and S_2, separated by $\Delta\mathscr{E} = 2eV$. For a constant dc voltage, there can only be an oscillating ac current at the Josephson frequency $f_J = 2eV/h$. But as in an atomic system, transfer of Cooper pairs across the junction must be accompanied by emission or absorption of photons at this resonant frequency.

Furthermore, for a very strong external ac voltage, multiple-photon effects are possible. In this case, a net dc supercurrent is compatible with photons at integral sub-harmonics of ω_J.

This picture might suggest that the ac Josephson effect can be used as a source of electromagnetic radiation across the entire spectrum. Unfortunately, this is not quite true; the picture breaks down at about the same point that the Cooper pairing breaks down, for $\Delta\mathscr{E} > \approx \mathscr{E}_g = 2\Delta$. For higher voltages, the response rolls off quickly. So it is not possible, for example, to apply 1 V across a Josephson junction and get out red light with $f \sim 500$ THz. Further analysis of the Josephson effect will be developed later in Chapter 5.

An ideal Josephson junction cannot conduct a current greater than I_c. In a real physical junction, there is typically a parallel dissipative channel associated with unpaired electrons that can be modeled most simply as a shunt resistor. This comes into play if a nonzero voltage is applied or with a current larger than I_c. This is the basis for the resistively shunted junction model. If an RSJ is biased with a constant current $I > I_c$, then the instantaneous voltage corresponds to a series of voltage pulses, as indicated in Fig. 1.5; the dc average voltage is shown in Fig. 1.10a. Each pulse with integrated value $\int V\,dt = \Phi_0$ corresponds to one fluxon crossing the junction and to the phase ϕ rotating by 360°. Therefore, a junction with a dc average voltage V corresponds to fluxons crossing at a rate $f = V/\Phi_0 = 2eV/h$, which again is the Josephson frequency.

As an example of Josephson junctions in a circuit, consider a superconducting loop with two identical RSJs, as shown schematically in Fig. 1.10b. This forms the basis of the device known as the dc-SQUID, or superconducting quantum interference device. If there is no magnetic flux in the superconducting loop, then the phase differences ϕ_1 and ϕ_2 across the two junctions must be identical, and the $I(V)$ relation would be simply double that of the individual junctions. However, if there is one-half of a flux quantum in the loop, this rotates the phase between the junctions by 180°, so that we must have $\phi_1 - \phi_2 = 180°$. This means that the supercurrents in the two junctions are always opposite to each other, corresponding to a net supercurrent of zero. This is equivalent to zero critical

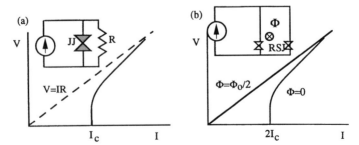

Figure 1.10. DC voltage-current characteristics of current-biased resistively shunted Josephson junction (RSJ) and dc SQUID (two parallel RSJs): (a) $V(I)$ for RSJ; (b) $V(I)$ for SQUID with flux in loop $\Phi = 0$ and $\Phi = \tfrac{1}{2}\Phi_0$.

current, or simply a resistor. The *I–V* curves for other values of Φ lie between these limits. A properly designed SQUID is sensitive to a small fraction of Φ_0.

1.6 APPLICATIONS OF SUPERCONDUCTING DEVICES

Superconductors are absolutely unique in many of their properties. A wide variety of superconducting devices have been proposed and developed, a few of which are already in commercial use. The discovery of high-temperature superconductors in the past decade has extended the practical range of some of these applications and enabled the development of new ones. However, the basic principles behind the operation of superconducting devices has remained very much the same. Of course, the discovery of superconductivity at room temperature would indeed be an amazing breakthrough. But even short of that unpredictable occurrence, it is likely that superconducting devices will play an ever-increasing role in the technology of the twenty-first century.

Superconducting devices tend to operate in one of three general regimes. There are those based on the zero resistance of the superconducting state, those based on the transition between the superconducting and the normal states, and those based on Josephson junctions. Below we give a brief survey of some of the different classes of applications, in order to provide a perspective as we develop the principles of superconducting devices in succeeding chapters.

In terms of zero resistance, the leading application is unquestionably the superconducting magnet. To produce magnetic fields greater than a few teslas, there is normally no reasonable alternative to using superconductors. This typically involves winding kilometers of wire into large solenoids and other geometries. Superconducting magnets are currently used primarily in medical and scientific instrumentation, particularly magnetic resonance imaging and particle accelerators. Other applications of bulk superconductors that focus on the low-frequency magnetic properties (which are really a consequence of zero resistance) include magnetic levitation, magnetic bearings, and magnetic shielding.

Most of the other applications we will discuss can be achieved using thin superconducting films, rather than wires, since the total currents needed are not quite so large. Superconductors exhibit extremely small resistance at rf and microwave frequencies, and resonant cavities, lossless transmission structures, and microwave filters are among the applications that are being developed.

Another class of applications relies on the transition between the superconductor and the normal metallic state. In an ideal superconductor, the resistive transition at T_c is extremely sharp, which provides the basis for a variety of thermally based detectors and switches. In particular, a superconductor held at T_c can be a very sensitive bolometer, or thermal detector of radiation. Similarly, a superconductor held at T_c can be switched from zero resistance to quite a large resistance with only a small exposure to thermal or optical energy. Furthermore, a transition from the superconducting to the normal state can be

14 PREVIEW

achieved not only with temperature, but alternatively with magnetic field or voltage.

But the Josephson junction has provided the basis for the widest range of superconducting devices. SQUID magnetometers are the most sensitive detectors of low-frequency magnetic fields available in any technology, with a sensitivity down to the femtotesla range. This makes SQUID arrays useful in localizing sources of magnetic fields in the brain, for example. At microwave frequencies, Josephson junctions (sometimes in arrays) have been used as detectors and sources of radiation, based on the ac Josephson relation $\hbar\omega = 2eV$. Limiting frequencies correspond to the superconducting energy gap, which approaches 1 THz. An important and unique application of Josephson junctions, based on the same equation, is the use of arrays of Josephson junctions to define the standard volt. As suggested in Fig. 1.9, an external microwave source can lead to a net dc supercurrent at $V = \hbar\omega/2e$. Since frequencies can be measured extremely accurately, this provides a much more accurate standard for voltage measurement than the old standard, which was based on an electrochemical cell.

The final application of Josephson junctions, and potentially the most significant, is to digital circuits, including ultimately a general-purpose computer. Any digital system requires a way to encode and to store a binary 0 or 1. Given flux quantization in a superconducting loop, it is natural to represent a bit by the absence or presence of a fluxon in a given loop. In terms of logic gates, two approaches have been used. In voltage-state logic, a Josephson junction or SQUID in the $V = 0$ state represents a 0, while the same device in the voltage state for $I > I_c$ is a 1. An alternative approach (single-flux quantum logic, or SFQ) has all of the junctions in the zero-voltage state, but with the binary information encoded in the presence or absence of discrete SFQ voltage pulses. In either approach, the intrinsic switching time for a Josephson logic gate is extremely fast, on the order of a few picoseconds. Integrated circuits with many thousands of such gates have been fabricated and demonstrate very fast digital computing, with extremely low power dissipation. If a number of technological hurdles can be overcome, a Josephson computer might be manufactured that is much faster than that possible using conventional semiconductor technologies.

1.7 OVERVIEW OF BOOK

The following chapters will address the subjects of this preview in more detail. In Chapter 2, we will deal with the consequences of zero resistance on dc and ac properties, and in Chapter 3, we discuss the magnetic properties of superconductors. Chapter 4 will describe the important aspects of real superconducting materials, including both low-T_c and the newer high-T_c materials. We will continue in Chapter 5 to discuss Josephson junctions and circuits containing them, and in Chapter 6 the application of such circuits to logic devices will be explored. We complete the survey in Chapter 7 with superconducting devices for the detection of electromagnetic radiation, and a final discussion of some future prospects for the development of superconducting circuits.

SUMMARY

- A material can superconduct only below a characteristic cryogenic critical temperature T_c. Recently discovered high-temperature superconductors (HTSs) exhibit 30 K $< T_c <$ 150 K; the earlier low-temperature superconductors (LTSs) have $T_c <$ 30 K and generally below 10 K.
- The superconducting state is a coherent quantum state of Cooper pairs of charge 2e, with a quantum phase θ, analogous in certain respects to a "macroscopic atom." This is responsible for quantization of magnetic flux in units of $\Phi_0 = h/2e$, and the ac Josephson effect whereby a dc voltage brings about ac current oscillations at frequency $f_J = 2eV/h$.
- A superconductor can be viewed as an ideal lossless inductor, which can store a circulating current indefinitely. Such induced currents are also responsible for magnetic repulsion and levitation.
- An ideal superconductor can also be viewed as a "flux insulator;" magnetic flux cannot cross a superconductor, except in quantized units of Φ_0. These "fluxons" are analogous to electrons or holes in a semiconductor.
- The leading bulk application of superconductors is the high-field magnet, with important uses in medical and scientific instrumentation.
- The Josephson junction, essentially a weak link between two superconductors, is the basis for most superconducting electronic devices, including sensitive SQUID magnetometers, ultrafast digital logic circuits, and the international voltage standard.

REFERENCES

A good introduction to the field of superconductivity may also be obtained in the following books.

M. Cyrot and D. Pavuna, *Introduction to Superconductivity and High-T_c Materials* (World Scientific, Singapore 1992).

R. W. Simon and A. D. Smith, *Superconductors: Conquering Technology's New Frontier* (Plenum, New York, 1988).

PROBLEMS

1.1. Superconducting solenoid. Consider a superconducting solenoid consisting of 10,000 turns of wire with a length of 30 cm and a diameter of 10 cm that is immersed in liquid helium. It takes a current of 100 A to reach the design magnetic field. Assume that the wire has a normal-state resistance of 0.1 Ω/m.

(a) Estimate the inductance L of this solenoid and the design magnetic field at full current.

(b) How long will it take to charge this up to a current of 100 A using a 1-V dc power supply?

(c) After charging to 100 A, the solenoid is placed in the persistent-current mode and the external source disconnected. How much energy is stored in this fully charged magnet?

(d) If the superconductor were to suddenly go normal, how quickly would it discharge? How much liquid helium would be boiled away in the process? (See Table C.1.)

1.2. Lossy superconductor and fluxon motion. Consider a superconducting element that "leaks flux" in the same way that a leaky capacitor leaks charge. This can be modeled as a resistance $R = 1$ pΩ in series with the superconducting inductor $L = 1$ µH.

(a) If this element is carrying a current $I = 100$ A, determine the average rate that fluxons cross the element.

(b) If this element were placed in an otherwise lossless superconducting loop with $L = 100$ µH, how long would it take the flux to leak out?

(c) Sketch the schematic of this circuit together with the dual circuit that represents the motion of flux in this circuit.

1.3. Room temperature superconductor? A hypothetical room temperature superconductor has critical temperature $T_c = 400$ K.

(a) Estimate the superconducting energy gap $\mathscr{E}_g = 2\Delta$ at room temperature (300 K).

(b) Up to what characteristic frequency (and equivalent wavelength) will this superconductor exclude electromagnetic fields?

(c) If a Josephson junction is made using this superconductor, up to approximately what voltage might it operate as an ideal Josephson element?

1.4. Magnetic levitation. A permanent magnet in the form of a disk 1 cm in diameter is levitating 2 mm above a superconductor plane. The magnetic field at the pole of the magnet is 0.2 T. The levitation is observed for a period of 24 h, with no indication of the height decaying (to at least 10% precision).

(a) If the magnet has a mass of 2 g, what is the levitation force?

(b) Estimate the magnitude of the screening currents needed to exclude the field and levitate the magnet.

(c) Estimate roughly the maximum resistivity of the superconductor given the imperceptible decay.

1.5. SQUID and Josephson junctions. A superconducting loop with area 1 cm² contains two Josephson junctions, each with critical current $I_c = 100\ \mu\text{A}$ and normal shunt resistance $R = 1\ \Omega$, biased as a dc SQUID.

(a) If the earth's magnetic field is 50 μT perpendicular to the loop, how many flux quanta Φ_0 are contained within the loop?

(b) Assume now that the SQUID is screened from the earth's field but permits smaller applied fields to couple in. How small a magnetic field would correspond to a flux quantum of $\tfrac{1}{2}\Phi_0$? Estimate the change in voltage ΔV for such a SQUID biased with a current just above $2I_c$ (see Fig. 1.10).

(c) If the measurement system has a noise level of $\approx 1\ \mu\text{V}$, estimate the minimum sensitivity of this SQUID to small changes in flux and magnetic field.

2
AC PROPERTIES AND SUPERCONDUCTING ENERGY GAP

In this chapter, we present the fundamental electrical properties of superconductors, focusing on circuit models of ideal lossless inductors. Section 2.1 begins with the important concept of kinetic inductance, which leads directly to an electromagnetic skin depth that is independent of frequency. Section 2.2 introduces a further refinement of the model to include a shunt resistance, which contributes a small rf surface resistance at high frequencies. We go on to apply this to low-loss superconducting rf transmission lines and filters in Section 2.3. The nature of the superconducting energy gap is briefly discussed in Section 2.4, with attention to how this gap leads to zero resistance at low temperatures. Finally, we show how this brings about highly nonlinear current–voltage characteristics of superconducting tunnel junctions.

2.1 KINETIC INDUCTANCE AND PENETRATION DEPTH

Kinetic Inductance

The usual self-inductance L in a circuit is associated with energy stored in the magnetic field produced by the electrical current. But there is another mechanism for storage of energy directly in the motion of the charge carriers, which gives rise to a "kinetic inductance." This kinetic inductance is present in normal metals as well as in superconductors, but it only becomes significant in superconductors when the series resistance goes to zero.

One can see the basis for this effect by considering the standard derivation of electrical conductivity in the presence of collisions. For normal electrons with a density n in the absence of electric field E, there is a wide distribution of velocities, but the average velocity (the drift velocity v) is zero. This is maintained by collisions with fixed ions that quickly relax any net nonzero velocity, with a characteristic collision time (or relaxation time) τ, typically of order 0.1 ps. Since an electric field accelerates the electrons, we have the combined equation

$$\frac{dv}{dt} = -\frac{eE}{m} - \frac{v}{\tau} \tag{2.1}$$

where m is the electron mass and $-e$ its charge. This can be rewritten in terms of the current density $J = -nev$ as

$$E = \left(\frac{m}{ne^2\tau}\right)J + \left(\frac{m}{ne^2}\right)\frac{dJ}{dt} = \rho J + \Lambda\frac{dJ}{dt} \tag{2.2}$$

This is schematically represented in Fig. 2.1a as an ideal voltage source in series with a resistor and an inductor. The first term is simply Ohm's law, with the usual resistivity ρ in Ohm-meters. Similarly, we can regard Λ as the "kinetic inductivity" of the conductor; for a carrier density $n \sim 10^{28}$ electrons/m^3, this corresponds to $\Lambda \sim 10^{-20}$ H-m. The frequency at which this inductivity becomes important is $R/L = \rho/\Lambda = 1/\tau \sim 10$ THz, so that the simple Ohm's law is valid across the entire common range of frequencies.

However, for the superconductor, if we let $\tau \to \infty$ ($\rho = 0$), then this crossover frequency is driven down to zero, and the kinetic inductance is of critical importance even for dc! We have commented that the fundamental charge carrier in a superconductor actually consists of a pair of electrons with charge $-2e$, but this does not affect the kinetic inductance, since $\Lambda = (2m)/(\frac{1}{2}n)(2e)^2 = m/ne^2$. One can also view this from energy considerations. The power (per unit volume) going into the superconducting electrons is $P = \mathbf{E} \cdot \mathbf{J} = d/dt(\frac{1}{2}\Lambda J^2)$. Because energy cannot be dissipated in a superconductor, this corresponds to an energy density $\mathscr{E} = \frac{1}{2}\Lambda J^2 = \frac{1}{2}nmv^2$ stored in the kinetic energy of the Cooper pairs. This is directly analogous to the usual term for energy storage in an inductor. It is important to reiterate here that this kinetic inductance does not replace the usual magnetic inductance; it is an additional term.

Transmission Line Picture

The picture of Fig. 2.1a is correct, but it is not complete, since we have left out the magnetic field H produced by the current and the effect of changing magnetic field on the electric field E. These relations are, of course, given by Maxwell's equations

$$\mathbf{\nabla} \times \mathbf{E} = -\frac{\partial \mathbf{B}}{\partial t} = -\mu_0\frac{\partial \mathbf{H}}{\partial t} \quad \text{and} \quad \mathbf{\nabla} \times \mathbf{H} = \mathbf{J} + \frac{\varepsilon\partial \mathbf{E}}{\partial t} \tag{2.3}$$

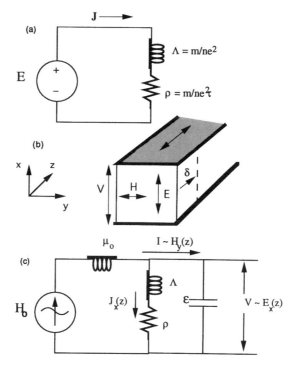

Figure 2.1. Circuit representation of electromagnetic fields in a conductor. (*a*) Relation of electric field *E* to current density *J*, including kinetic inductance Λ. (*b*) Geometry of plane electromagnetic wave propagating into the surface of a conductor and idealized transmission line that guides this wave. (*c*) Distributed circuit picture of idealized transmission line representing electromagnetic field equations in conductor, with current along the line representing magnetic field *H* and voltage across the line representing electric field *E*, penetrating a skin depth δ into the surface.

where $\varepsilon = \varepsilon_r \varepsilon_0$ is the permittivity of the medium and μ_0 is the permeability of free space. Consider a plane wave propagating in the z direction, as shown in Fig. 2.1*b*. For simplicity, consider a region with a square cross section (of side a) in the x–y direction, with **E** and **J** in the x direction and **H** in the y direction. The equations are then

$$-\frac{\partial E_x}{\partial z} = \mu_0 \frac{\partial H_y}{\partial t} \qquad -\frac{\partial H_y}{\partial z} = J_x + \varepsilon \frac{\partial E_x}{\partial t} \qquad (2.4)$$

One can conceptually place ideal electrodes on the top and bottom of this square and then model the system as a simple ideal transmission line in Fig. 2.1*c*, with a series distributed inductance per unit length $L = \mu_0$, a shunt capacitance per unit length $C = \varepsilon$, and a shunt admittance per unit length given by the J–E relation of Eq. (2.2) or Fig. 2.1*a*. (See Appendix A for a review of transmission lines.) Then, the wave in the medium will be carried by currents in and voltages between these ideal electrodes; in fact, $I = H_y a$ and $V = E_x a$. Furthermore, one can model the boundary condition with an ideal ac voltage or current source that specifies the value of E or H at the surface.

For a general transmission line (Fig. A.1) with series impedance per unit length Z and shunt admittance per unit length Y, the characteristic impedance of the line is $Z_0 = \sqrt{Z/Y}$ (the effective input impedance V/I of a long line), and the propagation constant is $\gamma = \sqrt{ZY}$ [the wave propagates as $\exp(-\gamma z)$]. For an insulator with $\rho \to \infty$, this gives $Z_0 = \sqrt{\mu_0/\varepsilon}$ and $\gamma = j\omega\sqrt{\mu_0\varepsilon} = j\omega\sqrt{\varepsilon_r}/c$ ($c = 1/\sqrt{\mu_0\varepsilon_0}$ is the speed of light in free space), which corresponds to the usual electromagnetic wave propagating with velocity $c/\sqrt{\varepsilon_r}$ through a dielectric medium with intrinsic impedance (E/H) of $377\,\Omega/\sqrt{\varepsilon_r}$. (We are using the engineering notation for complex numbers, where $j = \sqrt{-1}$.)

For a good conductor at typical frequencies, both Λ and ε can be neglected. (Note, however, that this picture also yields the plasma frequency $1/\sqrt{\Lambda\varepsilon}$, above which γ changes from imaginary to real and the conductor becomes transparent to electromagnetic waves. For most metals, this happens in the ultraviolet.) In that case, the admittance becomes simply $Y = 1/\rho$, and we have

$$\gamma = \sqrt{\frac{j\omega\mu_0}{\rho}} = \frac{1+j}{\delta} \tag{2.5}$$

where $\delta = \sqrt{(2\rho/\omega\mu_0)}$ is the skin depth, and

$$Z_0 = \sqrt{j\omega\mu_0\rho} = (1+j)\sqrt{\tfrac{1}{2}\omega\mu_0\rho} = Z_s = R_s + j\omega L_s \tag{2.6}$$

where $R_s = \rho/\delta$ is the surface resistance of the conductor and $L_s = \tfrac{1}{2}\mu_0\delta$ is the surface inductance, since $Z_s = E/H = V/I$ for this line. The significance of this is clear; the fields and currents are confined within an exponential decay length δ. The surface resistance corresponds to resistance from this top layer of the conductor, and the surface inductance reflects the penetration of magnetic field into the conductor on this same scale. Of course, at dc, $\delta \to \infty$ and $Z_s \to 0$, corresponding to currents and fields penetrating completely into the conductor.

Magnetic Penetration Depth

For the superconducting case, if we were simply to take $\rho = 0$ in Eq. (2.5), then we would obtain $\delta = 0$, and the fields and currents would not penetrate at all into the superconductor. In contrast, if we include the kinetic inductance term, we can replace the shunt resistor in the transmission line with a shunt inductor

22 AC PROPERTIES AND SUPERCONDUCTING ENERGY GAP

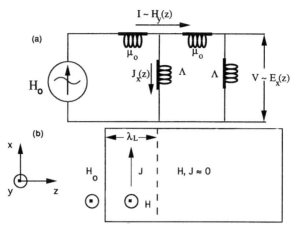

Figure 2.2. Electromagnetic field penetration into the surface of a superconductor. (*a*) Distributed circuit picture of currents in superconductor, with magnetic field H_0 at surface. (*b*) Cross-sectional view of screening currents within $\lambda_L = \sqrt{\Lambda/\mu_0}$ of surface.

Λ and arrive at the model of Fig. 2.2a. In the time domain, this model corresponds to the pair of equations

$$-\frac{\partial E_y}{\partial z} = \mu_0 \frac{\partial H_x}{\partial t} \qquad -\frac{\partial H_x}{\partial z} = J_y = \frac{1}{\Lambda} \int E_y \, dt \qquad (2.7)$$

which are, of course, Maxwell's equations for transverse electromagnetic waves in this perfectly conducting medium. In the frequency domain, the shunt admittance is then $Y = 1/j\omega\Lambda$, and we have

$$\gamma = \sqrt{\frac{\mu_0}{\Lambda}} = \frac{1}{\lambda_L} \qquad (2.8)$$

where

$$\lambda_L = \sqrt{\frac{\Lambda}{\mu_0}} = \sqrt{\frac{m}{\mu_0 n e^2}} \qquad (2.9)$$

is called the magnetic penetration depth (or London penetration depth) and is effectively the skin depth of the superconductor (not to be confused with a wavelength), and

$$Z_0 = j\omega\sqrt{\mu_0\Lambda} = j\omega\mu_0\lambda_L = j\omega L_s \qquad (2.10)$$

is the surface impedance of the superconductor. Note that the inclusion of the kinetic inductance term has led to penetration of the currents and fields a nonzero distance λ_L [$J, H \sim \exp(-z/\lambda_L)$, where z is the distance from the surface] independent of ω! This is in contrast to the resistive case, where the skin depth diverges at $\omega = 0$. The magnitude of λ_L in real metals is typically rather small, of order 100 nm. The superconductor still excludes a magnetic field, but the surface screening currents are spread out over a thickness of order λ_L, as one can see from the transmission line picture of Fig. 2.2a. Note that the total series inductance of order $\mu_0 \lambda_L$ is equal to the total shunt inductance Λ/λ_L. For any other distance, the combined total inductance would be larger.

Furthermore, the surface impedance does not include any resistance, but it does include an inductive term. It may not be immediately obvious from Eq. (2.10), but L_s includes equal contributions $\frac{1}{2}\mu_0 \lambda_L$ attributable to the kinetic inductance and to the magnetic inductance associated with the penetrating field. This is understandable in terms of the transmission line picture; half of the energy is stored in the series magnetic inductors, while the other half is stored in the shunt kinetic inductors.

Of course, a superconductor will exclude not only a time-varying magnetic field but also a static field, and can maintain a supercurrent indefinitely. It may seem odd to view this as merely the zero-frequency limit of an electromagnetic wave, since $E = 0$. However, it is quite consistent; for $\omega = 0$, the impedance $E/H = 0$ for the superconductor, so that a static magnetic field can be present in the absence of an electric field. Furthermore, the current density J is nonzero; since $J/E = 1/j\omega\mu_0\lambda_L^2$ and $E/H = Z_0 = j\omega\mu_0\lambda_L$, we have $J = H/\lambda_L$ for all frequencies, including dc. Both decrease as $\exp(-z/\lambda_L)$ as one goes into the surface (Fig. 2.2b).

It is also instructive to consider this in the time domain. Suppose we turn on a magnetic field near a superconductor. This will excite screening currents in the surface of the superconductor (up to a depth λ_L), which will act to screen out the magnetic field indefinitely, since there is no resistance. This is in contrast to a normal metal, for which resistive losses cause the eddy currents to decay, thus permitting the magnetic field to penetrate further into the conductor. This corresponds to a current going a large distance along the series lossless inductor in Fig. 2.1c.

An ideal classic superconductor always excludes a magnetic field from its interior. This is known as the Meissner effect, and we will discuss it further in the next chapter. It is important to understand that this transmission line picture does *not* automatically incorporate the Meissner effect. A uniform magnetic field penetrating into the superconductor (represented by a spatially constant dc current along the series inductor in Fig. 2.2) is perfectly consistent with this picture. If one started with these initial conditions, then removing the external field (setting $I = 0$ at the boundary) would result in a screening current (of depth λ_L) to *maintain* the internal magnetic field. Furthermore, a residual dc circulating current in any closed inductive loop in Fig. 2.2 is also consistent. All of these represent the possibility of trapped flux inside the superconductor. In fact, trapped flux does occur in superconductors under some conditions, but for now, we will assume that the initial state always has $H = 0$ and $J = 0$ in the superconductor. Then, based on our analysis, the magnetic flux will not penetrate the superconductor as long as the temperature is maintained below T_c.

Thin Superconducting Film

We can use the transmission line picture of Fig. 2.2 to consider a thin superconducting film of thickness $d < \sim \lambda_L$, but now we need to consider the boundary conditions on both sides, which in general may be independent. Adapting the relations for a general transmission line (Appendix A), we can express the spatial dependence of $H(z)$ and $J(z)$ in terms of the values H_1 and H_2 on the two sides (Fig. 2.3a). If we take $z = -\frac{1}{2}d$ at port 1 and $z = +\frac{1}{2}d$ at port 2, we have

$$H(z) = \frac{H_1 + H_2}{2} \frac{\cosh(z/\lambda_L)}{\cosh(d/2\lambda_L)} + \frac{H_2 - H_1}{2} \frac{\sinh(z/\lambda_L)}{\sinh(d/2\lambda_L)} \tag{2.11}$$

$$J(z) = \frac{H_1 + H_2}{2\lambda_L} \frac{\sinh(z/\lambda_L)}{\sinh(d/2\lambda_L)} + \frac{H_2 - H_1}{2\lambda_L} \frac{\cosh(z/\lambda_L)}{\cosh(d/2\lambda_L)} \tag{2.12}$$

Here we are following the sign convention that H is positive for I pointing to the right in Fig. 2.3b, which corresponds to sources that point up on the left and down on the right. Also, J is positive for shunt current pointing up. We can also relate the terminal characteristics of this length of transmission line using either the T or the Π lumped-element equivalent (Figs. 2.3c and d).

The boundary conditions and resulting field and current distributions will be distinctly different for several different physical situations, which we present in Fig. 2.4 for the limit where $d \ll \lambda_L$. If the field is confined on one side of the superconducting film (Fig. 2.4a), we have $H = H_0$ on one side but $H = 0$ on the other, corresponding to an open circuit. In this case, H decreases linearly from H_0 to zero, while $J = H_0/d$, much greater than the maximum screening current density $H_0 = \lambda_L$ for the thick-layer case. This corresponds to an input impedance $Z_{in} = j\omega\mu_0\lambda_L^2/d$, or a surface inductance $L_s = \mu_0\lambda_L^2/d$ that is enhanced by a factor of λ_L/d above that for the thick superconductor. This is exclusively kinetic inductance; the magnetic inductance component is reduced to only $\frac{1}{2}\mu_0 d$. More generally, we have

$$L_s = \mu_0\lambda_L\left[\tanh\left(\frac{d}{2\lambda_L}\right) + \frac{1}{\sinh(d/\lambda_L)}\right] = \mu_0\lambda_L\coth\left(\frac{d}{\lambda_L}\right) \tag{2.13}$$

The second case (Fig. 2.4b) where H changes sign on the two sides ($H_2 = -H_1 = -H_0$) is similar. This corresponds physically to the situation where the superconducting film is biased with a net (dc or ac) current. In the limit that $d \ll \lambda_L$, the current density J will be uniform, with $J = H_0/2d$, and H will change linearly from $+H_0$ to $-H_0$, passing through $H = 0$ at the center. The input inductance on each side corresponds to $\mu_0\lambda_L\coth(d/2\lambda_L)$, approaching

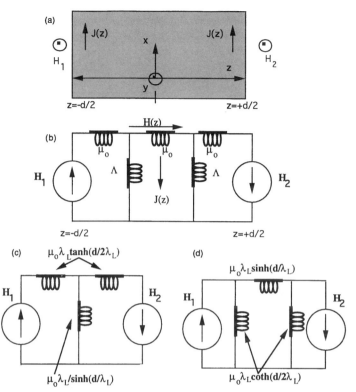

Figure 2.3. Representation of superconducting thin film of thickness d, centered around $z = 0$, with boundary conditions (magnetic field parallel to the surface) of H_1 on the left and H_2 on the right. (*a*) Geometry of currents and fields. (*b*) Inductive transmission line representation of thin film. Note that current sources representing positive H point up on the left and down on the right. (*c*) T lumped-element equivalent and (*d*) Π lumped-element equivalent of transmission line.

$2\mu_0\lambda_L^2/d$ in the thin-film limit. From the point of view of a current source feeding the film, the currents on the two sides are in parallel, so that the total surface inductance that would be measured would again be $\mu_0\lambda_L^2/d$ for $d \ll \lambda_L$. The total inductance would, of course, still include the magnetic inductance associated with fields outside the film, but the kinetic inductance can sometimes be dominant in a sufficiently thin film.

The third case (Fig. 2.4c) has $H = H_0$ on both sides of the film and corresponds to the film being immersed in an applied magnetic field H_0. For a bulk superconductor, of course, the field will be excluded except for a screening length λ_L on both sides, with screening current density $J = \pm H_0/\lambda_L$. In the thin-film limit, H is uniform and J changes sign linearly from $-H_0 d/2\lambda_L^2$ to $+H_0 d/2\lambda_L^2$, much smaller than in the case of the thicker film. More generally

$$H = \frac{H_0 \cosh(z/\lambda_L)}{\cosh(d/2\lambda_L)} \qquad J(z) = \frac{(H_0/\lambda_L)\sinh(z/\lambda_L)}{\sinh(d/2\lambda_L)} \qquad (2.14)$$

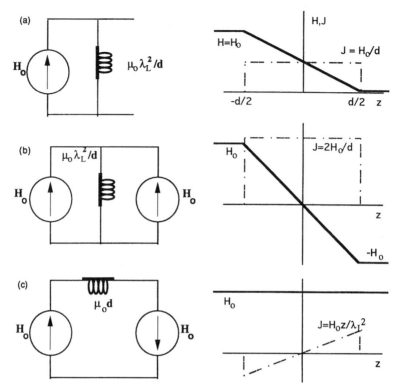

Figure 2.4. Field and current distributions (and effective inductances) for superconducting thin film with $d \ll \lambda_L$ for H_1 and H_2 given: (a) $H_1 = H_0$, $H_2 = 0$; (b) $H_1 = H_0 = -H_2$; (c) $H_1 = H_2 = H_0$.

Note also that the effective input impedance on the two sides is greatly reduced;

$$Z_{in} = j\omega\mu_0\lambda_L \tanh\left(\frac{d}{2\lambda_L}\right) \to \frac{j\omega\mu_0 d}{2} \quad \text{for } d \ll \lambda_L \quad (2.15)$$

In this limit, the kinetic inductance is barely excited; this is primarily the magnetic inductance associated with flux penetration.

Other Examples of Current Distribution

Let us consider several examples that illustrate the effects of kinetic inductance on the current distribution. Consider first a cylindrical superconducting wire of radius $a \gg \lambda_L$. If we supply a current to the wire, the current will flow only within λ_L of the surface of the wire, producing a circumferential magnetic field at the surface $H = I/2\pi a$ (see Fig. 2.5). This is similar to the skin effect in a normal wire at high frequencies, but in superconductors it is present at dc as well. The contribution to the inductance due to the penetration depth is $\mu_0\lambda_L/2\pi a$ per

2.1 KINETIC INDUCTANCE AND PENETRATION DEPTH

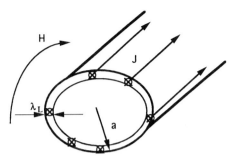

Figure 2.5. Current and field distribution in a cylindrical wire of radius $a \gg \lambda_L$, being fed by a current I. The currents and fields penetrate only a distance λ_L from the surface.

unit length. This is not typically a significant contribution to the total inductance of the wire, since the external magnetic inductance per unit length is $\approx (\mu_0/2\pi)\ln(b/a)$, where $b \gg a$ is the distance to the return current.

Consider next a parallel-plate transmission line with two superconducting electrodes of thickness d and width w separated by an insulating dielectric of thickness h with permittivity ε, as indicated in Fig. 2.6. The electromagnetic wave is now directed along the line, rather than into the superconductor, so that in the representation of this transmission line the surface inductance of the superconductor is in series with the external magnetic inductance between the electrodes (Fig. 2.6c). Neglecting edge effects, the fields are confined between the two electrodes, so that we have the boundary condition that $H = 0$ on the outer edge of the superconducting films. Within this approximation, this is the same as the microstrip geometry (Fig. 2.6b), with a superconducting film above a wider ground plane. Then, the capacitance per unit length is $C = \varepsilon w/h$ in the usual way, but the inductance per unit length now includes a component L_i due to the surface inductance as well as the usual external inductance L_e:

$$L = L_e + L_i = \frac{\mu_0 h}{w} + \frac{2L_s}{w} = \frac{\mu_0}{w}\left[h + 2\lambda_L \coth\left(\frac{d}{\lambda_L}\right)\right] \approx \frac{\mu_0 h}{w}\left(1 + \frac{2\lambda_L}{h}\right) \quad (2.16)$$

where the factor of 2 comes from the contributions of both electrodes in series (assumed to have the same values of λ_L and d) and the latter approximation assumes that $d \gg \lambda_L$. These values give rise to the characteristic impedance

$$Z_0 = \sqrt{\frac{L}{C}} = \sqrt{\frac{\mu_0}{\varepsilon}\frac{h}{w}}\sqrt{1 + \frac{2\lambda_L}{h}} \quad (2.17)$$

and the propagation constant

$$\gamma = \alpha + j\beta = j\omega\sqrt{LC} = j\omega\sqrt{\mu_0\varepsilon}\sqrt{1 + 2\lambda_L/h} \quad (2.18)$$

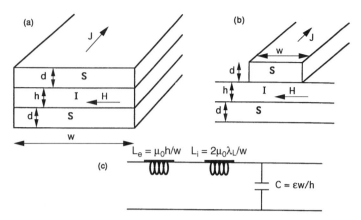

Figure 2.6. Superconducting transmission line, composed of superconducting electrodes S of thickness d separated by insulator I of thickness h and permittivity ε. (a) Parallel-plate transmission line. (b) Microstrip transmission line, equivalent to (a) if edge effects can be neglected. (c) Distributed circuit representation of lines in (a) and (b), with total inductance per unit length $L = L_e + L_i$ and capacitance per unit length C.

This is notable because it exhibits no attenuation ($\alpha = 0$) and no dispersion; the group velocity

$$v_g = \frac{1}{\partial \beta / \partial \omega} = \frac{c}{\sqrt{\varepsilon_r}} \bigg/ \sqrt{1 + 2\lambda_L/h} \qquad (2.19)$$

is independent of frequency ($c = 1/\sqrt{\varepsilon_0 \mu_0}$ is the speed of light in free space and $\varepsilon_r = \varepsilon/\varepsilon_0$ is the dielectric constant). This means that a pulse will maintain both its height and its shape on propagating down such a line. (Real superconducting transmission lines do have some losses at high frequencies, as described in the next section, but the losses can be extremely small.) The correction to the free-space velocity can be significant if $h < \lambda_L$, which is not hard to achieve using a thin-film dielectric. Further slowing of the wave can be obtained if the superconducting electrodes are thin films with thickness $d \ll \lambda_L$.

As a final example, consider a thin film of width $w \gg \lambda_L > d$ that is not close to a ground plane or other return path. From the one-dimensional analysis we have been using, one might expect that the current density would then be uniform through the film. However, although it may be uniform across the thickness, it is in general *not* uniform across the width. In particular, the current density peaks sharply at the edges of the film, where the magnetic field is perpendicular to the film and is also large (see Fig. 2.7). This is directly analogous to the surface charge distribution of a charged conductor of similar shape; the charge density would also be peaked at the edges. In this configuration, the field penetration at the edges of a film with $d \ll \lambda_L$ is governed by an effective penetration depth (also called the transverse penetration depth for a thin film) $\lambda_\perp = \lambda_L^2/d$ (Tinkham, 1996, p. 101). This problem does not have an analytic

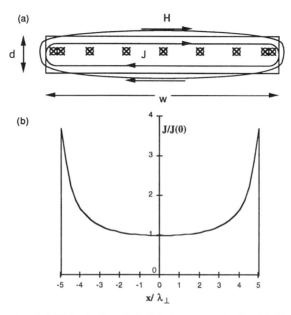

Figure 2.7. Current and field distributions in isolated superconducting thin film with thickness $d < \lambda_L$, with net transport current I. (a) Rough schematic of current density J (into the paper) and lines of magnetic field H circulating around it, showing concentration of perpendicular fields near the edges. (b) $J(x)$ from Eq. (2.20), for $w = 10\lambda_\perp$, where $\lambda_\perp = \lambda_L^2/d$ is the effective transverse penetration depth for a very thin film.

solution, but an approximate solution (Van Duzer, 1981, p. 107) is given by

$$J(x) = \begin{cases} \dfrac{J(0)}{\sqrt{[1-(2x/w)^2]}} & |x| < x_0 \\ \dfrac{7}{6}J(0)\sqrt{\dfrac{w}{\lambda_\perp}}\exp\left(-\dfrac{w/2-|x|}{\lambda_\perp}\right) & x_0 < |x| < \dfrac{w}{2} \end{cases} \quad (2.20)$$

with a connection between the two expressions at $\pm x_0 = \pm \frac{1}{2}(w - \lambda_\perp)$, where $J(0)$ is the current in the center of the film. This expression is valid in the limit that $w \gg \lambda_\perp$, where the second expression serves to cut off the divergence at the edge. This is shown in Fig. 2.7b for $w = 10\lambda_\perp$. This might correspond, for example, to $d = 10$ nm, $\lambda_L = 100$ nm, $\lambda_\perp = 1$ μm, and $w = 10$ μm. In the other limit that $\lambda_\perp > w$ (i.e., $wd < \lambda_L^2$), the current density is essentially uniform across the width, as well as through the thickness, with a surface inductance $L_s = \mu_0\lambda_L^2/d = \mu_0\lambda_\perp$. Alternatively, the presence of a nearby superconducting groundplane (which generates opposite screening currents) tends to suppress the peaking of currents at the edge, so we have a configuration closer to that in Fig. 2.6 with similarly uniform currents.

2.2 TWO-FLUID MODEL AND RF SURFACE RESISTANCE

The preceding analysis suggests that we can determine the kinetic inductivity Λ and magnetic penetration depth λ_L directly from well-defined microscopic parameters, but the situation is somewhat more complicated. For a typical metallic carrier density of order 10^{28} electrons/m^3, the free-electron values give $\lambda_L = (m/\mu_0 n e^2)^{0.5} \sim 50$ nm. This is the right order of magnitude but is a bit on the low side; 100 nm is more typical. There are two main factors that tend to increase the penetration depth: temperature and electron scattering. Furthermore, for some materials, such as the high-T_c cuprates, the anisotropic nature of the material gives rise to an anisotropic penetration depth. But if we regard λ as a semiempirical macroscopic parameter for a given superconducting sample (with the subscript L dropped), we can continue to make use of most of what we have been discussing. We will treat the effect of temperature first, since that is the most fundamental.

Two-Fluid Picture

Superconductivity is a low-temperature phenomenon; it exists only below a critical temperature T_c. But even below T_c, superconductivity does not exist at "full strength." The transition occurs gradually and continuously; in the parlance of thermodynamics, this is known as a "second-order phase transition." We can quantify this in terms of an effective superelectron density n_s, which represents that portion of the total electron density n that contributes to superconductivity. The remainder of the electrons $n_n = n - n_s$ are (to a first approximation) still "normal" (i.e., resistive). One can think of these as two interpenetrating electronic fluids, each of which can independently carry current. So we have

$$J = J_s + J_n = n_s e v_s + n_n e v_n = \left(\frac{1}{\Lambda}\right)\int E\, dt + \frac{E}{\rho} \qquad (2.21)$$

or equivalently,

$$E = \Lambda \frac{dJ_s}{dt} = \rho J_n \qquad (2.22)$$

which can be represented by the parallel kinetic inductivity $\Lambda = m/n_s e^2 = \mu_0 \lambda^2$ of the superconducting electrons and the normal resistivity $\rho = m/n_n e^2 \tau$ of the normal electrons, as indicated in Fig. 2.8a. [The two channels could alternatively be combined into an effective complex conductivity $\sigma(\omega) = 1/\rho - j/\omega\Lambda$.] For dc, of course, the parallel resistive channel will be shorted out by the lossless superconducting channel, so that our previous derivations of magnetic penetration should remain valid. But at higher frequencies, some of the current will flow through the resistor, leading to resistive loss in the "superconductor."

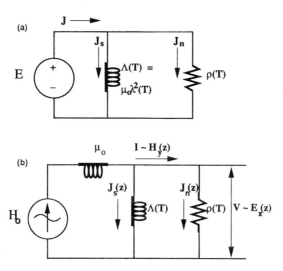

Figure 2.8. Circuit representation of two-fluid model for superconductor. (*a*) Relation between electric field and parallel currents J_s and J_n. (*b*) Transmission line picture for distributed electromagnetics of two-fluid superconductor.

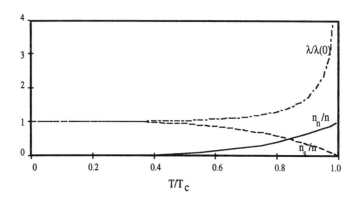

Figure 2.9. Conventional temperature dependence in the two-fluid model of superconducting and normal electron densities $n_s(T)$ and $n_n(T)$ and the magnetic penetration depth $\lambda(T)$, which diverges as $T \to T_c$.

Furthermore, these quantities are also now functions of temperature. The most common empirical form for the temperature dependence of n_s and n_n (see Fig. 2.9) is given by

$$n_s = n - n_n = n\left[1 - \left(\frac{T}{T_c}\right)^4\right] \tag{2.23}$$

This dependence, which goes smoothly from 0 at T_c to 1 at $T = 0$, is often called the "two-fluid" temperature dependence, although strictly speaking, the two-fluid model does not require any particular dependence on temperature. This predicts that the magnetic penetration depth diverges near T_c as

$$\lambda(T) = \frac{\lambda(0)}{\sqrt{1-(T/T_c)^4}} \approx \frac{\lambda(0)}{\sqrt{2(1-T/T_c)}} \qquad (2.24)$$

and this divergence is also in accord with other more microscopic theoretical approaches as well as with experimental measurements. Similarly, if we assume that the collision time τ is independent of temperature in this range, we have

$$\rho(T) = \rho(T_c)\left(\frac{T_c}{T}\right)^4 \qquad (2.25)$$

which diverges as $T \to 0$. This is probably a reasonable first approximation, although the microscopic BCS theory suggests that n_n should go to zero exponentially at low temperature rather than as a power law.

Effects of Scattering

Before going on to explore some of the consequences of the two-fluid picture, let us first comment on the effect of impurity scattering on n_s. One might initially suppose that such scattering affects only the normal electrons, since superconducting electrons are not subject to scattering. However, the situation is more complicated; in a "dirty metal"—a material with a short normal-electron scattering length (or mean-free path) l—the effective value of Λ turns out to increase in proportion to l [see, e.g., Orlando and Delin (1991, p. 535) and Van Duzer and Turner (1981, p. 290)]. More specifically, in the "dirty limit" for $l < \xi_0$ (the BCS coherence length, an intrinsic characteristic length to be discussed later at the end of the chapter), the dependence of Λ_d and λ_d take the approximate forms

$$\Lambda_d(T) \approx \Lambda_c(T)\frac{\xi_0}{l} \qquad \lambda_d(T) \approx \lambda_c(T)\sqrt{\frac{\xi_0}{l}} \qquad (2.26)$$

where the "clean limit" values λ_c and Λ_c are the same as those described earlier. This increase in $\Lambda = m/n_s e^2$ for the dirty limit is conventionally attributed to an effective decrease in n_s, although (in contrast to the thermal effect) this does *not* lead to a corresponding increase in n_n. So perhaps an alternative attribution in terms of an increased effective mass m_s might be more consistent. Furthermore, since l depends on the impurity concentration, λ and Λ may vary significantly among different samples of the same nominal material. However, if we express formulas in terms of the experimentally determined values of λ, we should have a self-consistent picture.

Since the magnetic penetration depth $\lambda(T)$ is a critical parameter for a superconductor that may vary depending on sample quality, it is important to be able to measure it directly. This can be accomplished by measuring the effective magnetization (or ac magnetic susceptibility) of a small sample, or more commonly, changes in the magnetization as a function of temperature. Alternatively, either rf impedance measurements may be made to determine the resonant frequency of a superconducting transmission line resonator (which depends on λ through the surface inductance $L_s = \mu_0 \lambda$, which in turn determines the phase velocity) or the kinetic inductance contribution to the total inductance of a stripline or film may be measured directly.

Rf Surface Resistance

Returning to the parallel LR circuit that describes the response of the two-fluid superconductor to an electric field, the inductivity Λ will dominate the behavior up to a very high frequency $\sim \rho/\Lambda = \mu_0 \lambda^2$. Taking typical values $\lambda \sim 100$ nm and $\rho \sim 10^{-7}$ Ω-m, we obtain a crossover frequency of order 10^{13} Hz. So if we take into account the actual value of $\lambda(T)$ for a given sample, the earlier results for current distributions and kinetic inductance will continue to hold through the entire microwave range. (Actually, the two-fluid model does not apply above a critical frequency $f_c = 2\Delta/h \sim 10^{12}$ Hz associated with the energy gap; at higher frequencies, normal-metal properties apply.)

But one parameter will be strongly affected by this shunt resistivity, the surface resistance, which is strictly zero only for dc. Many rf applications of superconductors depend on a very small surface resistance, since excess resistance could lead to attenuation on transmission lines and loss in resonant structures. It is therefore critical to know just how small the rf surface resistance is. We can obtain this directly from the transmission line picture for the characteristic impedance (Fig. 2.8b). For a thick superconductor ($d \gg \lambda$), the surface impedance takes the form

$$Z_s = Z_0 = \sqrt{\frac{Z}{Y}} = \sqrt{j\omega\mu_0/\sigma(\omega)} = \frac{j\omega\mu_0\lambda}{\sqrt{1 + j\omega\mu_0\lambda^2/\rho}} \quad (2.27)$$

If we take the "low-frequency" approximation $\omega \ll \rho/\mu_0\lambda^2 = n_s/n_n\tau \sim 10^{13}$ s^{-1}, this becomes

$$Z_s = R_s + j\omega L_s \approx j\omega\mu_0\lambda + \frac{\omega^2\mu_0^2\lambda^3}{2\rho} \quad (2.28)$$

This is to be compared with Eq. (2.13) for the surface impedance of a normal metal. Clearly, there is a wide range of frequencies (up through the gigahertz range) where the surface resistance of a superconductor is very small (see Fig. 2.10). Note that this ω^2 frequency dependence is very different from that for the surface resistance of a normal metal, which goes as $\omega^{1/2}$. So R_s starts much lower for a superconductor but rises faster. We can also examine the

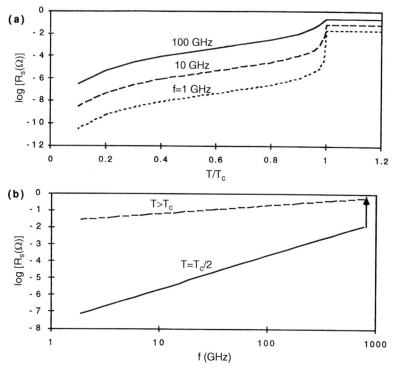

Figure 2.10. The rf surface resistance in the normal and superconducting states, based on the two-fluid model and conventional temperature dependence. Parameters used are $\lambda(0) = 100$ nm, $\rho(T > T_c) = 10^7$ Ω-m, and $f_c = 2\Delta/h = 700$ GHz, which might apply for a niobium thin film. (The film is assumed to be thicker than the relevant skin depth.) (a) Temperature dependence of R_s for values of $f = 1$, 10, and 100 GHz. (b) Frequency dependence, $\sim \omega^2$ for the superconductor and $\sim \omega^{1/2}$ for the normal metal. Note the jump up to normal state properties for $f > f_c$.

propagation constant

$$\gamma = \sqrt{ZY} \approx \frac{1}{\lambda} + \frac{j\omega\mu_0\lambda}{2\rho} = \alpha + j\beta \qquad (2.29)$$

where $\alpha = \text{Re}(\gamma)$ is the attenuation coefficient and $\beta = \text{Im}(\gamma)$ is the phase constant (also called the wave vector **k**). Given that $\beta \ll \alpha$ over the entire relevant frequency range for this model, our previous picture of fields and currents penetrating to a distance λ remains valid.

We can also examine the surface resistance for a thin film. Using the boundary condition that no radiation penetrates to the other side (Fig. 2.4a), in the limit that $d \ll \lambda$, we have

$$Z_s = \frac{1}{Yd} = \frac{1}{d\sigma(\omega)} = \frac{j\omega\mu_0\lambda^2/d}{1 + j\omega\mu_0\lambda^2/\rho} \approx \frac{j\omega\mu_0\lambda^2}{d} + \frac{\omega^2\mu_0^2\lambda^4}{\rho d} \qquad (2.30)$$

where this is in the same low-frequency approximation. Both L_s and R_s are enhanced by the same factor of λ/d with respect to the thick film, and R_s maintains the same ω^2 dependence.

This frequency dependence of $R_s \sim \omega^2$ is rather general; it will occur any time there is a large resistance in parallel with a superconducting inductor. For example, if there are small microscopic spots in a superconducting film that are resistive rather than superconducting (due to impurities or trapped flux), they too would be modeled as parallel resistances and would exhibit a similar dependence of the rf surface resistance. Furthermore, even radiation loss through a thin superconducting film can be modeled in the same way with a similar effect. So at very low temperatures ($T \ll T_c$), there are a number of reasons why a residual rf surface resistance may remain in real samples, despite the fact that the parallel $\rho \to \infty$ in theory.

There are a number of effects in the rf behavior of superconductors that cannot be explained using the classical two-fluid model as we have done here. First, as we have mentioned above, the two-fluid model completely breaks down when photons have enough energy to break a Cooper pair, that is, when $hf > 2\Delta$, the energy gap of the superconductor. These and other quantum corrections to the rf behavior of superconductors have been calculated using the Mattis–Bardeen formulas, based on the microscopic theory of BCS [see, e.g., Van Duzer and Turner (1981, p. 135)]. This approach also gives a correction to the low-temperature dependence of R_s, which should go approximately as $\exp(-\Delta/kT)$. On the other hand, the two-fluid model is still widely used to predict and extrapolate surface resistance of superconductors. This is partly because real superconductors (particularly the newer high-temperature superconductors) do not always follow the detailed predictions of Mattis–Bardeen, either because of nonideal samples or for more fundamental reasons. We will make further use of the two-fluid model as appropriate, since it provides both an intuitive picture and a good first approximation to the high-frequency behavior of practical superconducting materials.

The other key assumption we have made is that the rf surface impedance is independent of the rf power, which is true for low power but breaks down at higher power. In particular, the rf surface resistance of a superconductor tends to rise sharply beyond some critical level of rf power. This is essentially because at high power the critical current or critical field of part of the superconductor is being exceeded, driving it out of the fully superconducting state. This is in contrast to a normal metal, which shows constant surface impedance up to very high powers, at least until the surface of the metal starts to heat up significantly. The nonlinear and critical behavior of superconductors will be discussed in Section 3.2.

2.3 RF TRANSMISSION LINES AND FILTERS

Superconducting Transmission Lines

One class of applications of superconductors takes advantage of the small rf resistance for low-loss transmission lines and high-Q resonating elements.

Superconducting transmission lines not only have very low attenuation, but in addition exhibit low dispersion, which is equally important for the propagation of narrow pulses. High-Q resonating elements can be composed of either thin-film transmission line structures, or alternatively using bulk cavities. We will illustrate these phenomena using a simple microstrip transmission line (neglecting edge effects), as in Fig. 2.11. Here the resistance per unit length is $R = 2R_s(\omega)/w$ (where the factor of 2 comes from contributions of both top and bottom electrodes), and the inductance includes both the external magnetic inductance and the internal (surface) inductance of both superconducting electrodes. For the low-loss case, the attenuation factor per unit length is given by $\alpha = R/2Z_0$, where $Z_0 = \sqrt{L/C} = (377\,\Omega/\sqrt{\varepsilon_r})(h/w)\sqrt{1+2\lambda/h}$ is the characteristic impedance of the (lossless) line (for films with thickness $d \gg \lambda$). (There is also a small imaginary component to the characteristic impedance, $-jR/2\beta$, but this can normally be neglected.)

Consider an example with $Z_0 = 50\,\Omega$ and $w = 2\,\mu$m, which might represent an interconnect in an integrated circuit or multichip module. For $\varepsilon_r \approx 4$, this could be achieved with $h \approx 400$ nm, $\lambda \approx 100$ nm, and superconducting films with $d \approx 200$ nm. This also corresponds to a phase velocity $v = \omega/\beta = c/\sqrt{\varepsilon_r(1+2\lambda/h)} \approx 0.4c$, where $c = 3 \times 10^8$ m/s is the speed of light in a vacuum. For a frequency $f = 100$ GHz, Fig. 2.10 gives $R_s \approx 10^{-4}\,\Omega$ for the superconductor, which gives $R = 100\,\Omega$/m and $\alpha = 1$ Np/m. In other words, the signal amplitude would decay to $1/e = 0.37$ of its initial value in an attenuation length of 1 m. Compare this to a normal-metal system with the same geometry. If we have $\rho = 2 \times 10^{-8}\,\Omega$-m (comparable to copper) in a thin film 0.2 μm thick (of order the skin depth at 100 GHz), then we can approximate $R_s \approx \rho/d = 0.1\,\Omega$. This yields $\alpha = 1000$ Np/m, or an attenuation distance $1/\alpha$ of 1 mm instead of 1 m. So on the millimeter scale, the normal transmission line would be lossy while the superconducting line would exhibit negligible loss.

One can also view this in the time domain by considering the effect of a voltage pulse propagating down a transmission line (Fig. 2.12). Consider for simplicity a Gaussian pulse of the form $V(t) = V(0)\exp(-t^2/2\tau^2)$, where the effective width of the pulse is 2τ. The Fourier transform of this pulse is $V(\omega) \sim \exp(-\tfrac{1}{2}\omega^2\tau^2)$, which rolls off at a characteristic value $\omega = 1/\tau$. If we choose a 2-ps pulse with $\tau = 1$ ps, this contains frequency components up to

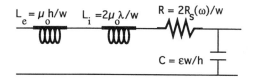

Figure 2.11. Distributed circuit model for microstrip transmission line neglecting edge effects (see Fig. 2.6) with superconducting electrodes separated by dielectric spacer, including resistive loss in superconductors.

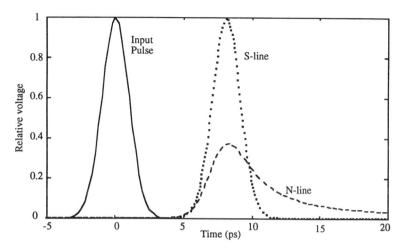

Figure 2.12. Propagation of 2-ps voltage pulse on transmission line with electrodes in superconducting or normal state. The initial Gaussian pulse is shown by the solid line on the left (around $t = 0$), together with the output pulse after propagation along 1-mm-length of a microstrip transmission line [either superconducting (dotted line) or normal (dashed line)], 2 μm wide with an insulating layer 0.4 μm thick (with $\varepsilon_r = 4$). The superconducting line properties are similar to those of a niobium line at $T = T_c$ (see Fig. 2.10), while the normal-line properties are those of copper at ambient temperature. The normal line exhibits both attenuation and dispersion, while the superconducting line shows neither.

$\omega \sim 10^{12}$ rad/s, or f up to ~ 160 GHz. So the propagation characteristics of a 2-ps pulse would be similar to the attenuation characteristics of frequencies on the order of 100 GHz, as described above. This is simulated in Fig. 2.12, where the resulting ouput pulse 1 mm down a transmission line is shown for electrodes in either the superconducting or normal states for parameters as given in the previous paragraph. In this simulation, the Fourier transform of the input pulse is multiplied by $\exp[-\gamma(\omega)l]$ for $l = 1$ mm (the thick-film formulas for γ are used) and then numerically inverse transformed back into the time domain. Note again that the pulse on the superconducting line is propagated without attenuation, while that on the normal line is reduced substantially.

Furthermore, there may also be effects of dispersion, that is, spreading of pulse components of different frequencies. Since the wave goes as $\exp[j(\omega t - \beta z)]$, if the phase velocity $v = \omega/\beta$ is a function of frequency over the relevant range, then different frequency components will propagate at different velocities, and a propagating voltage pulse will not maintain its shape. For a normal-metal electrode, if $d > \delta$ in order to reduce R_s, then the surface inductance $L_s = \frac{1}{2}\mu_0\delta$ will be dependent on frequency and will contribute to variations in $v = 1/\sqrt{LC}$. For the parameters given above, this is a significant effect, as shown in Fig. 2.12; it accounts for the asymmetry in the output pulse corresponding to the long tail for large times. This will be less significant for a thinner film with $d < \delta$, since then $L_s \approx \frac{1}{2}\mu_0 d$ is independent of ω. But then, the attenuation would be somewhat greater. For the superconducting line, on the other hand, $L_s = \mu_0 \lambda$ for all ω, so v will be constant until we get to very high frequencies, and this line will not exhibit significant dispersion. (There are other dispersive effects unrelated to the electrode material, e.g., modal dispersion in coplanar transmission lines, but we will not deal with them here.) Attenuation of picosecond pulses on the millimeter scale is not yet a problem for current digital semiconductor technology, but it is likely to become a concern as circuit speeds and the scale of integration increase. Superconducting transmission lines offer a clear advantage throughout this regime.

Superconducting Resonators

Superconductors are also widely used for high-frequency electromagnetic resonators and filters. The prototypical resonator is of course the lumped LC circuit, with a resonant frequency

$$\omega_0 = 2\pi f_0 = \frac{1}{\sqrt{LC}} \tag{2.31}$$

When constructed using normal-metal conductors, the resistive loss in both L and C can be incorporated in an effective series resistance R, leading to the series LCR circuit of Fig. 2.13a. Conventional inductors also typically use a high-permeability magnetic material, which is usually also somewhat lossy, frequently more so than the wires, due to a combination of eddy current and magnetic hysteresis loss; this also contributes to the effective series R. A resonator is often characterized by its response in the frequency domain. If we consider the magnitude of the effective impedance, we have

$$|Z(\omega)| = \left| j\omega L + R + \frac{1}{j\omega C} \right| = \sqrt{\left(\omega L - \frac{1}{\omega C}\right)^2 + R^2}$$

$$= R\sqrt{1 + Q^2\left(\frac{\omega}{\omega_0} - \frac{\omega_0}{\omega}\right)^2} \tag{2.32}$$

where

$$Q = \frac{\omega_0 L}{R} = \frac{\sqrt{(L/C)}}{R} \tag{2.33}$$

is the dimensionless quality factor of the resonator. In the high-Q limit where $Q \gg 1$, we have

$$|Z| = R\sqrt{1 + N^2} \tag{2.34}$$

where $N = 2(f-f_0)Q/f_0 = 2(f-f_0)/B$ is the number of half-bandwidths off resonance and $B = f_0/Q$ is the bandwidth of the resonator. This dependence is

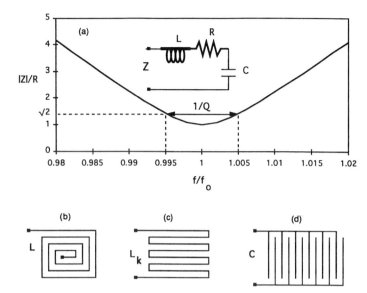

Figure 2.13. Lumped-element series *LCR* resonator, showing compact thin-film inductors and capacitors. (*a*) Frequency dependence of |*Z*| near resonance for $Q = 100$. (*b*) Multiturn square spiral inductor. (*c*) Long meander line for inductor using thin-film superconductor with large kinetic inductance L_k. (*d*) Interdigital capacitor patterned from single layer of thin film.

plotted in Fig. 2.13a (for $Q = 100$) as a function of f and shows a sharp minimum at f_0 with value R, with a characteristic width B at $|Z| = R\sqrt{2}$. The endpoints of this bandwidth are also called the half-power points, since half the maximum power will be dissipated at these points if the resonator is driven by a fixed voltage. Equivalently, in the time domain, the amplitude of the natural oscillation at f_0 will decay as $\exp(-\frac{1}{2}Bt)$, so that a fraction $2\pi/Q$ of the energy will leave the system in one period $1/f_0$. For some applications (e.g., very narrow bandwidth resonant filters) a very high value of Q is desirable. Superconductors, with their extremely small rf resistance, make this possible.

If the aim is to make a low-loss inductor and one goes to the trouble of using superconducting coils, then a magnetic material is normally *not* used; a larger number of turns around a hollow core are used instead, so that the only loss is due to the small surface resistance of the superconductor. A lumped inductor can certainly be made by wrapping wire around a cylinder, but alternatively the multiple turns of a flat spiral inductor can be patterned in a thin film (either rounded or rectangular; see Fig. 2.13b), since the current will only flow near the surface in any case. The measured inductance of a superconducting thin film will include a contribution from the kinetic inductance, particularly if $wd < \lambda^2$. In fact, for a very thin film with large λ and $d \ll \lambda$, we can have a situation where the inductance is completely dominated by the kinetic inductance $\mu_0 \lambda^2 / d$; values as large as 1 nH/square have been reported (Johnson and Kadin, 1997). In this case, a compact inductor can be made using a long meander line (Fig. 2.13c). For example, several centimeters of length of a 2-μm-wide line can be packed into a 100 μm square, leading to a total kinetic inductance of order 10 μH. However, in this thin-film limit, the surface resistance is also enhanced somewhat.

Low-loss dielectric materials for capacitors are easier to come by, so that superconducting capacitors can be made with, for example, aluminum oxide or silicon dioxide insulating layers. A parallel-plate capacitor can be made simply with two layers of a superconductor separated by a thin insulating layer. The thickness is typically greater than about 100 nm to avoid shorts or tunneling currents across the insulator. Alternatively, a design is sometimes used that consists of a single layer of superconductor patterned into a large number of closely spaced interdigital fingers (Fig. 2.13d).

As we go up to higher frequencies, eventually the device size becomes comparable to a significant fraction of a wavelength, and then we must use a distributed- rather than a lumped-element model. For devices on the centimeter scale, this typically occurs for microwave frequencies in the gigahertz range or above. In fact, in this frequency range we must take into account the interloop capacitance of inductors and the series inductance of capacitors; such lumped elements are typically self-resonant. So it makes sense to focus on the design of distributed elements, such as transmission lines, waveguides, and three-dimensional resonant cavities. The same concept of low-loss, high-Q resonators applies in all of these cases (see, e.g., Carr and McAvoy, 1991). We will illustrate the situation with a simple transmission line resonator.

Transmission Line Resonator

Consider a short length l of transmission line that is open on both ends so that waves reflect off the ends and form standing waves. The normal modes of such a resonator have nodes of the current (and antinodes of the voltage) at the two ends and correspond to the modes of a stretched string with an integral number of half-wavelengths along the length:

$$l = \tfrac{1}{2}n\lambda_n = \frac{nv}{2f_n} \tag{2.35}$$

where λ_n is the wavelength at the nth resonant frequency $f_n = \omega_n/2\pi$. Since $v = 1/\sqrt{LC}$, we can rearrange this equation to obtain an equation that is similar (but not identical) to Eq. (2.31) for the lumped resonator:

$$f_n = \frac{n/2}{\sqrt{(Ll)(Cl)}} \tag{2.36}$$

For example, if this is a microstrip with the parameters considered earlier, with $v = 1/\sqrt{LC} = 1.2 \times 10^8$ m/s, then a line $l = 1$ cm long will have resonances at

6 GHz and integral multiples thereof. This line need not be straight; it can be patterned into a more compact meander line as long as the curves are not too abrupt.

Strictly speaking, this picture of standing waves with alternating nodes of V and I is valid only for a completely lossless transmission line. In fact, for such a line, the effective impedance $Z_{\text{eff}} = V/I$ varies between zero at the nodes of V and infinity at the nodes of I. For a slightly lossy line, the picture is similar, except that I and V do not go quite to zero at the nodes, and $|Z_{\text{in}}|$ goes between a small (but nonzero) minimum and a large (but finite) maximum. Note in particular that the input impedance at the end of the line would correspond to one of these maxima.

Of course, a resonator is useful only if we can interact with it externally. In principle we ought to be able to measure the impedance at one of the ends with the other end open. Using the three-terminal lumped-element equivalent of Appendix A (Fig. A.5), this gives

$$Z_{\text{in}} = Z_0 \tanh(\tfrac{1}{2}\gamma l) + \frac{Z_0}{\sinh(\gamma l)} = Z_0 \coth(\gamma l)$$

$$= \frac{Z_0[\sinh(2\alpha l) - j\sin(2\beta l)]}{[\cosh(2\alpha l) - \cos(2\beta l)]} \tag{2.37}$$

For low loss, $|Z_{\text{in}}|$ oscillates between a minimum of Z_0 halfway between the resonances and a maximum of $2Z_0^2/Rl$ on the resonances. This is illustrated in Fig. 2.14 for the 1-cm superconducting transmission line of Fig. 2.12 at $T \approx \tfrac{1}{2}T_c$ with $Z_0 = 50\,\Omega$. Resonances at integral multiples of 6 GHz are clearly evident in the gross frequency scan, although a much finer frequency resolution (in Fig. 2.14b) is needed to measure the bandwidth and peak height accurately. For $T > T_c$ the attenuation would be so large ($\alpha \sim 10\,\text{cm}^{-1}$) that no resonances would be present.

We can also characterize this resonant behavior in terms of the Q factor from the bandwidth at the points where $|Z_{\text{in}}| = R_{\text{peak}}/\sqrt{2}$. As for the lumped-element resonator, $Q_n = f_n/B_n$, where the subscripts correspond to the particular resonance. We can relate Q to the transmission line parameters using the relation

$$Q_n = \frac{\beta}{2\alpha} = \frac{\omega_n L}{R} = \frac{\pi n Z_0}{Rl} = \frac{\pi n}{2} \frac{w}{l} \frac{Z_0}{R_s} \tag{2.38}$$

The values of Q_n decrease as we move to higher order resonances, reflecting the increase in $R_s \sim \omega^2$ from the two-fluid model. In many cases with real samples, the value of $R_s(f)$ may not be known accuraely; this provides a way to determine R_s from a measurement of Q. For the present example for $n = 1$, $B = 250\,\text{kHz}$ gives $Q = 25{,}000$, which in turn implies that $R_s < 1\,\mu\Omega$, consistent with Fig. 2.10 at $T = \tfrac{1}{2}T_c$. This analysis includes only the loss due to resistance in the electrodes. Any other source of loss, such as that due to dielectric loss in

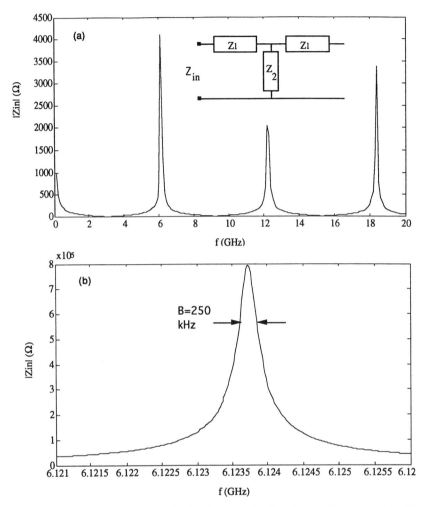

Figure 2.14. Resonance of superconducting transmission line. Parameters include $l = 1$ cm, $w = 2$ μm, $Z_0 = 50$ Ω, velocity $v = 1.2 \times 10^8$ m/s, and superconducting properties corresponding to Fig. 2.10 at $T = \frac{1}{2}T_c$. (a) Gross dependence of input impedance $|Z_{in}|$ on frequency, showing series of resonant peaks at multiples of 6 GHz. The inset shows the schematic model, with $Z_1 = Z_0 \tanh(\frac{1}{2}\gamma l)$ and $Z_2 = Z_0/\sinh(\gamma l)$. (b) Fine dependence of $|Z_{in}(f)|$ for first resonance from (a), showing $Q = 25{,}000$, corresponding to surface resistance $R_s < 1$ μΩ.

the insulator or radiation from the ends, would reduce the effective value of Q.

At low frequencies, we can measure impedance simply by connecting wires, supplying current, and measuring the voltage with a high-impedance voltmeter. It is not quite so simple at rf and microwave frequencies. Typically, a wave is directed to the resonator from an input transmission line with characteristic impedance Z_0, and the effective impedance Z of the resonator is determined by the voltage reflection coefficient at the input, often referred to as S_{11}:

$$S_{11} = \frac{V_r}{V_i} = \frac{(Z - Z_0)}{(Z + Z_0)} \qquad (2.39)$$

In general, this is a complex quantity (to account for phase shifts) and is typically measured with a network analyzer, which has an internal source that can sweep through a range of frequencies. If Z exhibits resonant behavior in its frequency dependence, then this will also be evident in S_{11}.

But we cannot accurately measure an impedance that is much greater than Z_0 using a "voltmeter" with a characteristic impedance of Z_0. The input line itself would load down the resonator, changing its apparent impedance characteristics. For this reason, a large impedance (such as a small capacitance) is normally placed in series with the resonator. Typically, we couple to the fringe electric fields at the end of the resonator across a gap in the microstrip. (An alternative method might be to couple inductively to the fringe magnetic fields near the center of the resonator.) In terms of the schematic of Fig. 2.15, the effective input impedance becomes

$$Z_{in} = \frac{1}{j\omega C} + Z_0 \coth(\gamma l) \qquad (2.40)$$

and this is the quantity that is substituted into Eq. (2.39) to determine the reflection coefficient S_{11}. We assume here for simplicity that the characteristic impedance Z_0 is the same for the input line and the resonator, although this need not always be the case.

We follow this through for the example above, the 50-Ω microstrip transmission line, 1 cm in length, very weakly coupled to a 50-Ω input line across a 0.3-fF capacitor (Fig. 2.15). We focus here on $|S_{11}|$ for the fundamental resonance around 6 GHz for several values of $T < T_c$. Here, there is a resonant dip rather than a peak, and the half-power points (for the calculation of B and Q) correspond to the frequencies where $|S_{11}|^2$ achieves half of its maximum dip. In the limit of weak coupling ($1/\omega C \gg Z_{peak}$), the value of Q inferred from this dip should be the same as that of the isolated resonator (and correction can be made for loading by the input line). In this limit, S_{11} is close to 1, so that these are the same frequencies where $|S_{11}|$ is itself at $\frac{1}{2}$ (rather than $1/\sqrt{2}$) of the maximum dip. We can use this method to calculate the intrinsic Q for the resonance. This provides a practical method to measure $R_s(f)$ for a superconducting thin film. In terms of the temperature dependence, the resonance shifts to lower frequencies as the kinetic inductance increases as we approach T_c. This effect can be used experimentally to measure $\lambda(T)$ if some of the other parameters are known.

44 AC PROPERTIES AND SUPERCONDUCTING ENERGY GAP

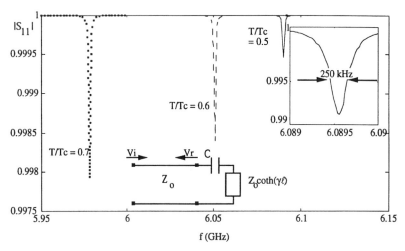

Figure 2.15. Resonant reflection from superconducting transmission line of Fig. 2.14 weakly coupled to 50-Ω input line across $C = 0.3$ fF for several temperatures (lower inset). The main plot shows the frequency dependence of the reflection coefficient S_{11} near the fundamental resonance for $T/T_c = 0.5, 0.6, 0.7$. The fine frequency dependence of S_{11} for $\tfrac{1}{2}T_c$ is shown in the inset at the top right with the bandwidth $B = f_0/Q$ noted. Note that the peak shifts to lower frequency for higher T due to the diverging kinetic inductance.

Two-Port Resonators and Filters

We can go one step further and make this transmission line resonator into a two port device by capacitively coupling another transmission line into the other end of the resonator, as indicated in Fig. 2.16. If we send an incident wave V_i down one line and measure the transmitted output V_t on the other, we can characterize the (voltage) transmission coefficient of the device, often known as $S_{21} = V_t/V_i$, where again this is generally complex to account for phase shifts. We assume here for simplicity that the coupling capacitors are the same on both ends. Solving the circuit equation from Fig. 2.16, we obtain

$$S_{21} = \frac{Z_0 Z_2}{(Z_0 + Z_1')(Z_0 + Z_1' + Z_2)} \tag{2.41}$$

where $Z_1' = Z_1 + 1/j\omega C$. In the weak coupling limit, when C is very small, this becomes

$$S_{21} \approx \frac{2j\omega C Z_0}{\sinh(\gamma l)} \tag{2.42}$$

which again will show resonant behavior, corresponding this time to a peak at resonance.

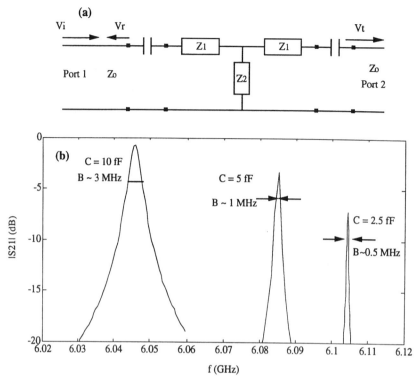

Figure 2.16. Transmission characteristics of simple superconducting bandpass filter. (*a*) Schematic of model, showing input and output lines (ports 1 and 2, respectively) capacitively coupled to microstrip resonator. Here $Z_1 = Z_0 \tanh(\frac{1}{2}\gamma l)$ and $Z_2 = Z_0/\sinh(\gamma l)$. (*b*) Frequency dependence of transmission coefficient $|S_{21}| = |V_t/V_i|$ in decibels near fundamental resonance [from Eq. (2.41)] for $T = \frac{1}{2}T_c$. Note that as C becomes larger the peak approaches 0 dB (full transmission), broadens, and shifts to lower frequencies.

The more general relation of $|S_{21}|$ for, Eq. (2.41) is plotted in Fig. 2.16 for several values of the coupling capacitor C, on a logarithmic scale in decibels, as is typical for filters:

$$S_{21} \text{ (dB)} = 20 \log_{10}|S_{21} \text{ (linear)}|. \tag{2.43}$$

For relatively large values of C, $|S_{21}|$ approaches 1 (or 0 dB), which is of course the maximum attainable for a passive linear device. The attenuation of the output signal relative to the input signal in the center of the band is often called the "insertion loss," and this approaches 0 dB for the largest values of C. Again, the bandwidth (3 dB below the peak) includes a component proportional to the resistive loss in the superconductor. But there is also a contribution due to energy coupled out of the input and output lines, and this becomes increasingly dominant for stronger coupling. From the resonator's point of view, any energy

46 AC PROPERTIES AND SUPERCONDUCTING ENERGY GAP

loss is equivalent and broadens the resonant peak. This accounts for the peaks that are much broader than the dips in the weakly coupled reflection measurement of Fig. 2.15. Also, the large values of C cause the resonance itself to shift somewhat to lower frequencies.

In contrast, if the same structure were made using normal-metal thin films, there would be no resonant peak at all, and the insertion loss would be at least $20\log(e^{-10}) = 85$ dB even for strong coupling. In other cases, we may want a filter with a wider bandpass but even sharper rolloffs at the edges. This can be obtained by using several coupled superconducting resonators to make a higher order filter. Filters with up to nine (or more) superconducting elements in series (a "nine-pole filter") have been demonstrated. Since the losses for each element add up, this would be unachievable using normal-metal thin-film transmission lines.

Another type of superconducting device is based on a very long thin-film transmission line and is known as a "tapped delay line" (Withers and Ralston, 1989). Up to several meters of such a line can be fabricated on a single 5-cm wafer using a spiral geometry. Given the very low loss of superconducting transmission lines, microwave-frequency signals may be propagated down such a line without excessive loss. This can provide the basis for a class of analog signal processing circuits with possible applications to radar technology. Chirp filters (which can be used for spectral analysis), correlators, and convolvers have been proposed and prototypes demonstrated at frequencies of several gigahertz or above. It is difficult to perform such real-time signal processing with digital circuits, although ultrafast superconducting digital circuits (Chapter 6) may offer an alternative.

2.4 SUPERCONDUCTING ENERGY GAP AND TUNNELING

Superconductors are metals with an energy gap. That may seem paradoxical, given the conventional understanding of metals and semiconductors. In order to clarify this, let us first review the basis for the behavior of electrons in metals and semiconductors [see, e.g., Kittel, (1996)], and show how superconductors fit into this picture.

Metals and Semiconductors

Atoms in both metals and semiconductors are typically arranged in crystals. The outer electrons in adjacent atoms are normally strongly overlapping, so that the electronic states extend over many atoms. However, the electrons are not entirely free; the wave nature of the electrons, combined with the crystalline organization of the fixed ions, prevents some electron states from propagating at all. This fact leads to the formation of discrete energy bands of extended electronic states separated by energy gaps, as illustrated, for example, in Fig. 2.17b, which shows the density of available electronic states $D(\mathscr{E})$ as

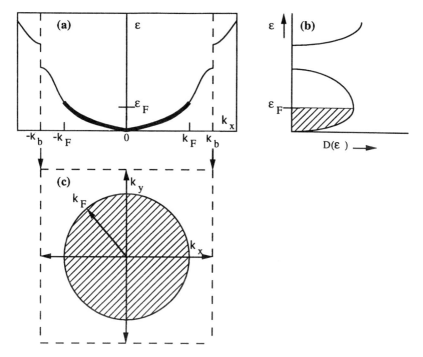

Figure 2.17. Electronic states in metal. (a) Energy versus electron wave vector $\mathscr{E}(k)$, showing states filled up to $\mathscr{E}_F$ and k_F, well below the energy gap. (b) Density of electronic states $D(\mathscr{E})$ in partially filled band. (c) Fermi sphere in k-space (shown in two dimensions) inside cubic Brillouin zone.

a function of energy. Electrons are subject to the Pauli exclusion principle, so that as electrons are added to the system, they fill up the lowest energy states up to the Fermi energy $\mathscr{E}_F$. In a metal there are only enough electrons to partially fill up the top band, whereas in a semiconductor or insulator (Fig. 2.18b) the valence band is completely occupied.

We consider first the metallic case (Fig. 2.17). In general, the electrons near $\mathscr{E}_F$ are the most important, and here these are far from the energy gap. They are thus largely unaffected by the lattice and hence behave essentially as free electrons, with only an overall constant shift in energy. It is customary (for metals) to take the zero of energy at the bottom of the band, so that we have $\mathscr{E} = \frac{1}{2}mv^2$ and $\mathscr{E}_F = \frac{1}{2}mv_F^2$, where v_F is the Fermi velocity. For typical electron densities of $\sim 10^{28}/m^3$, $\mathscr{E}_F$ is of order 1 eV, so that $v_F \sim 10^6$ m/s. This is much faster than one would expect from thermal excitation, as one can see by evaluating $\mathscr{E}_F/k_B \sim 10{,}000$ K. This is not the actual temperature of the electrons; in fact, they move this fast even at $T = 0$. Rather, they are forced to move this fast given the high density of electrons and the Pauli exclusion principle.

Since electrons are waves, it is customary to consider their wave vectors $k = mv/\hbar$ (using the de Broglie relation for free electrons), so that $\mathscr{E} = \hbar^2 k^2/2m$, and the states with lowest energy are also the ones with the smallest values of $|k|$. In Fig. 2.17c we sketch out the locations of the occupied states in "k-space," also known as reciprocal space, since the units of k are reciprocal distance. At low temperatures, they form a sphere in k-space (we show two dimensions for simplicity), called the "Fermi sphere," whose radius is $k_F = mv_F/\hbar \sim 1\,\text{Å}^{-1}$, which is related to the total density n of electrons in the band by the relation $k_F = (3\pi^2 n)^{1/3}$. At higher temperatures, there is only a slight smearing of the surface of the Fermi sphere at k_F.

We also show in Fig. 2.17c the boundary between the first and second bands, as indicated by the square surrounding the Fermi sphere (which would be a cube in three dimensions for a simple cubic crystalline lattice). This central zone (called the "Brillouin Zone" in solid-state physics) has the symmetry of the crystalline lattice, and electron states on this boundary form standing waves and cannot propagate in the crystal. The energy $\mathscr{E}(k)$ deviates significantly from the free-electron dependence for states close to this boundary and has a discontinuous jump—the energy gap—if one crosses the boundary. But for a metal the surface of the Fermi sphere is normally not close to this boundary.

Contrast this to an insulator or an intrinsic semiconductor at low temperature (Fig. 2.18). In this case, the valence band is completely filled, and there are no empty states with energies at $\mathscr{E}_F$, which lies in the middle of the energy gap. At higher temperatures, there may be some electrons thermally excited from the top of valence band into the bottom of the next higher "conduction band." The electron excitations in the conduction band need not be in the same region of k-space as the holes in the valence band, as suggested by Fig. 2.18c, appropriate for an indirect bandgap semiconductor such as Si.

Now consider what happens when we apply an electric field to either a metal or a semiconductor. For a metal, all of the electrons in the Fermi sphere move together along the field direction, limited only by collisions that relax the net shift δk of the center of the sphere:

$$\frac{d(\delta k)}{dt} = -\frac{eE}{\hbar} - \frac{\delta k}{\tau} \tag{2.44}$$

For a scattering event to relax the shifted distribution, it must take an occupied state on the leading edge with $k > k_F$ and scatter into an empty state on the trailing edge, as indicated in Fig. 2.19. There are many such electrons and empty states, so that such collisions are very likely, and τ is very small. The states deep inside the Fermi sphere do not scatter but are constrained by those on the outside; all electrons contribute to the current. So in steady state we have

$$J = ne\langle v\rangle = ne\hbar\frac{\delta k}{m} = \frac{ne^2\tau}{m}E = \sigma E \tag{2.45}$$

2.4 SUPERCONDUCTING ENERGY GAP AND TUNNELING

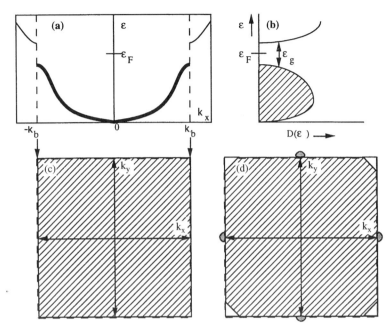

Figure 2.18. Electronic states in an intrinsic semiconductor. (*a*) Energy versus electron wave vector $\mathscr{E}(k)$ showing states filled up to zone boundary at k_b, just below the energy gap. (*b*) Density of electronic states $D(\mathscr{E})$ in filled valence band and empty conduction band. (*c*) Fully occupied valence band in *k*-space, filling entire Brillouin zone. (*d*) Electron and hole excitations in semiconductor at higher temperature, with hole states at corners of square and electron states at sides, indicative of an indirect-gap semiconductor.

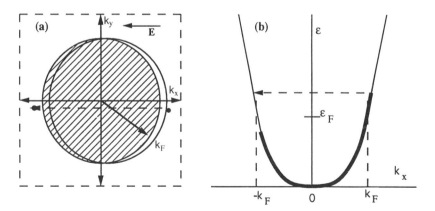

Figure 2.19. Electronic states in a metal in a electric field E. (*a*) The entire Fermi sphere shifts a distance δk opposite E limited by scattering across the sphere to empty states on the opposite side. (*b*) $\mathscr{E}(k)$ along the field direction showing shift in occupied states and the scattering from the leading edge.

giving the usual electrical conductivity for a normal ohmic metal. Note that the zone boundary does *not* move when the field is applied. This is because the zone boundary is determined by the spatial distribution of the fixed ions in the lattice, which remain fixed in the field. As long as the shift in the Fermi sphere does not bring it too close to the boundary, then the usual metallic behavior should hold.

Contrast this to the case of an insulator or cold semiconductor with the entire Brillouin zone occupied (Fig. 2.18c). In the absence of an electric field, there is no net current. But with an E-field applied, there can still be no net current, since the zone boundary remains fixed and the distribution cannot shift in k without a large increase in energy (states crossing the energy gap). In effect, the zone boundary acts as a "cage" for the electron states in k-space. Once the cage is full, all of the electronic states are trapped inside.

This is modified somewhat in a semiconductor at higher T when there is a density n_e of electron excitations in the conduction band and n_h of hole excitations in the valence band, which are equal for an intrinsic semiconductor. When an electric field is applied, there can be some redistribution of these excitations between the two sides since both types have nearby empty states. These contributions add together to a net current parallel to the field and a linear conductivity due to both electrons and holes, $\sigma = n_e e \mu_e + n_h e \mu_h$, where μ_e and μ_h are their mobilities ($e\tau/m$) for the two types. But this conductivity will be much smaller than in a metal, since the excited carrier densities are only a very small fraction of the entire population of the band.

Energy Gap in Superconductor

A superconductor displays aspects similar to both metals and semiconductors, but with some critical differences, as indicated in Fig. 2.20. We have a metal in which the band is only partially full, but in addition there is a small gap of width 2Δ (typically on the milli-electron-volt scale) right at the Fermi level. Note that, unlike the case of a semiconductor, the supercondcting energy gap does *not* occur at a Brillouin zone boundary, and the shape of the density of states at the gap edge (Fig. 2.20b) looks quite different from the semiconductor case. Furthermore, nothing much is visible in a sketch of the superconducting ground state in k-space to distinguish it from a metal. But if one looks more closely, we have a "frozen crust" at the surface of the Fermi sphere with a thickness of order $(\Delta/\mathscr{E}_F)k_F$ (we will discuss this in more detail below). This frozen crust not only locks in the electron states in the interior of the sphere but also prevents the electrons within the crust from scattering.

Now, if an electric field is applied to the superconductor, the entire Fermi sphere shifts, including the surface crust that defines the energy gap, as suggested in Fig. 2.20c. Furthermore, this shift is *not* limited by scattering, since such scattering is suppressed by the presence of this crust. This can also be understood in terms of the energy gap in the presence of the field in Fig. 2.20d. The electrons at the leading edge of the Fermi sphere cannot scatter to a state on the

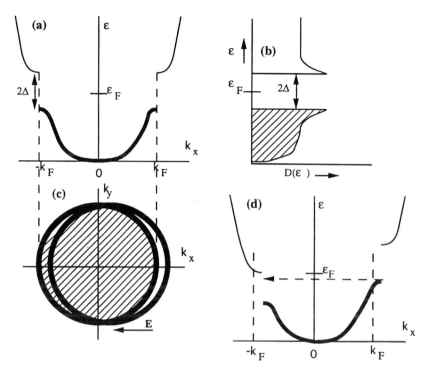

Figure 2.20. Electronic states in a superconductor. (*a*) Energy versus electron wave vector $\mathscr{E}(k)$ showing states filled up to energy gap at k_F. (*b*) Density of electronic states $D(\mathscr{E})$ showing filled states below gap and empty states above. (*c*) Fermi sphere (shown in two dimensions) surrounded by frozen crust (bold line) representing condensed superconducting pairs at Fermi surface. (*d*) $\mathscr{E}(k)$ for a shifted Fermi sphere showing that energy gap shifts with the Fermi surface. No relaxation by scattering is possible since there are no available states on the other side.

trailing edge, since no such state is available; it lies in the energy gap. (Note that this implies that the effective energy gap is suppressed by a sufficiently large current, as we will discuss in more detail later.) If there is no scattering, then the time-dependent term in Eq. (2.44) leaves us with the kinetic inductance of the superconductor and the rest of the phenomena that we discussed earlier.

So it seems that the key difference between the energy gap in a semiconductor and that in a superconductor is that in the former the gap is "pinned" to the fixed lattice, while in the latter it moves with the electrons. Of course, this does not explain why this should be the case. But in brief, the energy gap in a semiconductor is due to the interaction between an electron and the fixed lattice of ions, while the energy gap in a superconductor is due to the collective interaction among electrons around the Fermi surface.

Fermions and Bosons

As we pointed out earlier, superconducting electrons can be thought of as being bound in pairs, called Cooper pairs. This is important because particles in quantum mechanics can be classified in one of two distinct categories according to their "spin" (the intrinsic angular momentum), which is quantized in units of Planck's constant $\hbar$. There are spin-$\frac{1}{2}$ particles, such as the electron, which are known as "fermions." Then there are "bosons," with integral spin (0, 1, ...), such as the photon. Fermions are subject to the Pauli exclusion principle, which states that only two particles (with opposite spins) can be placed in any given energy level. This is the basis for atoms with many electrons; additional electrons must go into successively higher energy levels because the lowest ones are already "full." Without this rule, there would be nothing preventing all of the electrons from falling into the lowest energy level. But bosons are not subject to this principle; all of them *can* go into the lowest quantum energy level. Furthermore, the phases of the quantum wave functions in this ground state all add up coherently. One can think of a laser in these terms; all of the photons are in the same quantum state and are in phase with each other.

Something similar can happen in liquid helium. The standard helium atom (helium-4) is composed of two electrons and nucleus of two protons and two neutrons. All are spin-$\frac{1}{2}$ fermions, and an even number of such particles can only add up to an integer total spin—it turns out to be zero for both the electrons and the nucleus—making the helium atom a boson. At low temperatures (< 2.2 K), the atoms condense into a single ground state and become a "superfluid," which can flow with no dissipation or viscosity. The reason for the lossless flow is that this ground state can move as a whole and carry momentum but the atoms cannot individually dissipate energy without being raised to an excited state, and this takes a discrete amount of energy. In contrast, superfluidity does not occur (at least not until a much lower temperature) with the isotope helium-3 (with two protons and one neutron in the nucleus), which is a fermion.

Cooper Pairs

Electrons cannot condense into an electron superfluid (i.e., a superconductor) in this way, since as fermions they cannot all be in the same quantum state. This is why the Fermi sphere is required. However, if electrons with opposite spins pair up, then the resulting "particle" (which is known as a Cooper pair) has zero net spin, and those Cooper pairs are now effectively bosons that can condense into the ground state at low temperatures. [A similar Cooper pairing occurs at extremely low temperatures (< 3 mK) between helium-3 atoms, which are fermions, to form a superfluid.] We can then think of the superconducting energy gap as the energy needed to break one of these Cooper pairs in the collective ground state.

But how can electrons, which normally repel each other, form a bound pair? One mechanism, which forms the basis for the BCS theory of superconductivity

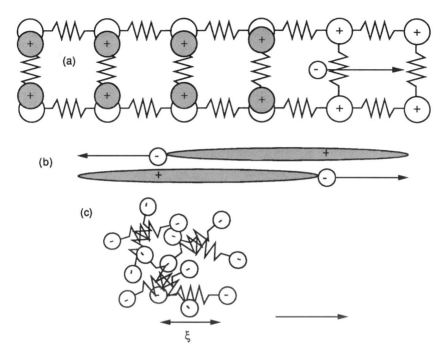

Figure 2.21. Cooper pairs in a conventional low-temperature superconductor. (*a*) Distortion of ionic lattice behind moving electron, leading to effectively positive tail. (*b*) Basis for Cooper pair: two electrons moving in opposite directions, each being attracted to the other's positive tail. (*c*) Illustration of strong overlap of Cooper pairs (on scale of coherence length ξ), leading to quantum wave functions with common phase θ.

(named after John Bardeen, Leon Cooper, and Robert Schreiffer, who received the Nobel Prize in Physics for this work), involves the interaction of rapidly moving electrons with an elastic lattice of much heavier positive ions. Consider how the lattice distorts near the path of a moving electron (Fig. 2.21). The positive ions near the electron are attracted to it, but by the time they have moved slightly toward the electron, the electron has moved far away. This creates a tube of net positive charge behind the moving electron (see, e.g., Weisskopf, 1981). Such a tube can be quite long, compared to the atomic scale. It is this positively charged tube, rather than the electron itself, that attracts a second electron. Furthermore, the second electron also leaves a wake of positive charge behind it. If two electrons are moving in exactly opposite directions, then each electron can be attracted to the positive wake of the other electron (Fig. 2.21b). This creates the possibility of a weakly bound state between the two electrons, which constitute the Cooper pair. This provides the basis for the so-called electron–phonon mechanism for superconductivity, although phonons (quantized lattice vibrations) do not appear explicitly in the picture. It is possible that other mechanisms may be more important for some superconductors (such as the high-temperature superconductors), but the end result, pairing of electrons (or holes) with opposite values of k, is likely to be very similar.

Superconducting Ground State

The ground state of a superconductor at $T = 0$ consists of a large number of these Cooper pairs. Each pair consists of two electrons with opposite k and opposite spin, with $|k| \approx k_F$ (from opposite sides of the Fermi surface), corresponding to a composite particle with total wave vector $K = 0$ and spin $S = 0$. Since they are effectively bosons, all of these Cooper pairs can go into the same quantum state, even if they overlap in space. In fact, the effective size of a Cooper pair (the superconducting coherence length ξ, to be discussed later) is much larger than the distance between them (at least for conventional LTS materials), so they are very strongly overlapping (see Fig. 2.21c). For this reason, the quantum wave functions of all of these pairs add up in phase, and the ground state of the superconductor can be characterized by a total wave function $\psi = |\psi|\exp(i\theta)$. This is a macroscopic quantum wave function, also known as the superconducting order parameter, which extends throughout the superconductor. The magnitude of this order parameter is typically defined so that $|\psi|^2 = n_s$, the effective superelectron density. Alternatively, it can also be shown that $|\psi|$ is proportional to the energy gap parameter Δ, so that a complex gap parameter $\Delta\exp(i\theta)$ is sometimes defined.

Note that this phase-coherent pair condensate is *not* explicit in the single-electron energy diagrams of Fig. 2.20, which are therefore incomplete. The frozen crust at the Fermi surface is an attempt to identify those electrons that are included in this pair condensate. But the pairs themselves act as if they all have a single energy $2\mathscr{E}_F$, rather than a spread of energies as suggested by the figure. Also, these pictures assume an isotropic Fermi surface and an isotropic energy gap, which is not always the case (particularly for HTSs).

We can think of the superconducting energy gap 2Δ as the binding energy of a Cooper pair within the superconducting ground state. It should also be no surprise that the thermal energy at T_c is expected to be similar in magnitude: $2\Delta/k_B T_c \approx 3.5$–4 in BCS theory. Although the net attractive interaction is related to lattice vibrations in this picture, Δ is much less than a typical quantum of vibrational energy, the phonon energy $\mathscr{E}_{\text{ph}} \approx hf_D \approx k_B \Theta_D$, where Θ_D is known as the Debye temperature of the material, typically on the order of 100–300 K. The BCS theory [see Tinkham (1996, Chap. 3) and Van Duzer and Turner (1981, Chap. 2)] shows that

$$k_B T_c \approx \tfrac{1}{2}\Delta(0) \approx hf_D \exp\left(-\frac{1}{\alpha}\right) \tag{2.46}$$

where α is a small dimensionless constant known as the electron–phonon coupling constant. This suggests that a large value of T_c will occur in a material with a large value of α, even approaching Θ_D. However, the value of α is normally in the range 0.1–0.3 (larger values tend to lead to an unstable crystalline structure), so that typically $T_c \ll \Theta_D$. This same constant α appears in the theory of electrical resistivity ρ of a normal metal due to collisions with phonons, characterized by an electron–phonon relaxation time $\tau_{e\text{-ph}}$. In particular, $\rho \sim 1/\tau_{e\text{-ph}} \sim \alpha$, so that metals that are good electrical conductors are normally poor superconductors, and vice versa.

Quasiparticle Excitations

In the weak-coupling (small-α) limit, the BCS theory gives simple expressions for the ground state and excitations. If we set the zero for energy at the Femi energy and define

$$\mathscr{E}_n = \frac{\hbar^2 k^2}{2m} - \mathscr{E}_F \approx \frac{\hbar^2 k_F}{m}(k - k_F) \tag{2.47}$$

and $D_n(0) = D_n(\mathscr{E}_F)$ as the density of states for the normal state at the Fermi energy, then we have

$$\mathscr{E}_s = \text{sign}(\mathscr{E}_n)\sqrt{\mathscr{E}_n^2 + \Delta^2} \tag{2.48}$$

$$D_s(\mathscr{E}_s) = \frac{D_n(0)|\mathscr{E}_s|}{\sqrt{\mathscr{E}_s^2 - \Delta^2}}, \tag{2.49}$$

as shown in Fig. 2.22. Note that this density of states diverges at $\mathscr{E}_s = \Delta$ but that this is an integrable divergence; the states that would be within the gap are simply pushed up or down in energy. Once can also estimate the "condensation energy" U_c, the reduction of the energy density of the superconducting state relative to the normal state at $T = 0$. It is *not* simply $\approx n\Delta$, since only the electrons within $\approx \Delta$ of the Fermi surface are really reducing their energy by $\approx \Delta$. The correct expression (from BCS theory) is reduced by $\sim \Delta/\mathscr{E}_F$:

$$U_c \approx \tfrac{1}{4} D_n(\mathscr{E}_F) \Delta^2 = \frac{3n\Delta^2}{8\mathscr{E}_F} \tag{2.50}$$

where the normal density of states for free electrons is $D_n(\mathscr{E}_F) = 3n/2\mathscr{E}_F$. A typical value is $U_c \sim 10 \text{ kJ/m}^3$.

Furthermore, the superconducting ground state is not sharp in k-space; the kinetic energy is a bit higher than the ground-state energy for normal electrons, but the potential energy associated with the attractive interaction more than makes up for this. As a function of $\mathscr{E}_n$ the fraction of occupied states in the

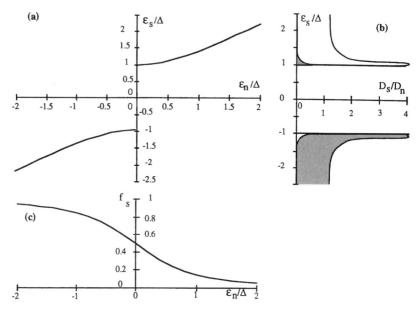

Figure 2.22. Semiconductor picture of superconducting ground state. (a) Energy gap 2Δ between filled lower band and empty upper band at Fermi energy. (b) Density of states in superconductor. For $T > 0$, a small number of electrons are excited from the top of the lower band (holes) to form electrons in the upper band. (c) Occupation fraction of electrons in superconducting ground state, showing "smearing" around Fermi surface even for $T = 0$.

superconducting ground state at $T = 0$ (Fig. 2.22c) is

$$f_s(\mathscr{E}_n) = \frac{1}{2}\left(1 - \frac{\mathscr{E}_n}{\sqrt{(\mathscr{E}_n^2 + \Delta^2)}}\right) \tag{2.51}$$

which looks similar to the thermal smearing in the excited state of a normal metal at T_c. With this in mind, the identification of an electron or a hole near the Fermi surface becomes somewhat fuzzy, so that these excitations are often referred to collectively as "quasiparticles." In fact, a better definition of a quasiparticle is a pairing state $(k, -k)$ that is only half-occupied. For $k \ll k_F$ the quasiparticle is definitely a hole, for $k \gg k_F$ it is definitely an electron, while for $k \approx k_F$ it exhibits a mixed character. Sometimes the term quasi-electron (or e-like) and quasi-hole (or h-like) are used to identify the two branches of the excitation spectrum.

Partly due to the mixed electron–hole character of low-energy excitations, an alternative "excitation picture" is sometimes used (Fig. 2.23), in which we focus on the positive excitation energy from the Fermi level. The energy gap is still 2Δ, since excitations must be created in pairs. At a temperature $T > 0$, there will be some thermal excitations across the gap, leading to the presence of both

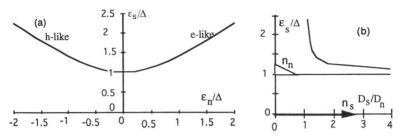

Figure 2.23. Excitation picture of superconductor. (a) Quasiparticle excitation energies, with quasi-holes on the left and quasi-electrons on the right. (b) Quasiparticle density of states (two branches combined). The ground-state Cooper pairs are effectively at $\mathscr{E}_F = 0$ and form the superelectrons n_s; thermal excitations above the gap together constitute the "normal electrons" n_n.

quasi-electrons and quasi-holes. Together, these quasiparticles constitute the "normal fluid" n_n of the two-fluid picture described earlier. The remaining electrons form the superfluid n_s and act as if they were all in a single quantum state at $\mathscr{E}_F$ per electron.

Since the superconducting energy gap is the result of a collective interaction among electrons near the Fermi surface, it should not be surprising that the presence of thermal excitations reduces the gap by removing electrons that would otherwise be contributing to the interaction. As T is increased, $\Delta(T)$ continues to decrease, finally decreasing sharply (but continuously) to zero at T_c. The overall dependence of $\Delta(T)$ from the BCS theory, which does not fit a single analytic expression, is shown in Fig. 2.24. This has the approximate form

$$\Delta(T) \approx 1.74\Delta(0)\left(1 - \frac{T}{T_c}\right)^{1/2} \approx \Delta(0)\left[1 - \left(\frac{T}{T_c}\right)^{3.3}\right]^{0.5} \qquad (2.52)$$

where the first expression is valid for $T > 0.9T_c$, and the latter is a simple empirical approximation over the entire range. This is similar in form to what we would expect from $\Delta^2 \sim n_s \sim 1 - (T/T_c)^4$ from the two-fluid picture.

Optical Properties

The gap in the excitation spectrum has some important implications for the properties of superconductors. As we have pointed out, it justifies some of the predictions of the two-fluid picture. But furthermore, it identifies where that picture will break down because of quantum effects. It is useful to compare the electromagnetic (optical) properties of a superconductor to those of a normal metal and a semiconductor. In all of these cases, a photon can provide the energy hf needed to raise an electron to a higher energy level if an empty state is available. An intrinsic semiconductor is transparent for $hf < \mathscr{E}_g$, since the filled band cannot screen out an electromagnetic field, and there are no available

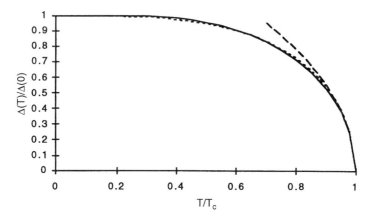

Figure 2.24. Temperature dependence of superconducting energy gap. Solid line, BCS theory; dashed line, $1.74(1 - T/T_c)^{0.5}$; dotted line, empirical approximation $[1 - (T/T_c)^{3.3}]^{0.5}$.

states to absorb the photon. A normal metal will reflect an incident wave, but it will also absorb some fraction, since there are many available states above $\mathscr{E}_F$. In contrast, a superconductor will be a perfect reflector of incident radiation for $hf < 2\Delta$, at least at low T when the density of excitations n_n is small. For $hf > 2\Delta$, the density of available states is again very large, and the optical properties should be very similar to those of a normal metal. The relevant critical frequency $f_c = 2\Delta/h$, typically ~ 1 THz, in the far infrared.

Tunnel Junctions

A superconducting device structure that is very important for both research and technology is the tunnel junction, a sandwich structure with an ultrathin insulating layer between two conducting electrodes one or both of which may be superconducting. This looks like a capacitor, but in addition, there can be a conductance across the insulating barrier even when it has no "holes." This is because an electron wave function does not immediately go to zero when it enters a classically forbidden region where its kinetic energy $\hbar^2 k^2/2m = \mathscr{E} - U$ is negative. Here U is the potential energy in the insulator, typically of order several electron-volts. This imaginary wave vector $k = -j\kappa$ gives a transmission coefficient across the insulator $T_t \sim \exp(-jkz) = \exp(-\kappa z)$ corresponding to exponential decay with decay constant

$$\kappa = \frac{[2m(U - \mathscr{E})]^{1/2}}{\hbar} \tag{2.53}$$

For typical values of energies, this gives a decay length $1/\kappa$ of about 1 Å. The electrical current flowing between the electrodes goes as T_t^2, so this effect falls off

2.4 SUPERCONDUCTING ENERGY GAP AND TUNNELING

very quickly indeed. Still, for a barrier thickness $d \sim 10$ Å, the factor $\exp(-2\kappa d)$ is large enough to give a significant current. This is commonly referred to as "tunneling," as if the electrons "tunneled through" the insulating barrier.

When we apply a dc voltage V between these electrodes, we can think of this as raising all the energies on one side by eV relative to the other, as shown in Fig. 2.25b. Now the possible transitions are straight across, from filled states on the left to empty states on the right. If there are no excited states ($T = 0$), the total tunneling current can be expressed in the form

$$I(V) \sim \int_0^{eV} d\mathscr{E}\ T_t^2 D_L(\mathscr{E} - eV) D_R(\mathscr{E}) \tag{2.54}$$

where L and R correspond to the left and right electrodes in the diagram [see, e.g., Tinkham (1996) or Van Duzer and Turner (1981)]. For the case where both electrodes are normal metals and assuming that $V \ll 1$ V (so that both T_t and D are essentially constant), we have

$$I_{nn}(V) = G_{nn}V \sim T_t^2 D_n^2(\mathscr{E}_F) eV \tag{2.55}$$

Perhaps not surprisingly, since we are dealing with normal metals, this acts like an ohmic resistor. This resistance is, of course, in series with the lead resistance, but the tunneling resistance can often be dominant. This is not in itself a distinctive device; if the insulating barrier has shorts across it, these will also generally be ohmic.

Things become more interesting for the case where one of the electrodes is superconducting (Fig. 2.26). Using the semiconductor picture of the superconducting density of states, we have

$$I_{ns}(V) \sim \int_0^{eV} d\mathscr{E}\ T_t^2 D_n(\mathscr{E} - eV) D_s(\mathscr{E}) = \frac{G_{nn}}{e} \int_0^{eV} d\mathscr{E} \frac{D_s(\mathscr{E})}{D_n}$$

$$= \begin{cases} 0 & |V| < \dfrac{\Delta}{e} \\ G_{nn} \sqrt{V^2 - \left(\dfrac{\Delta}{e}\right)^2} & |V| > \dfrac{\Delta}{e} \end{cases} \tag{2.56}$$

So for $eV < \Delta$, there is no tunneling current, since there are no available states to tunnel into; they lie within the energy gap. It is remarkable that the presence of a superconductor causes this device to act like an insulator, at least for low voltage. This is distinctly different from the behavior of a shorted insulating barrier, which would look more ohmic. It is also noteworthy that the superconducting density of states can be reconstructed by taking the differential conductance $dI/dV = G_{nn}D_s(eV)/D_n$. The rather sharp rise in current at $eV = \Delta$ is a direct consequence of the characteristic divergence in the density of states at

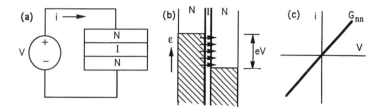

Figure 2.25. Normal/insulating/normal (NIN) tunnel junction. (*a*) Circuit schematic with dc voltage across junction. (*b*) Density of states on both sides of insulating barrier, showing electrons that are permitted to tunnel across. (*c*) *I–V* characteristic of NIN junction.

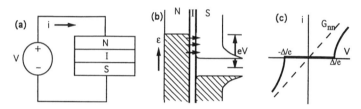

Figure 2.26. Normal/insulating/superconducting (NIS) tunnel junction. (*a*) Circuit schematic with dc voltage across junction. (*b*) Density of states on both sides of insulating barrier, showing electrons that are permitted to tunnel across. (*c*) *I–V* characteristic of NIS junction showing insulating behaviour for $|V| < \Delta/e$.

the gap edge. This rise will not be quite as sharp in a real material for $T > 0$; both factors will cause some smearing of the sharp rise. Still, such a tunneling junction provides one of the more common ways in practice to accurately measure the energy gap of a superconductor.

If one diverging density of states will give a sharp *I–V* characteristic, the convolution of two divergences can be expected to be even sharper. This is the case for a tunnel junction with two superconducting electrodes, sometimes called an SIS junction, illustrated in Fig. 2.27. Here,

$$I_{ss}(V) \sim \int_0^{eV} d\mathscr{E}\, T_t^2 D_s(\mathscr{E} - eV) D_s(\mathscr{E}) = \frac{G_{nn}}{e} \int_0^{eV} d\mathscr{E}\, \frac{D_s(\mathscr{E} - eV)\, D_s(\mathscr{E})}{D_n\, D_n}$$

$$= \begin{cases} 0 & |V| < \dfrac{2\Delta}{e} \\ \approx G_{nn} V & |V| > \dfrac{2\Delta}{e} \end{cases}$$

(2.57)

There is a finite jump in current at $V = 2\Delta/e$, essentially up to the normal resistance. (Theoretically, the jump is only up to $\frac{1}{4}\pi$ of the normal resistance, and the rise beyond that is continuous, but the distinction is normally insignificant.)

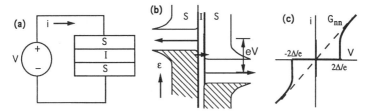

Figure 2.27. Superconducting/insulating/superconducting (SIS) tunnel junction. (*a*) Circuit schematic with dc voltage across junction. (*b*) Density of states on both sides of insulating barrier, including Cooper pairs at Fermi energy. (*c*) *I–V* characteristic of NIS junction showing insulating behavior for $|V| < 2\Delta/e$ and also Josephson current at $V = 0$.

Again, this exhibits insulating behavior for low voltage, distinctly different from a superconducting short, which would tend to look more ohmic. Thermal and real-junction effects tend to smear this a bit, but real SIS junctions show I–V curves that are about as sharp as any device known. This nonlinearity provides the basis for detection and mixing of high-frequency electromagnetic signals, as will be discussed in Chapter 7.

One additional feature of SIS junctions is the zero-voltage current shown in Fig. 2.27*c*. This is not due to a short across the insulating barrier, but rather is a manifestation of the Josephson effect associated with Cooper pairs tunneling across the barrier, as we will discuss further in Chapter 5. At this point, let us merely point out that we can think of this as due to the coupling of the pair states on the two sides for $V = 0$. The ground state at the Femi level $\mathscr{E}_F$ can contain any number of these "bosons" so that such coupling is possible.

Superconducting Coherence Length

There is a widely used phenomenological theory of superconductors, known as the Ginzburg–Landau (GL) theory, which is based on two characteristic distances [see, e.g., Tinkham (1996, Chap. 4) or Van Duzer and Turner (1981, Chap. 7)]. One is the magnetic penetration depth λ, which governs variations in fields and currents. The other is the superconducting coherence length ξ, which governs the spatial variation in the magnitude of the superconducting energy gap 2Δ, or equivalently of $\Delta^2 \sim n_s \sim |\Psi|^2$. In its simplest form, we can think of the coherence length as the characteristic size of a Cooper pair. One way to estimate this is to note that the superconducting ground state has an "uncertainty" in k-space of order $\delta k \approx (\Delta/\mathscr{E}_F)k_F$. By the quantum uncertainty principle, or equivalently by Fourier analysis, this implies a size in real space of order $\delta x \approx 1/\delta k = \hbar v_F/2\Delta$. The correct result for the BCS coherence length is

$$\xi_0 = \frac{\hbar v_F}{\pi \Delta(0)} \tag{2.58}$$

If we estimate typical values of $v_F \sim 10^6$ m/s and $\Delta(0) \sim 1$ meV, this gives $\xi_0 \sim 200$ nm, similar in magnitude to λ. However, like the case of $\lambda(T)$, the effective GL coherence length $\xi(T)$ can change both with temperature and with impurities, can also be anisotropic, and is generally taken to be a phenomenological parameter for a particular sample. Whereas λ increases as $\sqrt{(\xi_0/l)}$ in the dirty limit where the normal-electron scattering length $l < \xi_0$, the effective value of ξ decreases by the same factor. The temperature dependence of ξ and λ are similar; both diverge as $(1 - T/T_c)^{-1/2}$ close to T_c. These effects are summarized in Table 2.1.

It is also significant that we can write several of the key quantities of the superconducting state in terms of these characteristic lengths ξ and λ. For example, the condensation energy at $T = 0$ [Eq. (2.50)] can be expressed as

$$U_c(0) \approx \frac{\Phi_0^2}{130\mu_0\lambda_L^2\xi_0^2} \tag{2.59}$$

which is essentially the same as the more general GL expression at higher T in terms of the phenomenological parameters:

$$U_c(T) \approx \frac{\Phi_0^2}{158\mu_0\lambda^2(T)\xi^2(T)} \tag{2.60}$$

Note that this energy goes to zero at T_c, as one would expect.

Proximity Effect

Although $n_s \sim |\Psi|^2$ is normally constant inside a superconductor, it may vary near a surface, in particular at the boundary between a superconductor and a normal metal (an S/N interface). If the interface between them is clean, then some of the Cooper pairs can move out of S into N. They are certainly unstable in N, but the proximity of S means that there will be a nonzero density of pairs near the surface of N, on a scale of ξ_n, the effective coherence length in the normal metal. For the same reason, the density of pairs will be decreased at the surface of S, on the scale of ξ_s, as shown in Fig. 2.28. Since $n_s \sim \Delta^2$, a similar behavior will be exhibited by the energy gap. Thus, we can induce superconductivity near the surface of a normal metal, which is known as the "proximity effect." If we apply this to a thin-film N/S bilayer of thickness less than the relevant coherence lengths, superconductivity will be present as a sort of average of the two materials. A similar proximity effect can also be extended to interfaces between superconductors with different values of n_s.

It is also worth noting that in the case of a tunnel junction to a superconductor, it is really the top layer of the superconductor, on the scale of ξ_s, that is being characterized in Fig. 2.26. It is not uncommon that the surface of a superconductor may have a somewhat degraded T_c, corresponding to a weakened layer W.

SUMMARY

Table 2.1. Characteristic Lengths in Superconductors from GL Theory

	Clean Limit $(l \gg \xi_0)$	Dirty Limit $(l \ll \xi_0)$	$T \to T_c$
Magnetic penetration depth	$\lambda_L = (m/\mu_0 n e^2)^{1/2}$	$\lambda \approx \lambda_L(\xi_0/l)^{1/2}$	$\sim (1 - T/T_c)^{-1/2}$
Coherence length	$\xi_0 = \hbar v_F/\pi\Delta(0)$	$\xi \approx (\xi_0 l)^{1/2}$	$\sim (1 - T/T_c)^{-1/2}$

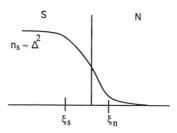

Figure 2.28. Proximity effect at superconducting/normal (SN) boundary *without* insulating barrier. Motion of Cooper pairs across interface leads to reduction in $n_s \sim \Delta^2$ on S side and induction of weak superconductivity on N side, both on the scale of the appropriate coherence length ξ from the interface.

Then, the junction might be identified as NIWS (instead of NIS), and the I–V curve of the junction would depend on the properties of the W layer and its thickness. This is particularly critical for those superconductors (such as high-T_c materials) where the coherence length is very short. So the true energy gap inside a superconductor may be different from that inferred using tunneling.

SUMMARY

- The superconducting kinetic inductance $\Lambda = m/\mu_0 n_s e^2$ determines the electromagnetic response of a superconductor; $\frac{1}{2}\Lambda J^2$ is the kinetic energy density of the superconducting electrons.
- The magnetic penetration depth $\lambda(T) = \sqrt{\Lambda/\mu_0} \sim 100$ nm is the effective electromagnetic skin depth of a superconductor for all frequencies from dc up to the gap frequency.
- In the two-fluid model, alternating current is carried in parallel by lossless superelectrons (Cooper pairs, described by Λ) and lossy normal electrons (excited quasiparticles, described by ρ_n).
- The transmission line model of the superconductor can be used to visualize and calculate the surface impedance of a superconductor; for a thick layer, this includes a surface inductance $\mu_0 \lambda(T)$ and a small surface resistance $R_s \sim \omega^2$.

64 AC PROPERTIES AND SUPERCONDUCTING ENERGY GAP

- The small R_s can be applied to low-loss transmission lines, resonators, and filters up to frequencies approaching the gap frequency $2\Delta/h \sim 1$ THz.
- A superconductor is a metal with an energy gap $\mathscr{E}_g$ that moves with the electrons, where $\mathscr{E}_g = 2\Delta \approx 3\text{-}4kT_c \sim 1\text{-}10$ meV at low temperatures, decreasing continuously to zero at $T = T_c$. This energy gap can also be viewed as the binding energy of electrons in Cooper pairs.
- Superconductors can be understood in terms of a microscopy theory (BCS) of pair formation and the energy gap, and also a macroscopic phenomenological theory (GL) of spatial variations of the coherent quantum wave function $\psi(r)$.
- A tunnel junction is a thin insulating layer between two metals. An SIS tunnel junction acts like a weak superconductor for $V = 0$, an insulator for $0 < V < 2\Delta/e$, and a resistor for $V > 2\Delta/e$.
- The superconducting coherence length ξ is the characteristic size of a Cooper pair and also determines the scale of the proximity effect in a superconductor/normal metal boundary.
- The superconducting condensation energy $U_c \approx \frac{1}{4}D(\mathscr{E}_F)\Delta^2 \approx \Phi_0^2/160\mu_0\lambda^2\xi^2$ is the reduced energy density of the superconductor compared to the normal resistive state.

REFERENCES

P. H. Carr and B. R. McAvoy, Eds., "Special Issue on Microwave Applications of Superconductivity," *IEEE Trans. Microwave Theory Techn.* **39**, 1445 (1991).

M. W. Johnson and A. M. Kadin, "Anomalous Current Dependence of Kinetic Inductance in Ultrathin NbN Meander Lines," *IEEE Trans. Appl. Supercond.* **7**, 3492 (1997).

C. Kittel, *Introduction to Solid State Physics*, 7th ed. (Wiley, New York, 1996).

T. P. Orlando and K. A. Delin, *Foundations of Applied Superconductivity*, Chaps. 2–4 (Addison-Wesley, Reading, MA, 1991).

M. Tinkham, *Introduction to Superconductivity*, 2nd ed., Chaps. 2 and 3 (McGraw-Hill, New York, 1996).

T. Van Duzer and C. W. Turner, *Principles of Superconductive Devices and Circuits*, Chaps. 2 and 3 (Elsevier, New York, 1981).

V. F. Weisskopf, "The Formation of Cooper Pairs and the Nature of Superconducting Currents," *Contemp. Phys.* **22**, 375 (1981).

R. S. Withers and R. W. Ralston, "Superconducting Analog Signal Processing Devices," *Proc. IEEE* **77**, 1247 (1989).

PROBLEMS

2.1. Kinetic inductance for normal metal. Consider a normal metal with resistivity $\rho \sim 0.1$ $\mu\Omega$-cm and electron density $10^{23}/\text{cm}^3$, which is similar to pure copper cooled to a low temperature.

(a) Calculate the kinetic inductivity Λ and determine at what characteristic frequency this will begin to dominate the electromagnetic response.

(b) For a direct current, determine the internal magnetic inductance L_m per unit length for a wire of radius a. [Hint: Determine the magnetic field distribution $H(r)$ inside the wire and use the definition that $U = \frac{1}{2}L_m I^2$ together with the magnetic field energy density $\frac{1}{2}\mu_0 H^2$.]

(c) Determine the kinetic inductance per unit length for this wire for a direct current. In what regime of wire radius a will the kinetic inductance be significant? Does this suggest that the kinetic inductance can be ignored for normal metals in practical situations?

2.2. Kinetic and magnetic inductance in superconductor.

(a) Consider a superconductor much thicker than the penetration depth $\lambda(T)$. Show that the surface inductance $L_s = L_m + L_k$, where L_m is the magnetic inductance associated with penetration of the magnetic field into the superconductor and L_k is the kinetic inductance associated with the currents in this region. Furthermore, show that $L_m = L_k$.

(b) Now consider a superconducting thin film with $d \ll \lambda(T)$. Determine L_m and L_k in this case, and compare their magnitudes.

2.3. Two-fluid model of superconductor. At a given temperature, a particular superconductor can be modeled as a kinetic inductivity $\Lambda = 5 \times 10^{-20}$ H-m, in parallel with a resistivity $\rho = 100$ $\mu\Omega$-cm.

(a) Using the transmission line picture, determine the characteristic impedance and the propagation constant γ of the equivalent line.

(b) For a frequency $f = 10$ GHz, identify the magnetic penetration depth λ, the surface inductance L_s, and the surface resistance R_s. Can you scale to R_s at 20 GHz?

(c) Is a thin film with $d = 20$ nm in the thin-film limit? What are L_s and R_s at 10 GHz in this case?

2.4. Superconducting transmission line. A parallel-plane transmission line consists of two thick superconducting films each 1 µm thick and 5 µm wide, with $\lambda = 0.1$ µm, separated by 0.2 µm of an insulator with $\varepsilon_r = 10$.

(a) Determine the low-frequency surface inductance of the superconducting planes.

(b) Determine the inductance per unit length L and the capacitance per unit length C, neglecting edge effects.

(c) Determine the characteristic impedance Z_0 and the wave velocity v on this transmission line.

(d) If the surface resistance of the superconductor at a given frequency is 0.1 mΩ, what is the resistance per unit length R on the line? What is the attenuation length $1/\alpha$ for waves propagating on the line?

2.5. Superconducting filter. A fixed length of superconducting transmission line can serve as a high-Q resonator or filter. A given line is 3 mm long, 100 μm wide, with a capacitance $C = 1$ nF/m, and it exhibits a series of resonances at multiples of 20 GHz. It is composed of superconducting electrodes with surface resistance $R_s = 1$ mΩ at 20 GHz.

(a) Determine the wave velocity v and the characteristic impedance Z_0 on the line.

(b) Determine the maximum Q and the minimum bandwith B for a 20-GHz resonator using this line.

(c) How would Q and B scale for higher order resonances assuming two-fluid behavior? Assume that the losses are exclusively from the superconductor and that the input and output coupling are not loading down the resonator.

2.6. Cooper pairs and superconducting coherence length. One can think of Cooper pairs as consisting of pairs of electrons, each with energies within $\sim \Delta$ of the Fermi surface, separated by a distance that is approximately a coherence length ξ. Estimate the number of overlapping Cooper pairs in a coherence volume $\sim \xi^3$ using numbers appropriate to LTS Pb and HTS YBCO. Use formulas for a free-electron metal.

(a) For Pb, electron density $n = 1.3 \times 10^{23}$/cm^3, $\Delta = 1.4$ meV, and $\xi = 800$ Å.

(b) For YBCO, $n \approx 2 \times 10^{21}$/cm^3, $\Delta = 30$ meV, and $\xi \approx 10$ Å.

(c) Do these results suggest that the basic picture of superconductivity may be different in LTS and HTS materials? Why?

2.7. Superconducting tunnel junction. Consider a tunnel junction between two metals with an insulator of thickness d.

(a) If the "barrier height" of the insulator is $U - \mathscr{E} = 1$ eV, determine the decay length $1/\kappa$ associated with tunnelling.

(b) Assume that a tunnel junction with area 1 mm^2 and insulator thickness $d = 30$ Å has a normal-state resistance $R_n = 1$ Ω. What value of d would maintain the same value of R_n for a junction with area 1 μm^2?

(c) Estimate the maximum dc Josephson current for $V = 0$ for the junctions of part (b) if $2\Delta = 3$ meV.

3

MAGNETIC PROPERTIES OF SUPERCONDUCTORS

In Chapter 2, we showed how supercurrents flow in a thin layer at the surface of a superconductor, preventing magnetic fields from penetrating into the bulk. However, practical superconductors often operate in a distinctly different regime, in which both current and magnetic field can be present throughout a superconductor. In this chapter, we will show how this is achieved via the presence inside the superconductor of quantized magnetic vortices (fluxons). This permits superconductors to operate up to much higher magnetic fields, leading to a variety of practical applications. We start in Section 3.1 with an analysis of diamagnetism associated with the Meissner effect in a type I superconductor. This is followed in Section 3.2 with an analysis of critical currents and critical fields that limit the Meissner state. The basic structure of vortices and the resulting "mixed state" in a type II superconductor are described in Section 3.3. The principle of duality between electric charge and magnetic flux is presented in Section 3.4, and this concept is developed to treat a vortex as a particle that transports flux, in much the same way that an electron transports charge. A moving vortex produces dissipation, so that true zero resistance in the mixed state requires that all of the vortices be held fixed. How such flux pinning can be achieved is described in Section 3.5, with important implications for the design of practical high-field superconducting magnet wire. Finally, a few applications of superconducting magnets and other power applications of superconductors are discussed in Section 3.6.

3.1 FLUX CONSERVATION AND DIAMAGNETISM

Conservation of Flux

One of the key consequences of zero resistance is that the magnetic flux inside a closed loop can never change. We can see this by first considering what

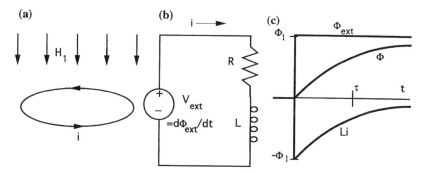

Figure 3.1. Response of resistive ring to application of magnetic flux. (*a*) Physical configuration, with $H = H_1$ applied perpendicular to plane of loop. (*b*) Lumped circuit equivalent of (*a*). (*c*) Contributions to flux Φ contained in loop, in response to application of $\Phi_{ext} = \Phi_1$ at $t = 0$. For a superconductor with $R = 0$, the induced circulating current will maintain $\Phi = 0$ indefinitely.

happens when one has a normal resistive wire in a circular loop of area A that can be modeled as a lumped resistor R in series with a self-inductance L (Fig. 3.1). If we start out with zero applied magnetic field and then increase the field to a value H_1 (perpendicular to the plane of the loop), the magnetic flux inside the loop will become $\Phi_1 = \mu_0 H_1 A$ after the initial transient. But we are more interested here in the transient response. By Faraday's law, while the flux is changing, there will be an induced electromotive force (emf) around the loop given by $V = -d\Phi/dt$. By linearity, we can separate both the emf and the flux into the sum of two parts, one due to the externally applied field and the other to the inductive response of the current:

$$\Phi = \Phi_{ext} + Li \qquad V = V_{ext} - L\frac{di}{dt} \qquad (3.1)$$

where $V_{ext} = -d\Phi_{ext}/dt$. The induced voltage $L\,di/dt$ is applied in the direction so as to induce a current that will oppose the change in the applied flux. We can model this with an ideal voltage source V_{ext} in series with the R and L in the loop (see Fig. 3.1), described by the circuit equation

$$V_{ext} = Ri + L\frac{di}{dt} \qquad (3.2)$$

Let us assume that the magnetic field is changed quickly, so that Φ_{ext} is a step function and V_{ext} is an impulse (delta function):

$$\Phi_{ext}(t) = \Phi_1 u(t) \qquad V_{ext}(t) = -\Phi_1 \delta(t) \qquad (3.3)$$

where $u(t)$ is the standard unit step function (0 for $t < 0$ and 1 for $t > 0$) and $\delta(t)$ is the standard unit impulse function (0 for $t \neq 0$, with a unit integral). The solution to the current response $i(t)$ is a simple standard problem in differential equations or Laplace transforms, which takes the form

$$i(t) = \left(-\frac{\Phi_1}{L}\right) \exp\left(-\frac{t}{\tau}\right) \quad (3.4)$$

where $\tau = L/R$ is the characteristic relaxation time. So the initial current is $i(0) = -\Phi_1/L$, which produces a flux $Li = -\Phi_1$. In other words, just after the change in the applied flux, the total flux is still zero:

$$\Phi(t) = \Phi_1 + Li = \Phi_1[1 - \exp\left(-\frac{t}{\tau}\right)] \approx \Phi_1\left(\frac{t}{\tau}\right)\cdots \quad (3.5)$$

This is equivalent to the usual eddy currents in a conductor that oppose penetration of any change in magnetic flux. Of course, for a normal metallic wire, the relaxation time is rather fast; taking $R \sim 1\,\Omega$ and $L \sim \mu_0 \sim 1\,\mu H$ (for a wire of order 1 m in length) gives $\tau \sim 1\,\mu s$.

In contrast, if a superconducting wire with $R = 0$ is used, then $\tau \to \infty$, and we have $i(t) = -\Phi_{app}/L$ and $\Phi(t) = 0$ for all times $t > 0$. The induced screening current will prevent the penetration of the applied magnetic flux inside the loop, and there is no resistance to cause the current to die out. This will be true regardless of how the applied field is changed; we will have $i(t) = -\Phi_{app}(t)/L$ for any variation in $\Phi_{app}(t)$. It is important to understand that for a superconducting wire the resistance is indeed zero, and not just a few orders of magnitude smaller than that in normal metals. Experiments have shown that such "persistent currents" do not diminish noticeably in times on the scale of a year or more. This implies that the resistance in a superconductor is at least 15 orders of magnitude smaller than in normal metals.

This applied flux can also be produced by a current in another part of the circuit, by mutual inductance (see Fig. 3.2). Then we would have $\Phi = Li + Mi_{app} = 0$, or $i = Mi_{app}/L$. This is, of course, the relation for an ordinary (ideal) transformer, but for the superconducting case it will continue to work at all frequencies, including both dc and high frequencies, because of the absence of any resistive loss.

Trapped Flux

In obtaining the relation that $\Phi = 0$ inside a superconducting loop, we have made one critical assumption: that $\Phi(t < 0) = 0$. This need not be the case (see Fig. 3.3). We could heat a loop of superconducting wire above its critical temperature T_c so that it exhibits a finite resistance. If we then apply a magnetic flux, it will quickly penetrate the loop as the screening current dies down. If we

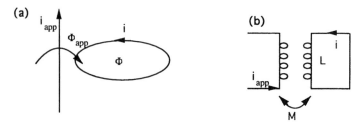

Figure 3.2. Inductive coupling of flux into superconducting loop. (*a*) Physical configuration. (*b*) Circuit schematic, with mutual inductance M between the two coils. The induced current $i = -Mi_{app}/L$ will act to maintain the total flux in the loop at the initial value.

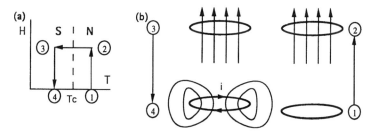

Figure 3.3. Flux trapping in superconduction loop. (*a*) Map of steps in H–T plane, showing separation into normal resistive (N) and superconducting (S) phases. (*b*) Configurations of field lines and current at each step in sequence showing persistent current and trapped flux in final step.

then cool down the loop below T_c, we have $R = 0$, and now there is a nonzero flux Φ_1 inside the loop. (Strictly speaking, as we briefly pointed out earlier, the flux in a superconducting loop is quantized. However, on the macroscopic scale, the flux is typically so much greater than the flux quantum that it can be regarded as a continuous variable.) This flux is now the conserved quantity; if we now remove the applied field, a screening current Φ_1/L will be set up to maintain Φ_1 inside the loop. To summarize this more general relation for a superconducting loop, we have

$$\Phi = Li + Mi_{app} + \Phi_{ext} + \Phi_1 = \Phi_1 \tag{3.6}$$

and its derivative, the loop equation for the voltage,

$$0 = L\frac{di}{dt} + M\frac{di_{app}}{dt} + \frac{d\Phi_{ext}}{dt} \tag{3.7}$$

Note that this circuit equation is consistent with either the presence or absence of trapped flux. It all depends on the initial condition at the instant when the loop is fully superconducting. The flux Φ_1 is effectively "trapped" inside the loop. The only way to change it is for the superconducting coil, or at least some part of it, to become resistive, normally by heating above T_c. Then the trapped flux could "escape" out of the ring, as the current dies out in an L/R time (as in Fig. 3.1).

The same principle can be applied to an N-turn loop such as a solenoid, provided we recognize that the total flux $\Phi_{tot} = NBA$, where A is the area of a single turn. In the charging circuit for a superconducting magnet, there is typically a short length of superconducting wire across the input leads of the magnet (Fig. 3.4). This creates a closed superconducting loop, so that the flux in the solenoid cannot be changed. But this superconducting short, known as a "thermal switch," is mounted on a small heater that can heat the section of wire above T_c while the rest of the superconductor is maintained below T_c. This thermal switch now becomes a resistor, and the flux in the superconducting solenoid can be changed by applying a small voltage V_0 across this resistor, in parallel with the inductive coil. This voltage is used to ramp up the current in the inductor at a constant rate V_0/L. When the desired operating current in reached, the heater current for the thermal switch is turned off. Then, the current in the power supply is reduced, while the superconducting solenoid maintains the same current, returning it through the now-superconducting shunt. In the end, the leads can be disconnected from the solenoid (to reduce thermal conduction from room temperature), and the solenoid will maintain in this persistent current state indefinitely (or as long as it is kept below T_c).

When there is flux Φ trapped in a superconducting loop with inductance L, there is also stored energy $\mathscr{E} = \Phi^2/2L$, with no loss of energy with time (since $R = 0$). This is also equivalent to $\frac{1}{2}Li^2$, where i is the circulating current, and also to the volume integral of the magnetic field energy $B^2/2\mu_0$. For example, for

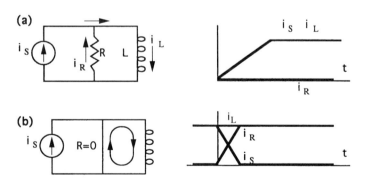

Figure 3.4. Charging of superconducting solenoid and persistent current state. (*a*) Schematic of charging circuit with thermal switch "on," creating resistive shunt, and associated time dependence of currents. (*b*) Fully charged solenoid with thermal switch "off," creating superconducting shunt and trapping flux in solenoid. The current source can then be ramped down to zero and the leads removed. Discharging the circuit requires following the same steps in reverse.

a long solenoid with area A and length l, with N turns, we have $L = N^2\mu_0 A/l$, $\Phi = NBA$, and $B = \mu_0 Ni/l$ in the interior of the solenoid. The stored energy can be quite substantial for a large solenoid with a large magnetic field. This forms the basis for "superconducting magnetic energy storage" (SMES). Energy can be extracted by temporarily driving a section of the coil normal, as in the thermal switch. The voltage across this section can be used to drive an external electrical load.

Let us consider a specific example with a solenoid with $L = 10$ H and a thermal switch with $R = 100\ \Omega$. If the operating current is 100 A, the energy stored when fully charged is $\frac{1}{2}Li^2 = 5000$ J. This can be charged using a voltage of 1 V, taking 1000 s to ramp up to full power. The thermal switch would divert only 10 mA of current and would dissipate only 10 J during charging. After turning off the thermal switch, the current in the supply can be quickly reduced, diverting the return current into the thermal switch. Note that we must carefully reverse these procedures to discharge the magnet. It is *not* good practice simply to turn the thermal switch back on; this would dump the entire 5 kJ into the switch in short time, $\sim L/R = 0.1$ s, catastrophically heating up a good part of the magnet and boiling off the associated cryogens.

Meissner Effect

There is one important phenomenon of superconductivity that is *not* explained simply by zero resistance. Consider here a solid bulk superconductor with no holes, rather than a closed loop as above. If such a superconductor is heated above T_c and a dc magnetic field is applied, then flux will eventually penetrate the conductor. If the conductor is now cooled below T_c with the external field still applied (see Fig. 3.5), what happens to the flux lines that lie inside the superconductor? Are they "frozen in place" (as is the case for the superconducting loop in Fig. 3.3), since magnetic flux cannot move through a perfect conductor? Actually, in real superconductors under ideal conditions, for moderately small magnetic fields, the magnetic flux lines are expelled from the superconductor at the same temperature and time as superconducting phase coherence is established. This is known as the "Meissner effect" and is a consequence of the macroscopic quantum wave function. In certain respects, therefore, a superconductor acts like a *diamagnetic* material: It opposes a magnetic field, not simply a change in magnetic field.

One way to think about the Meissner effect is that the superconducting state acts like a gas bubble that nucleates when the metal is cooled below T_c. This bubble quickly expands to fill the entire body, pushing aside the magnetic lines of force associated with the normal phase. This implicitly suggests (correctly) that there may be a maximum magnetic pressure that the bubble can withstand before collapsing. We will discuss this later in the context of the critical magnetic field.

It is also worth noting that we can construct a set of equations and corresponding transmission line model of a superconductor that do indeed

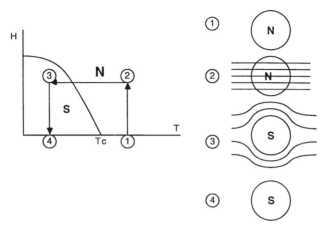

Figure 3.5. Illustration of Meissner effect, valid for ideal superconductor in moderately small magnetic field, whereby magnetic flux is expelled as a sphere is cooled into the superconducting state. (a) Map of steps in H–T plane showing separation into normal (N) and superconducting (S) phases. (b) Configurations of B-field at each step in sequence.

incorporate the Meissner effect. This involves dealing with the magnetic vector potential **A** defined such that

$$\mathbf{A} = -\int \mathbf{E}\, dt = -\Lambda \mathbf{J} \qquad \nabla \times \mathbf{A} = \mathbf{B} = \mu_0 \mathbf{H} \tag{3.8}$$

In essence, here **A** is the driving field for the supercurrent **J**. (In the superconductivity literature, this definition of **A** is called "the London gauge," and this set of equations is known as the "London equations.") This follows by integrating Eq. (2.7) once in time and defining the initial conditions. For a wave propagating in the z direction, we then have the equations

$$\frac{\partial H_y}{\partial z} = \frac{A_x}{\Lambda} \qquad \frac{\partial A_x}{\partial z} = \mu_0 H_y \tag{3.9}$$

If we consider a transmission line with $I = H_y a$ as before, but now take $V = -A_x a$, we get the picture in Fig. 3.6, with resistors having the same values as the inductors in Fig. 2.2. Clearly, this line is less intuitive (inductance terms are indicated by resistors), and the units are not even consistent. Furthermore, Z_0 no longer has the significance here as the surface impedance of the superconductor, so one must be rather careful in using this picture. However, the propagation constant of the line is still $\gamma = \sqrt{\mu_0/\Lambda} = 1/\lambda$, corresponding to penetration of fields and currents within λ from the surface. In addition, because there are only resistors in this picture, initial conditions cannot give rise to any trapped flux. In fact, the possibility of both flux expulsion and flux trapping

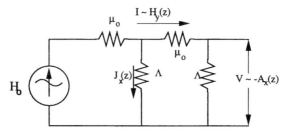

Figure 3.6. Resistive transmission line representation of London equations (3.8) in a superconducting plate in a magnetic field H_0, to be compared to the inductive line in Fig. 2.2. Here, currents along the line still represent $H(z)$, but the voltage across the line now represents the magnetic vector potential $A(z)$ (in the London gauge). Regardless of initial conditions, fields and currents are present only within a distance $\lambda = \sqrt{(\Lambda/\mu_0)}$ of the surface, consistent with the Meissner effect.

cannot be incorporated together in any classical model; they require the existence of a spatially varying quantum mechanical phase factor θ. [Equation (3.8) implicitly requires that θ is constant.] This will be discussed again in a later section.

I-Picture and M-Picture

On the macroscopic scale, at least for moderate magnetic fields, the magnetic flux density $B = 0$ inside a superconductor, as specified by the Meissner effect. There are two equivalent ways of viewing this situation. In the current viewpoint (the "I-picture"), the surface currents excluding this flux are explicitly calculated. Alternatively, in the magnetic viewpoint (the "M-picture"), the superconductor is regarded as a magnetic material, with an internal magnetization $\mathbf{M}$ that is responsible for the magnetic flux exclusion. In the literature on superconductivity, these two viewpoints are often used interchangeably, so it is important to understand how they are related.

In the I-picture, all magnetic fields are produced by currents using Ampere's law $\nabla \times \mathbf{H} = \mathbf{J}$ and $\mathbf{B} = \mu_0 \mathbf{H}$. For the superconductor, this means that an applied magnetic field is excluded by a surface screening current I_s, as indicated in Fig. 3.7a. In the M-picture, the currents are separated into the transport current $\mathbf{J}_t$, from an external power supply, and the "bound current" $\mathbf{J}_b$, associated with the equilibrium response of the material to a magnetic field. This leads to the field equations $\nabla \times \mathbf{H} = \mathbf{J}_t$, $\nabla \times \mathbf{M} = \mathbf{J}_b$, and $\mathbf{B} = \mu_0(\mathbf{H} + \mathbf{M})$, where $\mathbf{M}$ is the magnetization (the density of magnetic dipoles, i.e., microscopic current loops). It is important to realize that although $\mathbf{B}$ is the same in both pictures, they have different predictions for $\mathbf{H}$ inside the superconductor (see Fig. 3.7b). In a linear (nonhysteretic) magnetic material, $\mathbf{M} = \chi \mathbf{H}$, where χ is the (dimensionless) magnetic susceptibility, so that $\mathbf{B} = \mu \mathbf{H}$, where $\mu = \mu_0 \mu_r$ is the permeability and $\mu_r = 1 + \chi$ is the relative permeability. A typical linear magnetic material

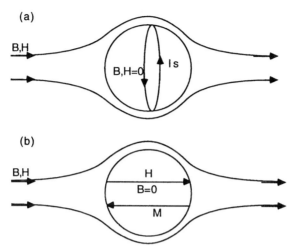

Figure 3.7. Exclusion of applied magnetic field from a superconducting sphere by screening current in I-picture (*a*) and by diamagnetic material in M-picture (*b*).

has both χ and $\mu_r \gg 1$. For a superconductor in the magnetic picture, $\mathbf{B} = 0$, so $\mathbf{M} = -\mathbf{H}$ and we must have $\mu = 0$ and $\chi = -1$, reflecting a fully diamagnetic response.

This fully diamagnetic response is reduced in a thin film with thickness $d < \lambda$. If the field H_0 is initially applied in the normal state for $T > T_c$, then it will be present on both sides of the film. Then, when the film is cooled into the superconducting state, the field will be only weakly expelled. Furthermore, the picture of uniform magnetization M is not valid. However, we can estimate the effective diamagnetic susceptibility of the film by integrating Eq. (2.14) to obtain

$$\chi = \frac{\langle M \rangle}{H_0} = \frac{\langle H \rangle}{H_0} - 1 = \frac{2\lambda}{d}\tanh\left(\frac{d}{2\lambda}\right) - 1 \to -\frac{d^2}{12\lambda^2} \qquad (3.10)$$

where $\langle H \rangle$ is the spatial average of the field inside the film. A different geometry, for example, a small spherical particle with radius $r \ll \lambda$ in a magnetic field H_0, would similarly yield $\chi \sim -r^2/\lambda^2$. The same would be true if we had a packed powder of small particles insulated one from another.

We note in passing that this dependence of χ is virtually the same as the formula for the quantum diamagnetism of nonmagnetic atoms. Almost any material that does not contain magnetic ions is very weakly diamagnetic due to lossless screening currents of electrons within each atom. Since $r \sim 1$ Å and $\lambda \sim 1000$ Å, typical values are $\chi \sim -10^{-6}$. This is another way in which a superconductor is like a macroscopic atom; what makes it special is that the lossless quantum screening currents are not limited to the atomic scale.

It may seem rather artificial to apply a model developed for ferromagnetic materials to a superconductor, which is apparently quite different. On the other hand, there are more similarities than differences. In particular, a magnetic material is composed of current loops on the atomic level. These are quantum mechanical currents, so that they cannot decay; only their direction can change. Consider a cubic array of $n \times n \times n$ microscopic current loops (each with current i, separation a, and magnetic dipole moment $\mu = ia^2$) all pointing in the same direction, as in Fig. 3.8a. This corresponds to a uniform magnetization $M = \mu/a^3 = i/a$. But note that all the internal currents cancel; we are left only with a surface current ni around the exterior of the material, as in Fig. 3.8b. This now looks a lot more like the situation of the superconductor in the I-picture, even though the screening currents are truly macroscopic only for the superconducting case. Apart from the ability of the superconductor to carry a lossless transport current and the fact that the magnetic susceptibility is positive for a magnet and negative for a superconductor, the two cases are formally rather similar.

It is standard to consider the effect of a superconductor placed in a uniform external magnetic field. The geometry is particularly simple for a long thin plate or rod with magnetic field H_0 applied parallel to the long dimension. Then, the external field is minimally perturbed, the effective magnetization is uniformly $M = -H_0$, and the circulating surface supercurrent per unit length is also H_0, oriented so as to produce a solenoidal field H_0 opposite to the applied field inside the superconductor.

Demagnetizing Factor

The situation is somewhat more complicated for other shapes and in general does not have a closed-form analytic solution. However, if the superconductor is

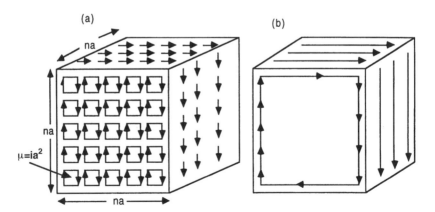

Figure 3.8. (a) Schematic of magnetic material consisting of cubic array of magnetic dipoles, each with dipole moment $\mu = ia^2$. (b) Equivalent macroscopic surface currents from (a).

in the shape of an ellipsoid and the field is applied along one of the principal axes, then we can use the concept of the "demagnetizing coefficient" N (between 0 and 1) to obtain a fairly simple solution. An ellipsoid is characterized by three such coefficients, which we can call N_x, N_y and N_z, where $N_x + N_y + N_z = 1$. Inside a body with magnetization along one of these directions, the effect is to produce a uniform reverse field NM so that the internal H-field is $H_{int} = H_0 - NM$, where H_0 is the applied magnetic field and N is the coefficient parallel to M. Since $M = -H_{int}$ for the superconductor, we can rearrange to obtain $H_{int} = H_0/(1-N)$.

The value of N has no simple form in general, but it simplifies in several important limits. First, for a sphere, $N = \frac{1}{3}$ in all directions by symmetry. Therefore, $H_{int} = 1.5H_0 = -M$. The external field is more complicated; it is a dipole field superimposed on the uniform applied field (Fig. 3.9a). However, since the tangential component of H is continuous across a boundary, the value of H_{ext} just outside the equatorial circle is also $1.5H_0$, corresponding to concentration of the magnetic flux that has been excluded from the sphere. The circumferential screening current per unit length is $I_s = M\sin\theta$, where θ is the angle from the polar axis parallel to the field.

Other simple ellipsoidal shapes include the oblate spheroid (shaped like a plate) and the prolate spheroid (cigar shaped) (see Fig. 3.9). For a field applied parallel to a long axis, $N \to 0$, so that $N \to 1$ for a field perpendicular to a plate and $N \to \frac{1}{2}$ for a field perpendicular to a rod. More generally, if $k \gg 1$ is the ratio of the long to the short diameter of the ellipsoid, $N \approx \pi/4k$ for a field parallel to the plate, and $N \approx \ln(k)/k^2$ for a field parallel to the rod.

For example, consider a magnetic field of 1 mT (10 G) perpendicular to a flat disk 10 μm thick and 1mm in diameter. To the extent that we can approximate this (roughly) as an oblate spheroid with major axis 1mm and minor axis 10 μm, $H_{int} = H_0/(1-N) \approx 2kH_0/\pi = 0.6\,\text{T}$. As noted above, this is also the real physical magnetic field just outside the corners of the disk, where the excluded field is most concentrated. This is particularly important when we consider (in the next section) that there is a critical magnetic field for a superconductor, above which it no longer excludes flux. The applied magnetic field H_0 may be well below the critical field H_c, whereas $H_{int} > H_c$. In this situation, the magnetic flux starts to penetrate the superconductor at the edges, where the external field is the highest.

Magnetic Shielding

In some circumstances, a superconductor may be used for magnetic shielding, taking advantage of this property of flux exclusion. It is important to understand how this compares to conventional magnetic shielding using a high-permeability magnetic material. As Fig. 3.10 indicates, in the magnetic material (with $\mu \to \infty$), the lines of flux are "sucked into" the material of the shield and thus are kept away from the region inside the shield. For the superconductor (with $\mu = 0$), on the other hand, the flux is excluded from the superconductor

78 MAGNETIC PROPERTIES OF SUPERCONDUCTORS

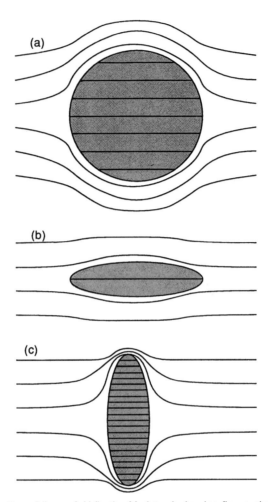

Figure 3.9. Illustration of lines of H (in the M-picture) showing flux exclusion from a superconducting ellipsoid in the presence of an applied field H_0. Inside the ellipsoid, $B = 0$ but $H = H_0/(1 - N)$, where N is the demagnetizing coefficient. (a) Sphere, with $N = \frac{1}{3}$ and $H = 1.5 H_0$ inside. (b) Prolate spheroid, with $N \ll 1$, $H \approx H_0$ inside, and only minimal perturbation to the field distribution outside. (c) Oblate spheroid, with $N \to 1$, $H \gg H_0$ inside, and concentration of field lines just outside sharp edges.

and also kept away from an object inside the superconductor shell. In both cases, demagnetization and nonlinear effects must be taken into account for a more complete picture.

3.2 CRITICAL CURRENTS AND FIELDS

As discussed earlier, the superconducting state forms and is stable at low temperatures, because it is lower in energy than the normal state. It can carry

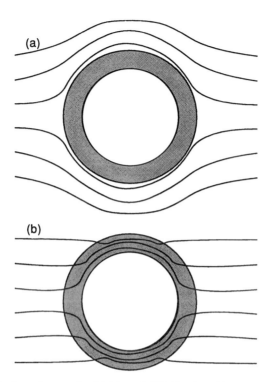

Figure 3.10. Illustration of magnetic shielding of B-field using either a superconducting shell (*a*) that excludes the flux or a high-permeability magnetic shell (*b*) that attracts the flux. In either case, an object in the center would be effectively shielded from the applied field.

lossless currents and screen out magnetic fields due to the energy gap. However, these currents raise the energy of the superconducting state until it is no longer stable. In this section we will demonstrate how this determines the critical currents and critical fields in a superconductor and some of the important implications. We will continue to assume in this section (unless otherwise noted) that the superconductor is in the Meissner state, with currents and fields only near the surface.

Critical Current in Thin Superconducting Film

Consider the case of dc current flow in a superconducting film with $d \ll \lambda$, in the absence of an external magnetic field, at $T = 0$. Apart from some current peaking at the edges, the current density should be uniform through the film. As we pointed out in Fig. 2.20d, the effective energy gap decreases as the Fermi sphere shifts. When the additional energy on the leading edge is equal to Δ, the top state below the gap has the same energy as the bottom state above the gap (at the trailing edge of the Fermi sphere), and Cooper pairs begin to break apart.

This determines the maximum current that the superconducting state can carry, which occurs for

$$\delta\mathscr{E} = \frac{\hbar^2 k_F}{m}\delta k = \Delta \tag{3.11}$$

which can be rearranged to yield

$$\delta k = \frac{\Delta}{\hbar v_f} = \frac{1}{\pi\xi_0} \tag{3.12}$$

This corresponds, in turn, to a critical current density

$$J_c = \frac{ne}{m}\hbar\,\delta k = \frac{\Phi_0}{\pi^2\xi_0\mu_0\lambda_L^2} \tag{3.13}$$

where ξ_0 and λ_L are the quantities as defined in Table 2.1. An alternative approach, which is more valid close to T_c, defines the critical current by setting the total kinetic energy density of the superconducting electrons n_s equal to the condensation energy [from Eq. (2.60)]:

$$U_k = \frac{1}{2}n_s m v_s^2 = \frac{1}{2}\mu_0\lambda^2 J_c^2 = \frac{\Phi_0^2}{158\xi^2\mu_0\lambda^2} = U_c \tag{3.14}$$

which yields

$$J_c = \frac{\Phi_0}{8.9\xi\mu_0\lambda^2} \tag{3.15}$$

This result is virtually the same as Eq. (3.13) above if we take the more general phenomenological values of these parameters, as given in Table 2.1. [The more complete GL calculation, which takes into account the fact that n_s (and λ) depend somewhat on J, replaces 8.9 with 16.3 (Orlando and Delin, 1991, p. 503).] Using typical values (for Nb) of $\lambda \approx 100$ nm and $\xi \approx 40$ nm, we can estimate $J_c \sim 3 \times 10^{11}$ A/m^2, an extremely large value. For example, in a wire with cross section $\approx \lambda^2$, this gives a total critical current $I_c > 100$ kA! Critical currents in real samples are somewhat smaller than this, as we will discuss later, but can still be very large. In addition, we can use Eq. (3.15) to show that the temperature dependence of J_c should go to zero as $(1 - T/T_c)^{3/2}$ near T_c and should level off below about $\frac{1}{2}T_c$.

Critical Magnetic Field

A similar picture holds for a thicker superconducting film or wire, where the current flows on the outside perimeter to a depth of λ. When the current density on the surface exceeds J_c, the superconducting state is destabilized. This will be true either for an imposed transport current or for a screening current in response to an applied magnetic field parallel to the surface. In either case, we can speak of a critical magnetic field H_c (or $B_c = \mu_0 H_c$) given by

$$H_c = J_c \lambda \approx \frac{\Phi_0}{10 \xi \mu_0 \lambda} \tag{3.16}$$

(the accurate numerical factor is $\pi\sqrt{8}$ from GL theory). This is also related to the condensation energy by

$$U_c = \frac{1}{2}\mu_0 H_c^2 = \frac{B_c^2}{2\mu_0} \tag{3.17}$$

This takes the form of an energy density associated with a magnetic field, and indeed we can think of $\frac{1}{2}\mu_0 H^2$ as the energy density needed to expel a magnetic field H from the interior of a bulk superconductor. When H exceeds H_c, this positive flux expulsion energy is greater than the negative condensation energy, and a superconductor in the Meissner state is no longer stable. Furthermore, $H_c(T=0)$ is often treated as a fundamental parameter of a given (bulk) superconductor, partly because the effect of impurities (on ξ and λ) cancels out. Typically $H_c(0) \sim 10^5$ A/m and $B_c(0) \sim 0.1$ T, or an order of magnitude higher for the high-T_c superconductors. This is *not* an extremely large magnetic field; a permanent magnet can produce a magnetic field of about 1 T. In fact, practical high-field superconductors can operate up to much higher magnetic fields.

As pointed out earlier, we can think of a superconductor in the Meissner state in analogy to a gas bubble, and we can make that analogy more quantitative here. In particular, the units of pressure, (newtons per square meter) are identical to the units of energy density (joules per cubic meter), and the condensation energy density $U_c = \frac{1}{2}\mu_0 H_c^2$ does indeed play the role of the internal hydrostatic pressure inside the bubble. Similarly, the magnetic energy density $\frac{1}{2}BH = B^2/2\mu_0 = \frac{1}{2}\mu_0 H^2$ is equivalent to a magnetic pressure acting on this bubble. When the magnetic pressure inward is greater than the superconducting pressure outward (i.e., for $H > H_c$), the bubble will collapse.

But it is important to realize that it is not the magnetic field itself that directly destroys the superconductivity. Rather, it is the kinetic energy of the screening currents. We can see this if we place a thin film with $d \ll \lambda$ parallel to a magnetic field H_0. As indicated in Fig. 2.4c, the screening currents have a maximum value $J = H_0 d/2\lambda^2$, which reaches the critical current density J_c at a critical field

$$H_{c\|} \approx \frac{J_c 2\lambda^2}{d} = \frac{H_c 2\lambda}{d} \gg H_c \tag{3.18}$$

The more accurate GL result is even larger than this (by a factor of $\sqrt{6}$). Note that Fig. 2.4c also indicates that the magnetic field fully penetrates the film in this limit. It turns out to be true more generally that a superconductor can coexist with a very large magnetic field only if the field penetrates into the superconductor, in apparent contradiction to the Meissner effect. This will be discussed further in Section 3.3.

Phase Diagram of Superconductor

For the time being, let us focus on a bulk superconductor that obeys the Meissner effect and therefore has the bulk critical field $H_c(T)$. A common empirical form for the temperature dependence (often grouped with the two-fluid dependence of n_s) is

$$H_c(T) = H_c(0)\left[1 - \left(\frac{T}{T_c}\right)^2\right] \qquad (3.19)$$

as shown in Fig. 3.11a. We can think of this as an effective "phase diagram" in the H–T plane. Below the parabola, the system will be superconducting; outside, it will be normal. This relation can also be inverted to yield a depressed critical temperature $T_c(H)$ in the presence of a magnetic field.

We can generalize this further by considering the presence of both an applied magnetic field H and a transport current density J. Since in this picture the key parameter is the local current density, which is the vector sum of the transport and screening contributions, this depends on the relative directions of J and H. The general effect is for the critical transport current J_c to be reduced to zero at H_c. For example, consider a narrow superconducting wire with a magnetic field applied parallel to the wire. The screening currents will flow circumferentially, perpendicular to the transport current that flows along the wire, so that the two

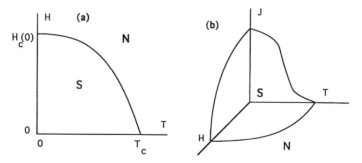

Figure 3.11. Phase diagrams for a type I superconductor. (a) Critical magnetic field $H_c(T)$ showing S phase below parabola and N phase above. (b) Superconducting critical surface as a function of T, H, and J.

current densities should add in quadrature (i.e., as the square root of the sum of the squares). Since the critical current is reached when the total current density reaches J_c, we have

$$J_c(H) = \sqrt{J_c^2(0) - \left(\frac{H}{\lambda}\right)^2} \approx J_c(0)\sqrt{1 - \left(\frac{H}{H_c}\right)^2} \quad (3.20)$$

We can combine this with the $H_c(T)$ plot to obtain a three-dimensional H–J–T phase diagram (Fig. 3.11b). If the operating point lies inside the paraboloid, the wire should remain superconducting.

Self-Heating of Superconducting Wire

What happens when the critical current of a superconducting wire is exceeded? Does it immediately go into the normal state? In general this is rather complicated, involving vortices and nonequilibrium effects. But in all cases, there is a voltage induced and power is dissipated. Given the extremely large currents, this power can be substantial, leading to self-heating of the wire. Unless special considerations are taken to remove this excess heat and/or to divert the current when a voltage appears, it is possible for the entire superconductor to be heated above T_c. Then, the heating may be sufficient to maintain the sample in the normal state, at least until the current is reduced below a lower I_{c2} (Fig. 3.12). Ironically, this heating effect is more prominent at lower temperatures, where I_c is larger and heat capacities are smaller. For these reasons, superconducting wires are often designed with low-resistance normal shunts such as copper or silver, which serve to divert both heat and current in the event that the critical current may be exceeded locally.

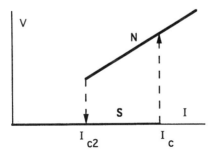

Figure 3.12. Typical $V(I)$ for a current-biased superconducting wire or film at a fixed "bath temperature." If I exceeds the critical current I_c, self-heating drives the wire into the normal state until I is reduced below a smaller return current I_{c2}.

Intermediate State

Thus far in dealing with critical fields, we have assumed a magnetic field parallel to the surface of a superconductor. More generally, we must consider demagnetizing effects (as in Section 3.1). For example, for a bulk superconducting sphere with factor $N = \frac{1}{3}$, the internal field in the M-picture is $H_{int} = H_0/(1 - N) = 1.5 H_0$. So if the applied field is $0.8 H_c$, then we have $H_{int} = 1.2 H_c$, and the sphere should be driven out of the superconducting state. But if the sphere becomes completely normal, then there is no longer any diamagnetism (after the initial transient) and $H_{int} = H_0 = 0.8 H_c$, at which point the sphere should become superconducting again. This paradox can be resolved only if the sphere is in a mixture of superconducting and normal domains, known as the "intermediate state," for applied fields between $H_{c1} = H_c(1 - N)$ and H_c. For a field applied perpendicular to a thin plate with $N \to 1$, this lower effective critical field approaches zero!

Looking more closely at the superconductor in this intermediate state (Fig. 3.13a), it should consist of two types of regions or domains: those with $B = 0$, which are still superconducting, and those with $B > B_c$, which are normal. (This is *not* the same as the mixed state with vortices, which we will introduce in the next section.) In that way, each domain should be stable. This can be achieved by having alternating N and S regions, in which the flux lines are concentrated in the N domains and excluded from the S domains, via the Meissner effect and screening currents at the surfaces of the S domains. The fraction of area (perpendicular to the field) in normal domains goes continuously from 0 at H_{c1} to 1 at H_c. So for this geometry, the H–T phase diagram of

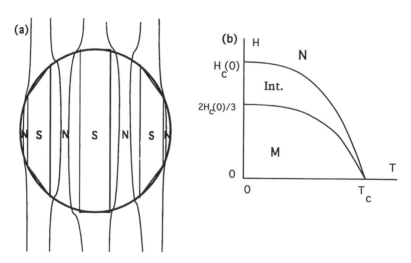

Figure 3.13. (*a*) Type I superconducting sphere in the intermediate state composed of alternating normal and superconducting domains. (*b*) Phase diagram showing three regimes: Meissner, intermediate, and normal.

a bulk superconductor separates into three regimes (Fig. 3.13b), Meissner, intermediate, and fully normal.

Normal/Superconducting Interface Energy

The domains in the intermediate state of a superconductor are analogous to magnetic domains in a ferromagnetic material. In both cases, the size of the domains is determined by a balance between the external magnetic field energy and the surface energy between the domains. The external field energy is minimized if the domains are very small. A positive surface energy, in contrast, will tend to make the domains larger. For ferromagnetic materials, the surface energy is associated with domain walls and is always positive. For a superconductor, the N/S interface energy can be of either sign, with profoundly different consequences. For a type I superconductor, where the N/S interface energy is positive, the domains in the intermediate state remain macroscopic. This is the classic superconductor that we have been discussing, where a domain is either in the Meissner state or in the fully normal state.

In contrast, for a type II superconductor, where the N/S interface energy becomes negative in large magnetic fields, the energetics favor breaking up the S and N domains into microscopically small dimensions, and the picture of well-defined N and S domains is no longer valid. The magnetic field penetrates in quantized vortices, and the N regions are tubes of radius ξ. On a macroscopic scale, the magnetic field penetrates uniformly through the entire material, so that the superconductor is no longer in the Meissner state. As we will show later in the chapter, this "mixed state" may or may not be properly superconducting. However, it has the advantage that the screening currents are greatly reduced, so that as in the case of the thin superconducting film, the effective maximum critical field can be very much larger than H_c. All practical superconducting materials are actually type II superconductors.

We can obtain some insights into what makes a superconductor type I or type II by considering the surface energy between an N and S domain in the intermediate state (Fig. 3.14). Inside the N domain, the magnetic field $H = H_c$ and $n_s = 0$. As one enters the S domain, n_s rises to its full value over a scale of ξ, while H falls to zero over a scale of λ. Compare this to the situation of a sharp N/S boundary. Some of the condensation energy at the surface of S is lost, corresponding to a positive surface energy $U_c \xi = \frac{1}{2}\mu_0 H_c^2 \xi$ per unit area. On the other hand, the field penetration reduces the energy by an equivalent field energy $\int \frac{1}{2}\mu_0 H^2 \, dx \approx \frac{1}{2}\mu_0 H_c^2 \lambda$. So the total surface energy is approximately

$$U_s = (\tfrac{1}{2}\mu_0 H_c^2)(\xi - \lambda) \tag{3.21}$$

If $\xi > \lambda$, then the surface energy is positive, and we have a type I superconductor. If $\xi < \lambda$, the surface energy is negative, and the superconductor is type II.

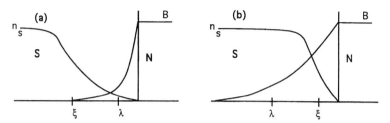

Figure 3.14. Interface between superconducting and normal domains in the intermediate state. (*a*) Type I superconductor with $\xi > \lambda$ and positive interface energy. (*b*) Type II superconductor with $\xi < \lambda$ and negative interface energy promoting formation of microscopic domains.

The more exact crossover from GL theory occurs at $\xi = \sqrt{2}\lambda$ (Tinkham, 1996, p. 134).

Two important consequences can be obtained directly from consideration of the interface energy. First, the interface energy will be negative for a type II superconductor even for fields below H_c, so that the Meissner state becomes unstable at a lower critical field H_{c1} (even apart from the demagnetizing factor). Second, once the S domains become smaller than λ, the magnetic field will largely penetrate the superconductor and the screening currents are relatively small. As suggested by the estimate above of the critical field for a thin film, this implies that for a type II superconductor the upper critical field H_{c2} can be much larger than H_c, so the phase diagrams shown thus far are incomplete.

All high-T_c superconductors are type II, as are all practical low-T_c superconducting alloys and compounds as well as the element with the highest T_c, Nb. But ironically, all other elemental superconductors are type I, which were the ones that were studied during the early years of superconductivity. Partly for this reason, it was initially believed that all superconductors were type I, effectively ruling out high-field applications. It was not until after type II superconductors and vortices were theoretically predicted (as a consequence of GL theory) that this conclusion was reexamined and practical high-field superconducting alloys developed. A more complete understanding of type II superconductors requires consideration of quantized vortices, which are the subject of the next section.

3.3 FLUXOID QUANTIZATION AND VORTICES

Macroscopic Quantum Wave Function

As we have discussed, a superconductor is characterized by a macroscopic quantum wave function that consists of a superposition of states of Cooper pairs, each with charge $-2e$. For a classical particle of charge $-2e$ and mass

$2m$ in an electromagnetic field, the energy and momentum relations are given by

$$\mathscr{E} = mv^2 - 2eV \qquad \mathbf{p} = 2m\mathbf{v} - 2e\mathbf{A} \qquad (3.22)$$

where $\mathbf{E} = -\nabla V - \partial \mathbf{A}/\partial t$ and $\mathbf{B} = \nabla \times \mathbf{A}$. The second term in the momentum equation may look odd, but it is needed for self-consistency. (This expression for $\mathbf{p}$ is sometimes known as the "canonical momentum.") For a quantum particle, the wave function is given by $\psi = |\psi|\exp(i\theta)$, where the phase function $\theta(\mathbf{r}, t) = \mathbf{k} \cdot \mathbf{r} - \omega t$ gives rise to waves with $\omega = -\partial\theta/\partial t = \mathscr{E}/\hbar$ and $\mathbf{k} = \nabla\theta = \mathbf{p}/\hbar$. (We assume here that the wave intensity $|\psi|^2 = n_s = $ const.) Focusing on the momentum equation, we have

$$\hbar \nabla\theta = 2m\mathbf{v} - 2e\mathbf{A} \qquad (3.23)$$

This is an important equation, from which both the Meissner effect and flux quantization can be obtained rather directly. We can also express it in terms of the supercurrent $\mathbf{J}_s$:

$$\mathbf{J}_s = -n_s e\mathbf{v} = -\left(\frac{\hbar}{2e\Lambda}\right)\nabla\theta - \frac{\mathbf{A}}{\Lambda} \qquad (3.24)$$

where $\Lambda = m/n_s e^2$ as before. If we assume that $\theta = $ const, then we have $\mathbf{A} = -\Lambda \mathbf{J}_s$, which was shown earlier to result in the Meissner effect: expulsion of magnetic flux by screening currents, except within λ of the surface. More generally, the term with the phase θ drops out if we take the curl of both sides (since the curl of a gradient is always zero), and we obtain the "second London equation":

$$\nabla \times \mathbf{J}_s = -\frac{\nabla \times \mathbf{A}}{\Lambda} = -\frac{\mathbf{B}}{\Lambda} \qquad (3.25)$$

which also leads to screening of magnetic flux from the interior of a superconductor. We can also see this by integrating Eq. (3.25) over a surface S within the superconductor, bounded by a closed path C (Fig. 3.15a):

$$\Phi = \int_S \mathbf{B} \cdot d\mathbf{S} = -\Lambda \int_S (\nabla \times \mathbf{J}_s) \cdot d\mathbf{S} = -\oint_C \mathbf{J}_s \cdot d\mathbf{l} \qquad (3.26)$$

where we have made use of a standard Stokes theorem transformation from the surface integral to the closed line integral. If there are no net circulating currents in the interior of a superconductor, there can be no magnetic flux Φ, and hence $B = 0$. This implicitly assumes that the superconductor has no holes in it, so that Eq. (3.25) is valid at all points inside the path.

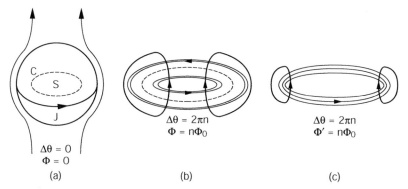

Figure 3.15. Fluxoid quantization in a superconducting loop: (a) $\Phi = 0$ inside closed superconducting path with no holes; (b) $\Phi = n\Phi_0$ inside superconducting path away from boundary, with a hole in the interior; (c) fluxoid $\Phi' = n\Phi_0$ for superconducting path with screening current.

Fluxoid Quantization

Of course, as we pointed out earlier, a superconducting ring can trap magnetic flux. Consider now a superconducting wire in the form of such a ring and a path that goes in the interior of the wire around the ring (Fig. 3.15b). Then $J_s = 0$, and we have

$$\Phi = \int B \cdot ds = \int A \cdot dl = \frac{\hbar}{2e} \int \nabla \theta \cdot dl = \frac{\hbar}{2e} \Delta \theta \qquad (3.27)$$

where $\Delta\theta$ is the change in the phase in going around the loop. Since θ must be single valued (modulo 2π), $\Delta\theta$ must be a multiple of 2π, so that we have

$$\Phi = \frac{nh}{2e} = n\Phi_0 \qquad (3.28)$$

where n is an integer. So not only is the magnetic flux in a superconducting ring conserved, but it is also quantized to one of a discrete set of values, which are multiples of the flux quantum $\Phi_0 = h/2e = 2.07 \times 10^{-15}$ Wb $= 2.07 \times 10^{-7}$ G-cm^2. For a macroscopic ring of ~ 1 cm, the B-field associated with a single flux quantum is very small, less than 1 µG, as compared with the earth's field of order 1 G, so this discreteness may not be noticed. On the other hand, this becomes more significant for microscopic rings, and furthermore, superconducting devices based on flux quantization can be designed to measure magnetic flux changes even much *smaller* than Φ_0.

We can generalize this result to the situation where the line integral includes regions where J_s may be nonzero. Consider, for example, a wire with cross section less than $\lambda(T)$, so that $J_s = I/S$ is uniform across the wire (Fig. 3.15c).

Then, from Eq. (3.24) we have

$$\Phi = \int \mathbf{B} \cdot d\mathbf{S} = \int \mathbf{A} \cdot d\mathbf{l} = n\Phi_0 - \Lambda \int \mathbf{J_s} \cdot d\mathbf{l} = n\Phi_0 - L_k I \quad (3.29)$$

where $L_k = \Lambda l/S$ is the kinetic inductance of the wire. We can rearrange this in the form

$$\Phi' = \Phi + L_k I = (L_m + L_k)I = n\Phi_0 \quad (3.30)$$

So it appears here that the *fluxoid* Φ', which includes both the usual magnetic flux and the "kinetic flux" associated with the kinetic inductance, is properly the parameter that is quantized. This is valid for either a type I or a type II superconductor.

It is also worth noting that fluxoid quantization is formally equivalent to the quantization of angular momentum in the Bohr model of the hydrogen atom. If we consider supercurrent flowing in a ring of radius r, then the condition $\Delta\theta = 2\pi n$ corresponds to angular momentum

$$\mathbf{L} = \mathbf{r} \times \mathbf{p} = r(\hbar\nabla\theta) = \frac{r\hbar(\Delta\theta)}{2\pi r} = n\hbar \quad (3.31)$$

exactly the result for the atomic case.

The Single Vortex

Consider now a superconducting cylinder of radius $R \gg \lambda$, with a cylindrical hole along its axis, of radius $a \ll \lambda$. This can have flux trapped in this central hole. From Eq. (3.24), if we take a line integral around the center at an intermediate radius r, we have

$$\Phi' = n\Phi_0 = \Phi(r) + \Lambda J_s 2\pi r \quad (3.32)$$

This is similar to the one-dimensional problem solved in Chapter 2 using a transmission line, except that this now has cylindrical symmetry. For $r \ll \lambda$, when the enclosed flux is still very small, we have

$$J_s \approx \frac{n\Phi_0}{2\pi r \mu_0 \lambda^2} \quad (3.33)$$

This is analogous to the current distribution in a fluid vortex, so that the same term is also used here. When r gets very small, this current density can get very large and can exceed the critical current density J_c for $r < \approx \xi$ [see Eq. (3.15)], which is possible in a type II superconductor with $\xi \ll \lambda$ (the large-κ limit).

Where this happens, the superconducting state is unstable, $n_s \to 0$, and the material acts like a normal non-superconducting metal. In fact, this normal "vortex core" of radius $\approx \xi$ acts just like a hole in the superconductor, even if there is no hole to begin with. This is the natural minimum size of a normal "domain" in a superconductor. If the N/S interface energy is negative (which is the case for type II superconductors, with $\xi < \lambda$), the magnetic flux will tend to be distributed among as many vortices as possible, so that $n = 1$ for each vortex. Each vortex is then a single "fluxon" containing one flux quantum Φ_0 and corresponding to $\Delta\theta = 2\pi$ around the normal core.

The structure of a single vortex is shown in Fig. 3.16. The phase function $\theta(\phi) = \phi$, the azimuthal angle in cylindrical coordinates, and is independent of the radial distance r. We take the "branch cut" at $\phi = 0$ and have a "hole" in the superconductor for $r < \xi$, which serves to cut off the singularity at $r = 0$. There is a current flowing in the ϕ direction corresponding to a magnetic flux Φ in the $+z$ direction. We may describe the superconductor surrounding the vortex core by modifying the inductive transmission line equivalent that we used earlier (Fig. 2.2), where both the series and shunt inductances are scaled by the circumference $2\pi r$ (Fig. 3.16b). The effective propagation constant remains $\gamma = 1/\lambda$, but the characteristic impedance is now $Z_0 = j\omega\mu_0\lambda(2\pi r)$, which depends on radial distance, so that the solution is no longer simple exponential decay (see below). We can think of the superconductor as initially being cut along the $\phi = 0$ line, and a voltage $V = \Phi_0\delta(t)$ (or any other function whose time integral is Φ_0) is applied across this line. This introduces a phase shift of 2π, so that now the cut can be reconnected.

An equivalent picture can be obtained by integrating in time to obtain the resistive transmission line of Fig. 3.16c, similar to that in Fig. 3.6 but with the same $2\pi r$ scaling of the resistances to account for the cylindrical geometry. Here, the "voltage source" is applying a flux Φ_0, and the voltage on the line represents $\Phi_0 - \Phi(r)$, where $\Phi(r) = \int \mu_0 H_z(r) 2\pi r \, dr = A_\phi(r) 2\pi r$ is the flux inside a path of radius r. Applying the transmission line equations, we have

$$\frac{\partial \Phi(r)}{\partial r} = \mu_0 2\pi r H_z(r) \tag{3.34}$$

$$J_\phi(r) = -\frac{\partial H_z}{\partial r} = \frac{\Phi_0 - \Phi(r)}{\Lambda 2\pi r} = \frac{\hbar}{2e\Lambda}\frac{2\pi}{2\pi r} - \frac{A(r)}{\Lambda} \tag{3.35}$$

where the first equation follows from the definition of flux and the second equation is consistent with Eq. (3.24) for $\Delta\theta = 2\pi$. These lead to a second-order differential equation

$$\frac{\partial^2 \Phi}{\partial r^2} - \frac{1}{r}\frac{\partial \Phi}{\partial r} + \frac{\Phi}{\lambda^2} = \frac{\Phi_0}{\lambda^2} \tag{3.36}$$

with the boundary condition that $\Phi(r = \xi) = 0$. The middle term on the left is the one that makes it more difficult to solve; the solution can be expressed in terms of "modified Bessel functions" (Orlando and Delin, 1991, p. 268).

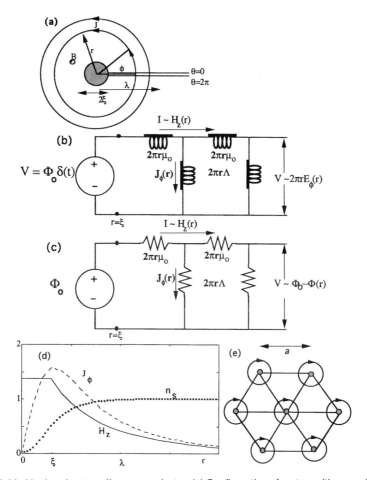

Figure 3.16. Vortices in a type II superconductor. (*a*) Configuration of vortex, with normal core of size $\approx \xi$, surrounded by screening current to a distance $\sim \lambda$. (*b*) Representation of currents and fields around vortex using inductive transmission line picture. (*c*) Vortex model using resistive transmission line. (*d*) Radial dependence of H_z, J_ϕ, and n_s near single vortex, in normalized units. (*e*) Triangular vortex lattice with lattice constant *a*, with each vortex core surrounded by screening currents.

However, both fields and currents still roll off essentially exponentially for $r \gg \lambda$. Approximate expressions for J and H are given below and illustrated in Fig. 3.16*d*:

$$H(r) \approx \begin{cases} \dfrac{\Phi_0}{2\mu_0 \lambda \sqrt{2\pi r \lambda}} \exp\left(\dfrac{-r}{\lambda}\right) & r \gg \lambda \\ \dfrac{\Phi_0}{2\pi\mu_0 \lambda^2} \ln \dfrac{\lambda}{r} & r \ll \lambda \end{cases} \quad (3.37)$$

92 MAGNETIC PROPERTIES OF SUPERCONDUCTORS

$$J(r) \approx \begin{cases} \dfrac{\Phi_0}{2\mu_0\lambda^2\sqrt{2\pi r\lambda}} \exp(-r/\lambda) & r \gg \lambda \\ \dfrac{\Phi_0}{2\pi\mu_0\lambda^2 r} & r \ll \lambda \end{cases} \quad (3.38)$$

For $r \ll \lambda$, the vortex core on the scale of ξ serves to cut off the divergence in J and B. The more accurate behaviors for J, H, and n_s are somewhat rounded at $r \approx \xi$.

Vortex Lattice

We have been describing the properties of a single vortex, but of course, when there is one vortex, there are likely to be many more. Since vortices are surrounded by screening currents, they will interact with one another on the scale of order λ or less. In fact, vortices of the same sense (with B in the same direction) repel each other. This has the effect of making the magnetic field penetration as uniform as possible. In ideal cases, the large number of parallel vortices form a two-dimensional lattice, generally a triangular lattice as illustrated in Fig. 3.16e. For such a lattice, the magnetic flux density $\langle B \rangle$ (spatially averaged over the vortices) can be related to the lattice constant a by the relation

$$\langle B \rangle = n_v \Phi_0 = \frac{2\Phi_0}{\sqrt{3}a^2} \quad (3.39)$$

where n_v is the density of vortices per unit area. This relation assumes a linear superposition of the fields of each vortex and that each one contributes a total flux of Φ_0. For large magnetic fields, there is very little diamagnetism (as discussed below), so that $\langle B \rangle \approx \mu_0 H$.

The vortex lattice in a type II superconductor can be imaged by several different techniques. With a magnetic field perpendicular to a plate, the magnetic field emanating from the surface will be somewhat larger near the vortex cores. This would attract fine magnetic particles, for example, forming the basis of a magnetic decoration method. Alternative methods include magneto-optic imaging and Lorentz electron microscopy.

Critical Fields in Type II Superconductor

A type II superconductor in the vortex state exhibits additional critical fields. Here, H_{c1} is the lower critical field at which it is energetically favorable for the first vortices to enter, and H_{c2} is the upper critical field, at which the last remnants of superconductivity disappear. We can estimate their magnitudes using simple arguments. First, once it is energetically favorable for vortices to enter, they will continue to enter until they begin to interact, which occurs at a distance $\sim \lambda$. So we can estimate $H_{c1} \sim \Phi_0/\mu_0\pi\lambda^2$. The more accurate GL result (for large $\kappa = \lambda/\xi$) is similar,

$$H_{c1} \approx \frac{\Phi_0}{4\pi\mu_0\lambda^2}\ln\kappa = \left(\frac{H_c}{\sqrt{2\kappa}}\right)\ln\kappa \tag{3.40}$$

and can be much smaller than H_c for large κ. Note also [from Eq. (3.37)] that the magnitude of H in the core of an isolated vortex is approximatly $2H_{c1}$.

For the upper critical field, given the normal cores of radius $\approx \xi$, we can estimate that at $H_{c2} \approx \Phi_0/\pi\mu_0\xi^2$ the entire sample will consist of such normal cores, and no superconductivity will be left. The more accurate GL result is very similar:

$$H_{c2} = \frac{\Phi_0}{2\pi\mu_0\xi^2} = \sqrt{2}\kappa H_c \tag{3.41}$$

This result is not limited to large κ and shows that $\kappa = 1/\sqrt{2}$ divides type I from type II. Apart from the weak factor of $\ln\kappa$, $H \approx \sqrt{(H_{c1}H_{c2})}$, that is, the geometric mean of the upper and lower critical fields.

Note that the GL parameter $\kappa = \lambda(T)/\xi(T)$ is only weakly dependent on temperature but is strongly dependent on the impurity level. In the dirty limit, $l \ll \xi_0$, which applies to some practical superconductors, $\kappa \approx \lambda_L/l$. This means that although H_c may be an intrinsic property of a given superconducting compound, H_{c1} and H_{c2} typically depend strongly on the impurity level. In particular, $H_{c2} \sim 1/l \sim \rho_n$, so that, ironically, the more resistive a material is above T_c, the better it is as a high-field superconductor below T_c. Consider, for example, the critical fields of superconducting Nb. If it is pure, both λ and ξ are about 40 nm, corresponding to $\kappa = 1$, making pure Nb marginally type II. In contrast, the alloy Nb–Ti has the same value of $T_c = 9$ K and $B_c = 0.2$ T but is very impure, with $\lambda \approx 300$ nm and $\xi \approx 5$ nm, or $\kappa \approx 60$. These numbers correspond to $B_{c2} \approx 0.3$ T for Nb, while Nb–Ti has $B_{c2} \approx 17$ T. This is why the dirty alloy is used as a high-field superconducting material while the pure metal is not. It is also worth remembering that some materials, in particular the high-temperature superconductors, are highly anisotropic, and this anisotropy is reflected in their critical fields as well.

Phase Diagram and Diamagnetism

Since κ is only weakly dependent on temperature, the temperature dependences of H_{c1} and H_{c2} are very similar to that of H_c, going roughly as $1 - (T/T_c)^2$. A phase diagram of a type II superconductor is shown in Fig. 3.17a, showing three different phases. For $H < H_{c1}$, the fully superconducting Meissner state is still stable. For $H > H_{c2}$, the material is fully in the normal resistive state. And between H_{c1} and H_{c2}, there is the *mixed* state (not to be confused with the *intermediate* state of Fig. 3.13), which consists of a microscopic mixture of "normal" vortex cores and superconducting regions with screening currents.

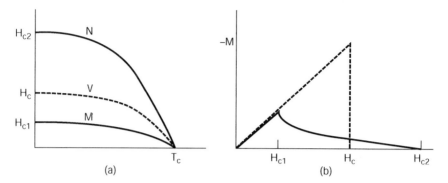

Figure 3.17. Critical fields and diamagnetism of equilibrium type II superconductor. (a) Temperature dependence of H_{c1} and H_{c2} showing stable regimes of Meissner, mixed, and normal states. (b) Magnetization $M(H)$, showing complete diamagnetism in Meissner state below H_{c1}, decreasing continuously to $M = 0$ at H_{c2}. Dependences $H_c(T)$ and $M(H)$ for a type I superconductor are shown in dashed lines for comparison.

It is clear that the mixed state of a bulk superconductor does not exhibit the full diamagnetism of the Meissner state, in which $B = 0$ and $M = -H$ (in the M-picture). In particular, we can define

$$M = \frac{\langle B \rangle}{\mu_0} - H = \frac{n_v \Phi_0}{\mu_0} - H \tag{3.42}$$

Just above H_{c1}, many vortices can suddenly enter, up to a density of order $n_v \sim 1/\lambda^2$, so we would expect M to drop sharply. For larger fields, as vortices become more closely spaced, M should drop more slowly, going to zero at H_{c2}, when the material consists of a dense array of vortex cores with no screening currents remaining (Fig. 3.17b).

There is an important assumption in the previous analysis, namely that the vortices can move around freely to achieve equilibrium with the applied magnetic field. However, for many practical type II superconductors, this is not always valid. More specifically, there are typically barriers to vortices crossing the surface as well as moving inside the superconductor. In fact, such effects that restrict the motion of vortices are essential if the mixed state is to exhibit zero resistance, as one would like for a practical high-field superconductor. These irreversible properties of type II superconductors will be discussed below in Section 3.5. But before proceeding with this, we will introduce in the next section the principle of duality in circuits and show how this can provide powerful insights into the effects of vortex motion in superconductors.

3.4 DUALITY AND FLUX FLOW

It is a well-known theorem in circuit theory that for every planar circuit, there is a dual circuit that satisfies the same circuit equations. It is perhaps not as widely appreciated that this dual circuit actually describes the transport of magnetic flux in the original physical system. Flux quantization in superconducting circuits brings an additional dimension to this picture. In particular, motion of a quantized fluxon (or vortex) in a superconductor is strongly analogous to motion of an electron or hole in a semiconductor, so that superconducting circuits and devices that are dual to those in semiconductors can be developed (Davidson and Beasley, 1979, Kadin, 1990).

Duality in Circuit

The construction of the dual of a given circuit involves transformation of both the network topology and the circuit elements, as summarized in Table 3.1. We illustrate this through a simple example of a series LCR resonator driven by a voltage source, as in Fig. 3.18a. This is described by the standard equation

$$V = RI + L\frac{dI}{dt} + \frac{Q}{C} \tag{3.43}$$

where $I = dQ/dt$. The topology of the dual circuit is shown in Fig. 3.18b, superimposed on the direct circuit (dashed line). Here we have located an element of the dual circuit on top of each element of the direct circuit but oriented transversely to the original element. Each node of the original circuit becomes a mesh in the dual circuit and each mesh a node. This turns a series connection into a parallel connection. The ground node transforms to the "open mesh" at infinity, that is, the outside of the circuit, and vice versa (all duality transformations are bidirectional).

In terms of the transformations of the elements in the dual circuit, these are indicated in Fig. 3.18c. The voltage source transforms into a current source, so that an ideal 1-V ac source becomes a 1-A ac source with the same time dependence. The dual element to a resistance R is a conductance G with the same value, that is, a 5-Ω resistor in the direct circuit becomes a 5-S conductor (a 0.2-Ω resistor) in the dual circuit. A 1-μH inductor transforms into a 1 μF capacitor. The dual circuit has become a current-driven parallel LCR resonator, described by the following circuit equation:

$$I = GV + C\frac{dV}{dt} + \frac{\Phi}{L} \tag{3.44}$$

Table 3.1. Transformations and Substitutions in Construction of Dual Circuit

Elements in dual circuit $\perp$ elements in direct circuit.
Exchange node $\leftrightarrow$ mesh, series $\leftrightarrow$ parallel, and open circuit $\leftrightarrow$ short circuit.
Substitute $G \leftrightarrow R$, $L \leftrightarrow C$, $I \leftrightarrow V$, $Q \leftrightarrow \Phi$.

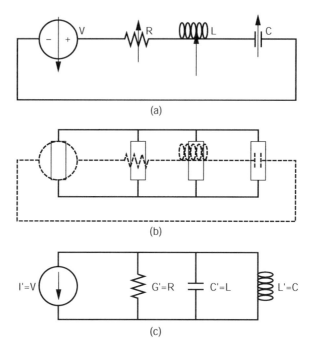

Figure 3.18. Dual-circuit transformation of series LCR resonator. (*a*) Direct circuit, driven by voltage source. The hatched arrows show the flux path. (*b*) Construction of network topology for dual circuit. (*c*) Dual circuit, current-driven *LCR* resonator, which is also the physical circuit for flux motion in (*a*).

where $V = d\Phi/dt$ from Faraday's law. If we switch the roles of current and voltage (and implicitly, also charge Q and flux Φ), Eq. (3.44) for the dual circuit is exactly the same differential equation as Eq. (3.43) above for the direct circuit.

The dual circuit is apparently physically distinct from the direct circuit, with a different topology and different element values. However, from an alternative point of view, the dual circuit describes the transport of magnetic flux in the direct circuit. This is consistent with the construction of Fig. 3.18*b*, since flux moves transversely to a voltage drop and hence to the flow of current. We can think of flux lines moving in the dual "fluxonic" circuit, in direct analogy to electrons moving in an electronic circuit. Of course, magnetic flux lines must ultimately form closed loops, but within the plane of the circuit, we can treat the intersection of a flux line as if it were a "point particle." It is also important to distinguish the flux transport circuit from a "magnetic circuit" that is often used in the analysis of static magnetic fields. In the latter case, the magnetic circuit follows parallel to the static magnetic flux loop. A flux transport circuit, in contrast, follows the motion of dynamic flux lines in a plane perpendicular to the lines themselves.

We can illustrate these principles by comparison of Fig. 3.18a and Fig. 3.18c. Here the voltage source acts to pump flux into the loop, that is, it acts as a "current source" for flux. This flux in the loop can leak out via one of three parallel channels. It cannot leak across a solid wire, since by assumption there can be no voltage drop. The resistor acts like a linear flux leak, where the rate of flux leakage (the voltage) is inversely proportional to the resistance. Hence we have a flux conductor, where the conductance G has the same value as the resistance R. Similarly, the inductor is a flux storage element, which can be regarded as a "flux capacitor" (with apologies to the movie *Back to the Future*). A capacitor can be viewed as storing energy in flux motion (i.e., voltage) in the same way that an inductor stores energy in charge motion. It may seem odd to regard a static electric field in a capacitor (with no magnetic field) as due to steady-state transverse motion of flux, but it is nonetheless consistent. Furthermore, both the resonant frequency $\omega_0 = 1/\sqrt{LC}$ and the Q-factor are the same for both circuits, as they must if they describe aspects of the same resonator. This exercise in switching from a charge representation to a fluxonic representation does not actually add anything new in this case, but development of such a facility will prove useful in understanding superconducting devices.

As a second example, consider an ideal lossless transmission line (Fig. 3.19), which might be constructed using superconducting electrodes. If a current pulse $I(t)$ is launched on this line, it will propagate to the right with velocity $v = 1/\sqrt{LC}$, transporting a charge $Q = \int I\,dt$ along with it. Of course, we can also

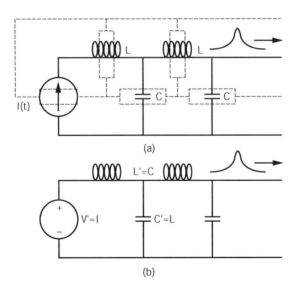

Figure 3.19. Duality transformation for a lossless transmission line. (*a*) Line with current pulse. Constructions for dual circuit are shown as superimposed dashed lines. (*b*) Dual line to (*a*), with voltage pulse.

multiply by the characteristic impedance to obtain the voltage pulse $V(t) = Z_0 I(t)$, which transports a net flux $\Phi = \int V \, dt$ down the line. This should be consistent with the picture we obtain by applying duality to this transmission line. And indeed, as shown in Fig. 3.19b, the dual to an LC transmission line is another LC line with the same velocity. Interestingly, both charge and flux are conserved in such a signal. A voltage pulse containing one flux quantum Φ_0 for a transmission line with $Z_0 = 50 \, \Omega$ would also contain a charge of about 250 electrons.

Duality and Vortices in Superconductors

A bulk superconductor in the Meissner state acts as a "flux insulator" in that no flux can cross it. But vortices in the mixed state of a superconductor are analogous (dual) to charge carriers in a semiconductor (see Table 3.2). Each moving vortex can transport a flux Φ_0, giving rise to a voltage V, in much the same way that an electron can transport a charge e to produce a current I. This also creates a consistent picture of both flux conservation and flux quantization in a superconducting loop, analogous to the quantization of charge on a charged object. The total flux in the loop must be an integral number of flux quanta, which can only change if and when one or more vortices move across the superconductor.

Furthermore, a uniform transport current J exerts a force on a vortex that is analogous to the electrostatic force $e\mathbf{E}$ on an electron. The force on a vortex (per unit length) is

$$\frac{\mathbf{F}}{l} = \mathbf{J} \times \mathbf{\Phi}_0 \qquad (3.45)$$

where $\mathbf{\Phi}_0$ points in the direction of the magnetic flux density $\mathbf{B}$ in the center of the vortex. This force is perpendicular to both the magnetic field and the current. This follows from the standard $\mathbf{J} \times \mathbf{B}$ transverse Lorentz force on a current element, and the same term is often used to describe the force on a vortex. This is also the same force that gives rise to the usual attraction between two parallel

Table 3.2. Duality between Electrons in Semiconductor and Fluxons in Superconductor

	Semiconductor	Superconductor
Background	Insulator	Flux insulator
Carriers	$Q = \pm e$	$\Phi = \pm \Phi_0$
Force	$F = QE$	$F = J \times \Phi$
Current	$J = \sigma E$	$E = \rho J$
Conductivity	$\sigma = n e \mu_e$	$\rho = n_v \Phi_0 \mu_v$
Mobility	$\mu_e = e\tau/m$	$\mu_v = \rho_n/B_{c2}$

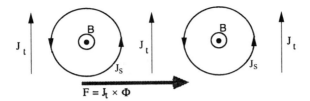

Figure 3.20. Lorentz force on a vortex due to a transport current J_t. The magnetic field points out of the paper. Note that current–current interactions on both sides of the vortex tend to push the vortex to the right, consistent with the net $J \times \Phi_0$ force. Note also that adjacent vortices will repel, since overlapping currents are antiparallel.

currents and the repulsion between two antiparallel currents. In Fig. 3.20 we show a microscopic view of uniform transport currents and counterclockwise screening currents near a vortex with B pointing out of the page. The force due to the transport current on the right of the vortex is attractive and so tends to pull the vortex to the right. On the left-hand side, the transport current repels the vortex screening current, again pushing the vortex to the right. These add together and account for the net $J \times \Phi_0$ force pointing to the right.

This picture also shows why adjacent vortices (with fields pointing in the same direction) should repel each other: Their overlapping currents are in opposite directions and hence repel. For the same reason, two vortices with opposite polarities (sometimes called a vortex and an antivortex) will attract one another and can actually recombine (or annihilate) when they come together. Treating a vortex as a particle analogous to an electron or hole in a semiconductor, this makes perfect sense. Like charges repel, and opposite charges attract and can recombine.

How does a vortex respond to a constant Lorentz force? In analogy with carriers in a semiconductor, there are three possibilities:

1. The vortex is bound to a localized state and does not move significantly, despite the force. This would correspond to an electron bound to an impurity or trap state.
2. The vortex accelerates without loss, as an electron in a perfect conductor.
3. The vortex moves at a constant "drift velocity," limited by resistive loss, as an electron in a metal or semiconductor.

In real superconducting materials, either condition 1 or condition 3 predominates. "Supervortex" behavior does not normally occur, although it may be present in certain Josephson junction arrays at low temperatures. We will deal with the important implications of "pinning" of vortices on localized defects in the next section. For the remainder of the present section, we will focus on the free flow of vortices with resistance.

Unipolar Fluxonics

For an electron in a semiconductor, we can define an effective mobility such that $v = \mu E$ and a conductivity $\sigma = ne\mu$, assuming carriers of a single sign. In the vortex case, by duality, the effective driving field is the transport current density J, and the effective vortex conductivity is actually an electrical resistivity, so that

$$\rho_v = n_v \Phi_0 \mu_v \approx B \mu_v \tag{3.46}$$

where $\mu_v = v/J$ is the vortex mobility. Here, we have also assumed that all the vortices have the same sign and are produced by an applied magnetic field. A closely related parameter that is widely used is the vortex viscocity η, defined so that $F = J\Phi_0 = \eta v$ or $\eta = \Phi_0/\mu_v$. A proper derivation of the value of μ_v is rather complicated, but we can obtain a simple approximate result if we assume that μ_v is independent of the vortex density, and goes over continuously into the normal state for $B > B_{c2}$. At B_{c2}, we have $\rho = \rho_n$, so that $\mu_v \approx \rho_n/B_{c2}$, and Eq. (3.46) becomes

$$\rho_v = \frac{n_v \Phi_0 \rho_n}{B_{c2}} = \rho_n \left(\frac{B}{B_{c2}} \right) = n_v 2\pi \xi^2 \rho_n \tag{3.47}$$

This formula is generally called the "Bardeen–Stephen relation" for vortex flow. This would be expected to apply for magnetic fields between B_{c1} and B_{c2} (Fig. 3.21).

For a significant magnetic field and density of vortices, this resistance is lower than that in the normal state but still much too large for a "good superconductor." This should not be surprising; mobile carriers in an insulator turn it into a semiconductor, so mobile flux carriers would also be expected to lead to dissipation. The key word here is "mobile." If we can prevent the carriers from moving in response to the force, then we can recover the lossless state. For a semiconductor, this means that all of the carriers must be localized on traps or other defects. In the same way, if all vortices are fixed on pinning sites, then a transport current can simply flow around them, leading again to a true zero-resistance state.

One way to interpret Eq. (3.47) is to note that $\rho_v/\rho_n = n_v 2\pi \xi^2$, which is (up to a factor of two) the fraction of area covered with normal vortex cores of radius $\approx \xi$. But this explanation is not quite convincing, since one would expect that the current could simply flow around the vortex cores even if they are moving. But moving flux implies that $V > 0$, which means that an electrical power IV is going into the vortex system. In steady state, this power must be dissipated and will show up as resistance. On a more microscopic level, we have a normal core $\sim 2\xi$ moving through a superconductor. This requires breaking Cooper pairs on the leading edge of the core and re-forming them on the trailing edge; both of these are irreversible processes that correspond to dissipation and hence a nonzero resistance R.

3.4 DUALITY AND FLUX FLOW

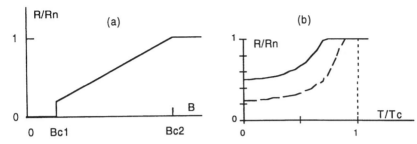

Figure 3.21. Resistance of superconductor in vortex state with flux flow. (a) Dependence on magnetic field B for fixed temperature $T < T_c$. (b) Dependence on temperature T for (left to right) $B = \frac{1}{2}B_{c2}, \frac{1}{4}B_{c2}$, and $B < B_{c1}$.

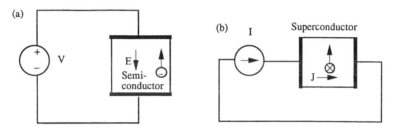

Figure 3.22. Unipolar carrier flow in semiconducting and superconducting cases. (a) Voltage across n-type (electron-doped) semiconductor. (b) Dual of circuit in (a), corresponding to vortex flow in superconductor with B pointing into the paper.

In certain respects, applying a magnetic field perpendicular to a superconductor is analogous to doping a semiconductor; both create the possibility of a larger number of "free" carriers of a single polarity. In the semiconductor case, as shown schematically in Fig. 3.22a, if we apply ohmic contacts on two ends and apply a voltage between these, then the carriers will flow from one electrode to the other, each carrier corresponding to a charge transfer (or integrated current pulse) of e. The dual of this circuit is shown in 3.22b, which represents a current-driven superconductor in the vortex state due to the presence of a large perpendicular magnetic field. Although the average magnetic field is static, the vortices are moving up under the influence of the Lorentz force. Each vortex enters the film on the free edge on the bottom and leaves on the top and corresponds to an integrated voltage pulse (parallel to the current) of Φ_0. Under some circumstances, there may be a surface barrier to vortex entry into the superconductor, which is perhaps analogous to the Schottky barrier that can sometimes form on semiconductor contacts.

Flux Flow Transistor

We can take this analogy further, by developing a device that is dual to the field-effect transistor (FET) in a semiconductor surface layer (Nordman, 1995).

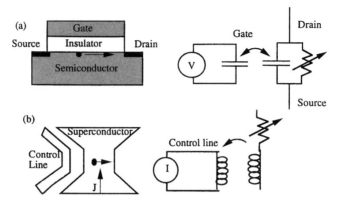

Figure 3.23. Field-effect devices in semiconductor and superconductor. (*a*) Layout of insulated-gate field-effect transistor with *n*-type semiconducting channel between source and drain and equivalent circuit of capacitively controlled variable conductance. (*b*) Superconducting flux flow transistor, essentially the dual of the FET in (*a*), with current in control line inductively modulating the resistance across the vortex flow channel of the superconductor.

As shown in Fig. 3.23*a*, a voltage applied to an insulated gate electrode acts to vary the effective electron density n in the n-channel of the inversion layer; this is the (electric) field effect. We have indicated the circuit schematic for this using a "mutual capacitance" between the two parts of the circuit. This change in n, in turn, modulates the conductivity $\sigma = ne\mu$ between the source and the drain electrodes. Under appropriate conditions, this can provide the basis for a voltage gain.

A superconducting device that is in certain respects a dual to the FET is shown in Fig. 3.23*b*. This is known as the superconducting flux flow transistor (SFFT), or sometimes the vortex flow transistor (VFT). Here, current in an inductive control line (also typically superconducting) produces vortices in a "vortex channel" in the center, the active region of the device. As in the FET, this modulates the resistance across this vortex channel. In principle, this too could give rise to voltage gain, although a large gain has sometimes proven difficult to achieve in practice. One problem is that the magnetic field that is attainable using a control line such as in Fig. 3.23*b* is rather small. For example, if we consider a control current of 100 mA (which can be carried by a superconducting line on the micrometer scale) and an active region 2 μm away, the field in the active region is $B \approx \mu_0 I/2\pi r = 10$ mT $= 100$ G. Such a small field will be able to inject vortices only if the active region contains Josephson junctions or other weak links.

Vortex motion can sometimes be quite fast in superconducting thin films, particularly on the micrometer scale of devices. Since the vortex traversal time limits the response time of the SFFT, this is an important consideration. Typical values for a high-T_c superconductor might be $\rho_n \sim 10^{-5}$ Ω-m and $B_{c2} \sim 100$ T, yielding $\mu_v \sim 10^{-7}$ m^3/A-s. Taking a current density $J \sim 10^{10}$ A/m^2, we obtain a velocity $v \sim 10^3$ m/s, or a traversal time ~ 1 ns for a 1-μm wide line. This suggests that devices of this type could perform into the microwave range, which has indeed been observed.

Bipolar Fluxonics

We have been dealing with situations in which the vortex density is reasonably uniform, or at least has vortices of only a single polarity. It is also possible for the perpendicular magnetic field to change sign across a superconducting film, which would require the presence of vortices of both polarities. This is analogous to having both electrons and holes in a semiconductor and suggests the possibility of bipolar fluxonic devices.

Let us first deal with the case of a superconducting film without an externally applied magnetic field. Then, the only vortices will be produced by the self-magnetic field of the transport current, which is maximum at the two edges and of opposite sign. Above a certain critical current, vortices (and antivortices) can enter the superconductor at both edges and will move toward the center under the influence of the transport current (Fig. 3.24). A vortex will continue to move until it either recombines with one of the opposite polarity or reaches the other side. Vortex motion of both polarities contribute to the resistance and to Joule heating in the superconductor. Under certain circumstances, this vortex motion can even lead to runaway self-heating above the critical temperature, as described earlier. This situation is perhaps analogous to the breakdown process that may occur in an intrinsic semiconductor under a high voltage. Electrons would enter from one electrode and holes from the other, moving in opposite directions across the semiconductor until they either recombine with a carrier of the opposite sign or reach the other side.

Another possibility is an "intrinsic superconductor" with thermally excited vortex–antivortex pairs, analogous to thermally excited electrons and holes in a small-gap intrinsic semiconductor. The effective energy gap for these vortex excitations is *not* the same as the energy gap 2Δ for breaking a Cooper pair. We

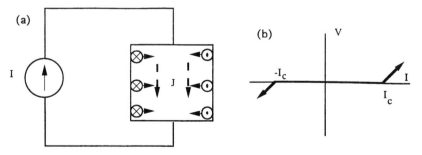

Figure 3.24. (*a*) Vortex configuration just above breakdown of current-biased superconducting strip in absence of externally applied magnetic field. (*b*) Corresponding $V(I)$ for strip.

can estimate $\mathcal{E}_{gv}$ as the condensation energy corresponding to the presence of two vortex cores $\sim \xi$ in radius and d in height, where d is the thickness of a superconducting film. From Eq. (2.60), we have

$$\mathcal{E}_{gv} \sim 2\pi\xi^2 d \frac{\Phi_0^2}{158\,\mu_0\lambda^2\xi^2} = \frac{\Phi_0^2 d}{25\,\mu_0\lambda^2} = \frac{\Phi_0^2}{25\mu_0\lambda_\perp} \qquad (3.47)$$

where $\lambda_\perp = \lambda^2/d$ is the effective transverse penetration depth for very thin films. This is generally much larger than 2Δ, in which case thermal excitation of vortices can be neglected, since then the thermal energy kT will be much less than $\mathcal{E}_{gv}$, except perhaps very close to T_c where $\lambda_\perp$ diverges. But if we take a very thin film with $d = 10$ nm and a moderately large value of $\lambda \sim 1$ µm, then we obtain $\mathcal{E}_{gv} \approx 10$ meV. A more complete analysis (Beasley, Mooij, and Orlando, 1979) has shown that there is a phase transition, called the vortex-unbinding transition (or Kosterlitz–Thouless transition) at a temperature T_v such that

$$T_v = \frac{\mathcal{E}_{gv}}{k} \approx \frac{1\,\text{cm-K}}{\lambda_\perp} \qquad (3.48)$$

For the example here, $\lambda_\perp \sim 100$ µm, which yields $T_v \sim 100$ K. Below T_v, there would still be some excited vortex–antivortex pairs, but they would remain bound (and eventually recombine) and therefore not contribute to resistance. In contrast, above T_v, the vortex and antivortex are effectively free and hence will contribute to resistance. Practically speaking, this effect is a concern for some high-temperature superconductors and not only in ultrathin films. This is because these materials consists of weakly coupled superconducting layers on the 1-nm scale, as will be discussed further in Chapter 4.

Alternatively, we can have an externally applied magnetic field that changes polarity across the film. This is analogous to doping a semiconductor p-type in one region and n-type in another. The simplest such semiconducting structure is the *pn* junction, consisting of a sudden change from p-type to n-type. In this situation, even without an applied voltage, some of the electrons and holes will diffuse and recombine in the "depletion layer" near their interface. The remaining space charge (associated with the dopant ions) creates an internal electric field that prevents further interdiffusion and recombination (Fig. 3.25a). It is this internal field that is responsible for asymmetric diode behaviour upon application of an external voltage. Application of an opposing electric field that exceeds this internal field ("forward biasing") leads to a large opposite flow of both electrons and holes across the depletion layer, that is, a large current. In contrast, application of an E-field in the same direction as the internal field ("reverse bias") just reinforces this depletion layer, giving very little current. This is a simplified picture of the operation of a *pn* junction diode, which in turn provides the basis for bipolar junction transistors.

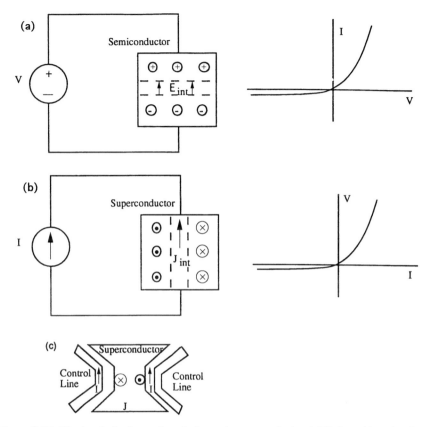

Figure 3.25. Bipolar diodes in semiconductor and superconductor. (*a*) Voltage-biased *pn* junction diode showing internal field E_{int} in depletion layer and $I(V)$. (*b*) Dual circuit to (*a*) showing bipolar fluxonic diode with internal current J_{int} in depletion layer in center and $V(I)$. (*c*) One configuration for field doping of diode using current in two control lines.

A superconducting structure that is dual to the *pn* junction diode (Kadin, 1990) would consist of an external magnetic field that reverses polarity sharply (Fig. 3.25*b*). Leaving aside for the moment how we might produce such a *B*-field, the vortices near the interface would be expected to recombine, leading to a depletion layer analogous to the one in the semiconductor diode. This would require an internal self-current between the two vortex domains, keeping them apart. (The necessary return current for an unbiased device would be around the exterior of the superconductor and does not affect the argument here.) If this structure is now biased with a current along the direction of the self-current (reverse bias), this will tend to move the vortices further apart, increasing the width of the depletion region and giving no net steady-state flux flow voltage. In contrast, a current in the reverse direction (forward bias) will lead to "minority carrier injection" across the depletion region, and these minority vortices will either recombine with the majority vortices or continue to traverse to the edge of the film, thus producing flux flow resistivity. The vortices that recombine or leave are replaced by new ones that enter from the edge, thus maintaining the overall flux distribution. Although the details of this may be different from the *pn* junction diode in Fig. 3.25a, we would expect a $V(I)$ relation for the "fluxonic diode" qualitatively similar to the $I(V)$ dependence for the *pn* junction diode.

How can one obtain the required flux profile for this bipolar fluxonic diode? One approach is to use two control lines on either side of the active region (Fig. 3.25c), with currents flowing in the same direction. The magnetic field profile does change sign in the center, although perhaps not as sharply as one might prefer. Other bipolar fluxonic devices, including various three-terminal transistors, have also been proposed, and some have been demonstrated. Some of these have been carried out using vortices in Josephson junctions, which we will discuss in Chapter 5. Although the real situation is generally more complicated than this simple picture would suggest, some of these results have indicated promising devices that exhibit gain.

Finally, it is worth emphasizing that this duality between flux and charge must be modified when we move to three-dimensional geometries. We can see this clearly by considering current flowing axially in a cylindrical wire. The magnetic field is circumferential, and above the critical current, a "vortex ring" will form on the periphery of the wire. This would be driven inward by the Lorentz force of the current, and if we maintain cylindrical symmetry, it will shrink down to zero radius and then annihilate (Fig. 3.26). This process will yield an integrated flux flow voltage along the wire of one flux quantum Φ_0 and is the analog of the recombining vortex–antivortex pair in Fig. 3.24. Similarly, the three-dimensional analog of the thermally excited vortex–antivortex pair is a thermally excited vortex ring, with a characteristic excitation energy [cf. Eq. (3.47)] of order

$$\mathcal{E}_{gv} \sim U_c (\pi \xi^2)(2\pi \xi) \sim \frac{\Phi_0^2 \xi}{8\mu_0 \lambda^2} \tag{3.49}$$

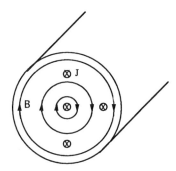

Figure 3.26. Collapsing vortex rings in cylindrical superconducting wire with axial transport current but no applied magnetic field. As each vortex ring shrinks to zero radius at the center, it is replaced in steady state with another entering around the periphery.

Such a ring would normally tend to collapse soon after production, since the vortex sections on opposite sides are opposite in polarity and will attract one another. However, in the presence of a large transport current, a properly oriented vortex ring can expand outward, again giving flux flow resistance. If we take $\lambda \sim 1$ μm and $\xi \sim 1$ nm, values that might be appropriate to some high-temperature superconductors close to T_c, we obtain a characteristic excitation energy ~ 30 K in temperature units. So even in this regime, by using this charge–flux duality carefully, we can continue to develop intuition about vortices in superconductors.

3.5 FLUX PINNING IN LARGE FIELDS

Although the previous section focused on vortex motion, for most purposes vortex flow should be avoided. Practical high-field superconductors contain a high density of defect sites that act to restrain, or "pin," vortices, so that they do not move when a current is applied. In this way, there is no resistive loss, and we have restored true superconductivity. However, the same pinning sites also prevent currents and fields from redistributing easily within a superconducting wire when the applied field or current is changed. So we want to permit necessary "flux creep" while avoiding catastrophic effects from "flux jumps" that cause runaway heating. This required stability can be achieved using a composite of microscopic superconducting and normal filaments, as will be discussed below.

Surface Pinning

Our previous discussions of the vortex lattice and flux flow in a type II superconductor have implicitly assumed that vortices can freely enter (and exit) the superconductor. In principle, we can apply a magnetic field $H > H_{c1}$ for $T > T_c$ and then cool down to a lower temperature. In this situation, the flux will remain inside the superconductor as an array of vortices, rather than being expelled as in the Meissner state. However, a more common situation is to start with a superconductor in zero field at $T < T_c$ and then to increase the field (or current). Then, vortices must enter the superconductor from the outside. Furthermore, even after the nucleation of a vortex core at the superconducting surface, there may still be a force attracting it to the surface, thus keeping it from penetrating further.

We can see the basis for such a surface barrier using the method of images to analyze a vortex near a plane boundary, as in Fig. 3.27. If we place an image antivortex on the opposite side of the boundary, then by symmetry there will be no net current perpendicular to the boundary, as required. As we pointed out earlier, a vortex and anitvortex will attract each other, so that the real vortex will tend to stay next to the boundary in the absence of a large transport current. However, the picture is not quite complete. In addition to the vortex itself, there

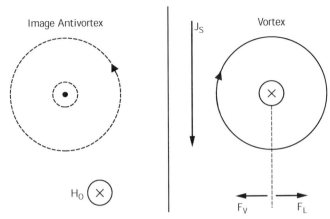

Figure 3.27. A vortex just inside a superconducting surface is attracted to its "image antivortex" on the other side but is repelled by the surface screening current J_s. The net effect is for the vortex to be pinned at the surface for an applied field H_0 less than the surface barrier field H_s.

is a large magnetic field outside the superconductor that will be partly screened by surface currents in the superconductor. These surface currents, as indicated in Fig. 3.27, are in a direction that tends to force the vortex away from the surface. For an applied field H_0 less than the "surface barrier field" H_s, there is net attraction to the surface; for $H > H_s$, repulsion is dominant. A detailed analysis (Orlando and Delin, 1991, p. 340) shows that H_s is given in the large-κ limit by

$$H_s \approx \frac{\Phi_0}{4\pi\mu_0\xi\lambda} = \frac{H_c}{\sqrt{2}} = H_{c1}\left(\frac{\kappa}{\ln \kappa}\right) \gg H_{c1} \qquad (3.50)$$

For H_0 between H_{c1} and H_s, then, we would expect that no vortices would penetrate into the superconductor, and hence there would be no flux flow and zero resistance, despite what we described earlier. If we increase H to a value greater than H_s, then vortex entry occurs unimpeded, and the superconductor may exhibit some resistance. However, if we now decrease H_0 to below H_s, the vortices already inside are not expelled. Instead, they can continue to flow in the superconductor, contributing to resistance, until they are swept out by the transport current. This sort of hysteretic behavior, which depends on past history as well as the present state, is characteristic of practical superconductors in the mixed state.

The situation is actually more complicated than this; a field that has components perpendicular to the surface, or a surface with roughness on the scale of λ, would cause vortex penetration at a field somewhat smaller than H_s. Furthermore, even if the vortex is bound to the surface, a sufficiently large transport current will bring about a Lorentz force $J \times \Phi_0$ that can cause "depinning" of the vortex and initiate flux flow resistance.

Pinning on Defects

Pinning of a vortex will occur not only on an edge but also on a variety of holes and other defects in a superconductor. Consider, for example, a cylindrical hole or normal metal of radius $\sim \xi$ parallel to the vortex. The vortex will tend to locate with its core superimposed on this hole, since $n_s = 0$ in the center of the vortex anyway. Here, the vortex does not have to give up any condensation energy. We can estimate the force binding the vortex to this hole by considering the amount of energy (work) W that would have to be supplied to move the vortex out of the hole a distance $\Delta x \sim 2\xi$ from its equilibrium position (see Fig. 3.28) as the condensation energy [from Eq. (2.60)] in a cylindrical volume equal to the core:

$$W \approx U_c\,(\pi \xi^2 l) \approx \frac{\Phi_0^2 l}{50\mu_0 \lambda^2} \tag{3.51}$$

so that the pinning force per unit length F_p/l would be

$$\frac{F_p}{l} \sim \frac{W}{l\Delta x} \approx \frac{\Phi_0^2}{100\,\xi\,\mu_0\lambda^2} \tag{3.52}$$

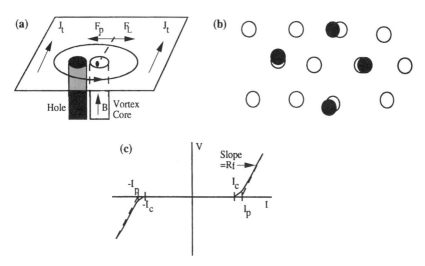

Figure 3.28. Vortex pinning in type II superconductor. (*a*) Ideal pinning site for vortex in a superconductor showing vortex core displaced a distance 2ξ from cylindrical hole by Lorentz force of applied current. (*b*) Top view of a vortex lattice pinned by a random array of pinning sites. The vortices are the open circles, and the pinning sites are the solid circles. (*c*) $V(I)$ relation for type II superconducting film in uniform B, with pinning current I_p and flux flow resistance R_f identified, together with critical current $I_c < I_p$.

If a transport current J_t is applied perpendicular to the vortex, this will exert a Lorentz force $F_L = J_t \times \Phi_0$ against the pinning force. Therefore, the vortex will remain stationary unless $F_L > F_p$, that is, the critical depinning current density J_p should be (see also Orlando and Delin, 1991, p. 371)

$$J_p \sim \frac{\Phi_0}{100 \xi \mu_0 \lambda^2} \tag{3.53}$$

If we estimate $\lambda \sim 300$ nm and $\xi \sim 5$ nm, then we obtain $J_p \sim 3 \times 10^{10}$ A/m^2, a very large value.

We can think of J_p as an upper estimate of the critical current density J_c of a type II superconductor in the mixed state, assuming that every vortex is pinned by an optimum defect. For a larger current density, flux flow resistance would be present. It is worth comparing this to the critical current density for breaking Cooper pairs, from Eq. (3.15). The depinning current has the same parametric dependence, but is about a factor of 10 smaller. Since this is a maximum for optimum pinning, the value of J_p in a practical superconductor is typically one to two orders of magnitude smaller still than this ideal value. Even so, values of depinning current density $J_p \sim 10^9$ A/m^2 are commonly achieved in practice for both low-T_c and high-T_c superconductors. This corresponds to ~ 1000 A in a wire 1 mm in diameter, assuming uniform current flow.

Pinning in real superconductors is further complicated by a number of issues. First, one might initially think that every vortex must be individually fixed to a pinning site, but this is usually not necessary. Because vortices in large fields are generally within a vortex lattice that deforms locally toward pinning sites (Fig. 3.28b), they tend to move not as individual vortices but as correlated bundles. If a fraction of the vortices are individually pinned, it may well be enough to pin the entire lattice. On a macroscopic level, then, we tend to focus not on the individual pinning forces, but rather on the average density of pinning forces (per unit volume)

$$f_p = J_p \times B \tag{3.54}$$

where $B = n_v \Phi_0$ is the macroscopic average density of vortices. The resulting value of the depinning current density J_p will include a statistical average of both the reduced effectiveness of individual pinning sites and the relative sparseness of such sites.

Let us assume that we have a macroscopically uniform distribution of such pinning sites and a uniform magnetic field B in the superconductor. Then we can identify two regimes. First, for $J < J_p$, the vortex lattice is immobile and the voltage is zero. For $J > J_p$, the vortex lattice can move, subject to both pinning forces and resistive drag forces. The net force per unit length on the vortices is

$$\frac{F}{l} = J\Phi_0 - J_p \Phi_0 - \eta v \tag{3.55}$$

where v is the vortex velocity and $\eta = \Phi_0/\mu_v$ is the vortex viscocity that we discussed earlier. In steady state, we now have

$$v = \mu_v (J - J_p) \qquad (3.56)$$

and the dependence of the electric field on current density,

$$E = \rho_v (J - J_p) \qquad (3.57)$$

This is illustrated in the $V(I)$ relation in Fig. 3.28c. The presence of pinning provides a current offset I_p to the flux flow resistance for both positive and negative currents. For currents $|I| > I_p$, the differential resistance dV/dI is the same constant flux flow resistance that we described earlier in the absence of pinning. Of course, we have also assumed here that the temperature in the conductor remains fixed. Once the large value of I_p is exceeded, the amount of Joule heating IV may be so large as to drive the superconductor into the normal state.

There is an ambiguity in defining the critical current I_c (or current density J_c). The actual $V(I)$ will curve slightly above the "average critical current" I_p for two reasons. First, the pinning strength f_p may not be entirely uniform; the voltage would start to increase at the lower end of the statistical distribution. Second, thermal fluctuations can cause some vortex motion even if the static force is slightly less than the pinning force. This provides the basis for thermally activated "flux creep." So the "true" value of I_c (where V is first nonzero) may be slightly below the "average" value determined by I_p. From a practical point of view, I_c is often defined by the current at which V (or more properly, E) is below some low threshold, for example, 1 μV/cm. A lower threshold typically yields a somewhat lower value of I_c.

Critical State Model

In the example above, we assumed that B and J were uniform in the superconductor, but the presence of strong flux pinning means that this is generally not the case. In fact, the field and current distributions depend strongly on the past history of the superconductor. For example, consider a superconducting slab of thickness $d \gg \lambda$ in a parallel magnetic field (Fig. 3.29), where the superconductor was initially cooled down in zero field, in the absence of transport current, so that no flux has penetrated. As the applied field is increased (neglecting H_{c1} and the surface barrier to flux penetration), the effective magnetic pressure $\frac{1}{2}\mu_0 H^2$ increases, which increases the force on vortices near the surface. From Ampere's law, we have $dH_z/dx = -J_y$, so that a large screening current density is necessary to sustain a sharp gradient in H. If the magnitude of the field gradient $|dH/dx|$ is initially greater than the depinning current J_c, then magnetic flux will penetrate further, until the gradient decreases to just below J_c. At this point, the flux is in the "critical state" in the entire region in which the flux

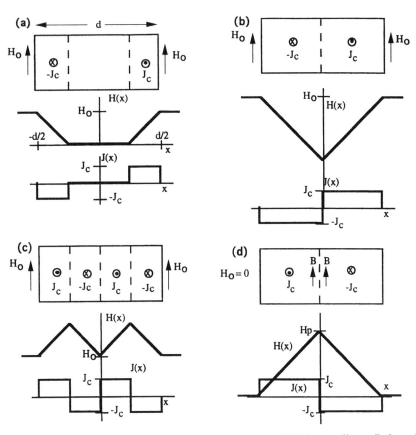

Figure 3.29. Critical state model for flat superconducting slab of thickness d in applied parallel magnetic field H_0 with constant J_c (Bean model). The penetration field is $H_p = \frac{1}{2}J_c d$. The slab was initially cooled in $H_0 = 0$; then H_0 is increased to $H_{max} = 2H_p$, then decreased back to 0. (a) Initial configuration and spatial dependence of H and J for increasing $H_0 < H_p$. (b) Configuration, $H(x)$ and $J(x)$ for $H_0 = H_{max}$. (c) $H(x)$ and $J(x)$ for decreasing $H_0 = H_p$. (d) $H(x)$ and $J(x)$ for H_0 decreasing to 0. Note flux and supercurrents trapped in the critical state.

has penetrated, and the flux distribution is stable. Hence, there is no further vortex flow, and the resistance is zero.

The details of such a critical state depend on $J_c(B)$. In the simplest model (sometimes called the "Bean model" of the critical state), J_c is independent of B, so that the profile of $H(x)$ exhibits a constant slope (see Fig. 3.29). The penetration depth of the field is then $x = H_0/J_c$, with a constant screening current density J_c within this layer of opposite sign on the two sides of the slab. Some flux reaches the center of the slab at a characteristic penetration field $H_p = \frac{1}{2}J_c d$. For larger fields, the current distribution remains the same, and the field distribution is a triangle, where the center lags behind the edges by H_p. We are most interested here in the regime where $H_0 \gg H_p$.

Now consider what happens when we decrease H_0. Any change in flux must first occur in the outer layers of the slab. The screening currents in these outer layers reverse sign, even though the external magnetic field has not reversed. For fields $H_0 < H_{max} - 2H_p$, the field profile is again a triangle. When H_0 has been reduced to zero, there is still a substantial amount of flux trapped inside the slab. We have neglected the lower critical field H_{c1} here, but this will have little effect, since the local trapped values of H are typically much higher than H_{c1}. Even if we reverse H_0, we still cannot completely eliminate the trapped field.

One can take the spatial average of the fields in Fig. 3.29 to obtain $\langle B \rangle$ and $\langle M \rangle = \langle B \rangle / \mu_0 - H_0$. In Fig. 3.30, we have plotted these versus H_0 to obtain a hysteresis loop, between H_{max} and $-H_{max}$, starting with a "virgin" sample with no flux. This is analogous in many respects to the hysteretic B–H loop of a permanent magnet material, where domain walls are pinned to defects such as grain boundaries. Apart from the initial magnetization and the region just after dH_0/dt is reversed, the curves in Fig. 3.30 consist of straight lines. These lines correspond to Figs. 3.29b and d, where the magnetic profile is a triangle and J is constant on each side. Here, $\langle M \rangle = \pm \frac{1}{2} H_p = \frac{1}{4} J_c d$, which corresponds to the "remanent magnetization" of a permanent magnet. This relation is sometimes used to determine the critical current $J_c = 4\langle M \rangle d$ using only a noncontact measurement of the magnetic moment ($\langle M \rangle V$) in an applied magnetic field. A similar relation can be obtained for sample shapes other than a slab in a parallel magnetic field. For example, a cylinder of radius R would yield $\langle M \rangle = \pm \frac{1}{3} H_p = \frac{1}{3} J_c R$, or $J_c = 3 \langle M \rangle / R$.

Just as in the ferromagnetic case, the presence of hysteresis has important implications for ac applications. That is because the area inside the hysteresis loop, $\oint \langle B \rangle dH = \int \mu_0 \langle M \rangle / dH \approx \mu_0 J_c d H_{max}$, corresponds to energy per unit volume dissipated in a cycle. For high-frequency applications, this vortex motion gives unacceptably high losses, but at low frequencies (including, in some cases, 50–60 Hz ac power frequency) the losses can still be quite small. Even for dc applications, this is relevant for the rate of ramping up or down of the field. Note also that this dissipation can be minimized by making the slab thickness (or wire radius) as small as possible.

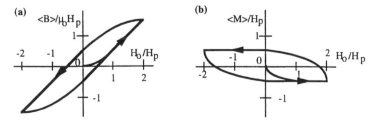

Figure 3.30. Magnetic hysteresis loops in critical state model (with constant J_c) for superconducting slab of Fig. 3.29 in parallel magnetic field H_0 up to $H_{max} = 2H_p = J_c d$, starting with $B(x) = 0$ in the initial state. Here $\langle B \rangle = [\int B(x) dx]/d$ across the slab and $\langle M \rangle = \langle B \rangle / \mu - H_0$, and the arrows indicate the direction H_0 is changing.

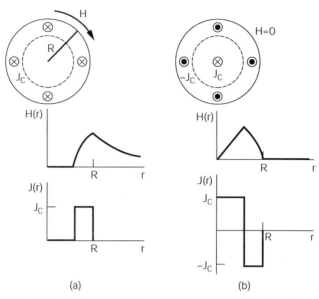

Figure 3.31. Critical state due to current I in a cylindrical superconducting wire of radius R with no applied magnetic field. In the initial state with $I = 0$, there is no flux inside the wire. The current is first increased to $I_c = J_c \pi R^2$ and finally ramped back down to $I = 0$. (a) The $H_\phi(r)$ and $J_z(r)$ during initial increase of I. (b) Dependence of J and H after decreasing I to zero. Note trapped flux similar to that in Fig. 3.29.

As a second example, consider the dc current distribution in a cylindrical wire of radius $R \gg \lambda$ with a transport current I and only the self-field of the current. For a type I superconductor, the current would be restricted to a thin layer $\sim \lambda$ on the exterior. For a normal metal, the current density would be uniform. For a type II superconductor, let us assume that, initially, $I = 0$. As the current is ramped up, the axial current density J and the circumferential B are initially on the outer layer of the wire (see Fig. 3.31). As I is increased, the vortex rings move inward, driven by the current density in the critical state with value J_c. [Note that the slope of $H(r)$ is not necessarily constant in cylindrical geometry.] The maximum current is achieved when $I = \pi R^2 J_c$ and the current is flowing uniformly in the critical state. But similarly to Fig. 3.29, when the current is reduced, J actually changes sign on the outside of the wire. When I is reduced to zero, the wire is still in the critical state, with counterflow between the outer and inner portions of the wire. More generally, one has both a transport current in a wire and a screening current due to magnetic fields produced by other parts of a solenoid. This makes detailed calculations more complex, but the general picture of the critical state and flux trapping should remain valid.

Note that we can also use this picture to estimate the largest magnetic field we can produce with a superconducting solenoid having a given J_c. The maximum field H in the interior that can be sustained in the critical state is $J_c d$, where d is

3.5 FLUX PINNING IN LARGE FIELDS

here the total thickness of the superconducting windings. (This also follows from simple dimensional analysis, but the critical state picture provides more insight.) For example, if $J_c \sim 10^8 \, \text{A}/\text{m}^2$ and the windings are 4 cm thick, then the maximum field in the center is $\sim 4 \, \text{MA}/\text{m} \sim 5 \, \text{T}$. If we want to create a larger field, then we must either have a large value of J_c or build the magnet on a larger scale. So even if we have a superconductor with $H_{c2} \sim 100 \, \text{T}$, we cannot create a table-top superconducting solenoid with H approaching 100 T unless we have J_c exceeding $10^9 \, \text{A}/\text{m}^2$.

In fact, our simple assumption that J_c is independent of B is not valid at large fields. A more accurate approximation for fairly large B is that the pinning force $f_p = J_c B$ is a constant corresponding to a constant magnetic "pressure gradient" $\nabla(B^2/2\mu_0) = B\nabla H = BJ_c$. Thus, we have $J_c \sim 1/B$. This does not invalidate our analysis above, based on constant J_c, but the details of the critical state will be different. For example, a hysteresis loop of $\langle M \rangle$ versus H would exhibit not flat parallel lines as in Fig. 3.30 but rather something like the peaked curves of Fig. 3.32. Assuming the $H_{\max} \gg H_p$, the applied field is close to $\langle B \rangle$, and we can estimate $J_c(B) \approx 4M(B)/d$ for a flat slab in a parallel magnetic field, or $\approx 3M(B)/R$ for a cylinder of radius R with an axial field.

Note that a constant value of $f_p(B)$ implies a different scaling relation for superconducting magnets of a given material than a constant $J_c(B)$ (Tinkham, 1996, p. 178). In particular, for a given value of f_p we have

$$B_{\max} = \sqrt{(2\mu_0 f_p d)} \tag{3.58}$$

or conversely, the winding thickness should go as $d \sim B^2$. Since the windings typically cover most of the cross section of a solenoid and most of the mass of a superconducting solenoid lies in the windings, this would indicate that the mass should scale approximately as B^4 for a given length, which is reasonably accurate for some real superconductors. In contrast, a field-independent J_c would suggest a much weaker dependence of the mass $\sim B^2$.

As we pointed out earlier, there is magnetic pressure $B^2/2\mu_0$ on the vortices, but if these are pinned, this pressure is transmitted through to the superconducting wires themselves. This pressure can be enormous in large magnetic fields; for

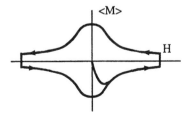

Figure 3.32. Magnetic hysteresis loop $\langle M \rangle$ versus applied field for superconductor in critical state, similar to that in Fig. 3.30 but with $J_c \sim 1/B$ in large fields.

116 MAGNETIC PROPERTIES OF SUPERCONDUCTORS

$B = 12$ T, this pressure amounts to 10^8 N/m^2, or about 1000 atm. In the case of a solenoid, this will tend to cause the windings to bow outward. Since many superconductors are sensitive to strain or even susceptible to brittle fracture, rigid mechanical support structures must be built into the design of high-field magnets.

Stability and Normal-Superconducting Composite Wires

A critical problem in high-current, high-field superconductors is the need to maintain electrothermal stability. This follows from the fact that J_c decreases with increasing T, while in the critical state, virtually the entire wire operates at current densities just below J_c. A small local increase in T therefore creates a local instability in the critical state, requiring a rearrangement in vortices known as a "flux jump." But this flux jump produces local heating, which will decrease J_c further, bringing about more flux jumps and further heating. This can produce an avalanche effect, known as "quenching" the superconductor, whereby the entire conductor is driven into the normal state with $T > T_c$.

We can see this if we consider the flux distribution in a strip of superconductor in the critical state, with critical current J_c, thickness d, width w, and $H_0 \gg H_p$, as in Fig. 3.33. If J_c is reduced by a small amount $\delta J_c = \delta T |\partial J_c/\partial T|$ over a section of the strip of length l, then the magnetic flux trapped in each half of the slab is reduced by

$$\delta \Phi = \frac{\langle \delta B \rangle l d}{2} = \frac{\mu_0 \langle \delta M \rangle l d}{2} = \frac{\mu_0 (\delta J_c d/4) l d}{2} = \frac{\mu_0 d^2 l \delta T |\partial J_c/\partial T|}{8} \quad (3.59)$$

The energy dissipated δQ by this flux jump is

$$\delta Q = 2(\int IV dt) = J_c\, dw\, \delta \Phi = \frac{\mu_0 d^3 wl\, \delta T |\partial J_c/\partial T|}{8} \quad (3.60)$$

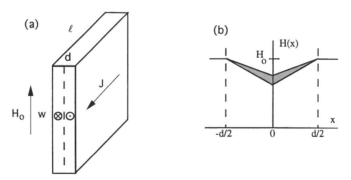

Figure 3.33. Superconducting slab in critical state with $H_0 \gg H_p$ before and after flux jump corresponding to decreased J_c. (a) Configuration of field and currents. (b) The flux change $\delta\Phi$ is proportional to the area between the two triangular $H(x)$ profiles.

But this heat dissipation will in turn create an additional temperature increase $\delta T'$ given by

$$\delta T' = \frac{\delta Q}{C\,dwl} \quad (3.61)$$

where C is the heat capacity per unit volume in the superconductor, assuming that there is not time for any of this heat to be conducted away (the "adiabatic condition"). If $\delta T' \ll \delta T$, then this sequence of heating steps should converge, avoiding thermal runaway. This corresponds to a requirement that

$$d < \left(\frac{8C}{(\mu_0 J_c |\partial J_c/\partial T|)}\right)^{1/2} \quad (3.62)$$

For a typical low-T_c superconductor at 4 K with $J_c \sim 10^9$ A/m^2, $|\partial J_c/\partial T| \sim J_c/T_c \sim 10^8$ A/m^2 K, and $C \sim 1$ kJ/m^3 K, this gives $d < \sim 100$ μm. A similar result would also apply for a cylindrical wire. Such a fine wire would be "adiabatically stabilized" against catastrophic flux jumps but could carry only ~ 10 A of current.

We can safely carry a larger current in a superconducting wire if we embed it in a good normal conductor. Copper and silver are typically used for low- and high-T_c superconductors, respectively. In either case, the pure normal conductor has a much higher thermal conductivity than the superconductor, and in addition has a lower electrical resistivity than the flux flow resistivity during a flux jump. For dc operation while the superconductor remains superconducting, the parallel resistive shunt has absolutely no effect. But if a flux jump occurs, the heat can be quickly diverted to the normal sheath, and from there to the cryogen bath, before catastrophic heating occurs. In fact, a sufficiently thick normal layer can locally carry the entire current of the wire, permitting the core to cool down and restore superconductivity. Based on this principle of "cryostatic stabilization," low-T_c wires have been made with superconducting cores ~ 250 μm in diameter that can carry ~ 50 A and are encased in copper with about 10 times the cross-sectional area of the superconductor (see Fig. 3.34). Of course, this reduces the effective current density of the composite conductor down to $\sim 10^8$ A/m^2.

If we want to further increase the current-carrying capacity of superconducting wires, we can go one step further to a multifilamentary composite containing many fine superconducting filaments encased in a normal metallic matrix (Fig. 3.34). But it is essential here to make sure that a flux jump within one superconducting filament does not catastrophically heat up the entire cross section. We can obtain a higher effective current density than that permitted by cryostatic stabilization described above by taking advantage of the principle of "dynamic stabilization." The basic concept is that if any locally excess heat in a superconducting filament can be diverted away faster than magnetic flux can

Figure 3.34. Typical comparative cross sections for high-field superconducting wire with adiabatic stability (*a*), single-core composite with cryostatic stability (*b*), and multifilamentary wire with dynamical stability (*c*). The dark regions represent superconductor and the lighter regions a normal metal such as copper.

move between superconducting filaments, then a catastrophic flux jump can be suppressed. The analysis of this is complex, but an argument can be made in terms of characteristic diffusion constants (Tinkham, 1996, p. 187).

It is well known that heat spreads by diffusion, with a thermal diffusivity.

$$D_{th} = \frac{\kappa}{C} \tag{3.63}$$

which has units m²/s. Here C is the heat capacity per unit volume and κ is the thermal conductivity. It is perhaps not quite as widely appreciated that magnetic flux also moves via diffusion in a conducting medium. This electromagnetic diffusion constant is

$$D_{em} = \frac{\rho_n}{\mu_0} \tag{3.64}$$

where ρ_n is the metallic resistivity and D_{em} also has units of m²/s. This can also be applied to flux flow in a superconductor if ρ_{ff} is used and if pinning is not significant. Of course, in a superconductor with strong pinning, magnetic flux will not move at all. For either type of diffusion constant, a characteristic time for diffusion across a distance L is $t \sim L^2/D$.

Note that D_{th} is proportional to the thermal conductivity, which for a metal is generally proportional to the electrical conductivity. So a good conductor will have a large value of D_{th} and a small value of D_{em}. Conversely, a poor conductor will have a small value of D_{th} and a large value of D_{em}. Fortunately, these are exactly the properties of the copper matrix and the superconducting filaments in the flux flow state. Estimates have shown that $D_{em}(SC) \sim D_{th}(Cu) \sim 0.1$ m²/s, while $D_{th}(SC) \sim D_{em}(Cu) \sim 10^{-4}$ m²/s. So the bottleneck for heat flow is the superconducting filaments themselves, while the bottleneck for flux flow is the copper.

In order to obtain dynamic stability, we would like to isolate the superconducting filaments magnetically but to couple much of the heat generated out to the copper. Since $D_{th}(SC) \sim D_{em}(Cu)$, it is straightforward to design a composite with filaments of radius R and interfilament distance $L > R$ such that $R^2/D_{th}(SC) < L^2/D_{em}(Cu)$. In addition, we would like to ensure that the remaining heat in the superconductor does not cause a runaway temperature increase, as in the earlier requirement for adiabatic stability. The complete analysis for typical low-T_c superconductors leads to a filament size less than ~ 50 μm in diameter, with about twice as much Cu as superconductor. The current capacity of each filament is small (~ 2 A), but the entire composite can have hundreds of filaments. A rough indication of the comparative scales in the different schemes of stabilized superconducting wire is given in Fig. 3.34.

There is one final requirement that is important for ac or transient applications of long, multifilamentary composite superconducting wire. In this case, the magnetic flux must distribute through the normal metal as well as through the superconductor, which tends to couple the filaments magnetically, reducing their stability. It has been shown that this problem can be ameliorated by twisting the wire, so that a given filament rotates its position relative to the others in a helical fashion, with a pitch less than about 10 cm for typical parameters.

We will leave discussion of how such multifilamentary composites are made for the next chapter on superconducting materials. But it is worth noting here that wire for a superconducting magnet requires carefully tailored structures over the entire range of physical parameters. This starts out on the atomic level with the proper atomic structure to give the optimum T_c. We may also need controlled atomic impurities to obtain a large value of H_{c2} and an appropriate distribution of small defects on the 10–100-nm scale to act as pinning sites to yield a large J_c. Superconducting microfilaments may have diameters on the 10-μm scale, and the wire itself is typically ~ 1 mm across. The wire may be twisted on the centimeter scale and wound around a solenoidal form on the 10-cm scale, together with support and cooling structures. And the total continuous length of the wire must be many kilometers long. All of these must be optimized simultaneously in order to achieve a practical superconducting magnet. Perhaps remarkably, this has been accomplished, at least with low-T_c materials, and is now a standard commercial product.

3.6 MAGNET AND POWER APPLICATIONS

A variety of applications have been developed that build on the high-field, high-current properties of superconductors. Some of these are already in service using LTS materials; others are under development for HTS materials. As listed in Table 3.3, a number of these involve the production or manipulation of large dc magnetic fields. Other applications take advantage of low-loss ac properties or of diamagnetic repulsion. The first applications were to scientific research, but more recently, superconductors have started to make inroads into medical diagnostics, electric power systems, materials processing, and transportation

Table 3.3. Large-Scale Applications of Superconductors

Superconducting magnets
  Magnetic resonance imaging (MRI)
  High-energy physics
  Nuclear fusion
  Magnetic separation
  Superconducting magnetic energy storage (SMES)

AC motors and generators
Transformers
Power transmission cables
Magnetic levitation and frictionless bearings
MagLev trains

(Sheahen, 1994; Grant, 1997; Montgomery, 1997). These are discussed in more detail below. Small-scale electronic applications will be described elsewhere.

Conventional Magnet Technology

Before discussing the applications of high-field superconducting magnets, let us first describe briefly the features and limitations of conventional magnet technology. One particularly common configuration is the iron-core electromagnet, which we indicate schematically in Fig. 3.35a. Here, a coil of N turns of wire (each with current I) is wound around an iron core with an air gap of width d. In the magnetic circuit picture, we can regard NI as the magnetomotive force, or mmf (analogous to the electromotive force of a voltage source in an electrical circuit), that generates magnetic flux lines that then cross the gap and return on the other side of the core. The iron core is analogous to wires leading to a circuit element, and the gap is analogous to a resistor. In the ideal situation, almost the entire mmf drops across the gap, leading to a magnetic field in the gap

$$H = \frac{NI}{d} \qquad B = \frac{\mu_0 NI}{d} \tag{3.65}$$

The quantity analogous to the resistance in an electrical circuit is the "reluctance"

$$R = \frac{l}{\mu A} \tag{3.66}$$

where l is the length of the element parallel to the flux lines, A its cross-sectional area, and μ the permeability. The flux Φ around the loop is analogous to the

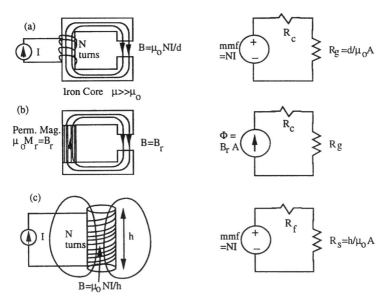

Figure 3.35. Technologies for magnet design. (*a*) Iron-core electromagnet with schematic magnetic circuit. The field of interest is in the gap of the core; the reluctance of the gap $R_g \gg R_c$ of the core. (*b*) Permanent magnet with soft iron core to conduct and focus flux to air gap. No external power supply is necessary. (*c*) Hollow-core solenoid commonly used for superconducting magnets with high magnetic field inside solenoid. After charging to full current, a superconducting coil can be operated in the persistent-current mode with no external power supply. For a long cylinder, the internal reluctance of the solenoid $R_s \gg R_f$ of the external fringe fields.

electrical current, so that

$$\Phi = \frac{NI}{R_{\text{tot}}} \tag{3.67}$$

where R_{tot} is the total series reluctance around the magnetic circuit. Due to the very large linear permeability of the iron ($\mu \gg \mu_0$), the reluctance R_c of the iron core should be small compared to the series reluctance $R_g = d/\mu_0 A$ of the air gap (and the parallel leakage reluctance of the surrounding air). This provides great flexibility in designing and distributing the coils and the iron core. However, all magnetic materials exhibits a saturation magnetization M_s, that is typically limited to less than 2 MA/m, corresponding to a saturation induction $B_s = \mu_0 M_s \sim 2$ T. For fields larger than this, the permeability of an iron core decreases and approaches μ_0, and the flux lines are no longer even confined within the iron. Then, this entire approach breaks down. Therefore, one cannot generally use an iron-core electromagnet to produce a magnetic field $B > \approx 2$ T.

Similar considerations are present in systems with a permanent magnet. A permanent magnet can be characterized by a remanent flux density

$B_r = \mu_0 M_r$, even in zero applied field. For typical high-performance permanent magnet materials, $B_r \sim 1$ T. As shown in Fig. 3.35b, we can approximate a permanent magnet as a flux source $\Phi = B_r A$, analogous to a current source, which can replace the drive coils of Fig. 3.35a. This has an advantage in operating without any power supply, but the field cannot be easily tuned. Furthermore, one can taper the iron core to concentrate the magnetic flux slightly, but there is still an effective limit of about 2 T for the magnetic flux density in the air gap.

Superconducting Solenoids

For these reasons, for magnetic fields larger than ~ 2 T, an iron core cannot be used to conduct or focus a magnetic field and is generally not part of the system. Furthermore, a permanent magnet is not of much use. Instead, we can only use coils of wire located near the region of concentrated magnetic field. The simplest and most common configuration for high-field superconducting magnets is the hollow solenoid, as indicated in Fig. 3.35c. Here, the flux is concentrated in a uniform field inside the solenoid and returns via the fringe fields outside. For a long cylinder of length h and area A, the reluctance of the cylinder $R = h/\mu_0 A$ is much larger than that of the fringe fields, which can be neglected. Then the flux inside the cylinder is

$$\Phi = \frac{NI}{R} = \frac{\mu_0 NIA}{h} \tag{3.68}$$

and the internal field is uniform with value

$$H = \frac{NI}{h} \qquad B = \frac{\mu_0 NI}{h} \tag{3.69}$$

In principle there is no limit on how large a field can be obtained, provided we can apply a large enough voltage to overcome wire resistance and also provide enough cooling to carry away the enormous amount of waste heat. With superconducting wire, however, neither of these is a problem.

Let us go through a sample design exercise for a superconducting solenoid (Fig. 3.36a). Assume that we want to produce an axial field of 2 T over a cylindrical volume 1 m in diameter by 2 m long. (This is comparable to some magnets used for magnetic resonance imaging, as discussed later.) Assuming also that this is being constructed using a wire that can carry 100 A at a current density $J = 1 \times 10^9$ A/m^2; this corresponds to a conductor diameter of 0.36 mm. The required magnetic field is $H = 1.6 \times 10^6$ A/m, which for $I = 100$ A and $h = 2$ m leads to $N = 32{,}000$ turns. If we estimate the wire diameter (including insulation) to be 0.5 mm, then each layer will contain 4000 turns, and the entire 32,000 turns can be achieved with eight layers, with a total winding thickness of only 4 mm. But the total length of wire needed is about 100 km! We can estimate

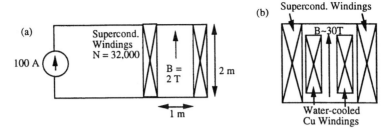

Figure 3.36. High-field solenoid design. (a) Superconducting solenoid with 2-T field inside cylinder 1 m in diameter by 2 m long obtained with 32,000 turns of wire carrying 100 A. (b) Simplified schematic of ultra-high-field hybrid magnet showing outer cryogenically cooled superconducting windings and inner water-cooled copper windings. Inside the inner coil, fields of 30–40 T can be achieved.

the self-inductance

$$L = \frac{N\Phi}{I} = \frac{\mu_0 N^2 A}{h} = 500 \text{ H} \tag{3.70}$$

so that one could linearly ramp such a lossless inductor up to full current by applying 10 V for 5000 s (a bit over one hour), as suggested in Fig. 1.1. All of this power VI goes into magnetic energy storage, giving a total energy at full power

$$\mathscr{E} = \tfrac{1}{2} L I^2 = 2.5 \text{ MJ} \tag{3.71}$$

(This is the same energy one obtains by multiplying the magnetic energy density $\tfrac{1}{2}BH = B^2/2\mu_0$ by the volume of the cylinder.) We can then switch to the persistent-current mode, removing the leads, and the magnet will continue to store this energy without loss indefinitely.

We should also consider the same magnet in the normal state for $T > T_c$. If we estimate a resistivity $\rho = 1$ μΩ-cm $= 10^{-8}$ Ω-m (pure Cu at room temperature has $\rho = 1.7$ μΩ-cm), then the resistance of the entire 100-km length is $R = 10$ kΩ. If we try to conduct 100 A in this wire, this requires a voltage drop of $IR = 1$ MV and dissipates $I^2 R = 100$ MW! This is clearly not practical and illustrates the great advantage of superconductivity. The charging circuit described in the previous paragraph needs to deliver a maximum of 1 kW of power.

The forces involved in such a solenoid are rather large. As we discussed earlier, magnetic energy density corresponds to pressure. The stored 2.5 MJ lies in a volume of 1.6 m³, corresponding to an energy density or pressure of 1.6 MN/m²; this is the same as would be obtained by calculating $B^2/2\mu_0$. This in turn corresponds to 16 atm, or a total outward force on the windings of about 10^7 N, equivalent to about 1000 tons of force. This cannot be supported by the windings themselves and requires a reinforced metal framework. This becomes even more critical for larger magnetic fields (10 T or above), since the magnetic pressure goes as B^2.

Low-temperature superconducting magnets have become widespread for producing large magnetic fields. The largest field thus far achieved in a superconducting solenoid is 20 T using Nb_3Sn cooled to 2 K. This is close to the upper critical field H_{c2} of the superconductor, when the critical current starts to drop sharply. However, it is worth noting that the largest dc magnetic fields in the world (available in a few high-magnetic field research centers), ~ 30 T or greater, are even larger than this and are produced using normal resistive copper. But the need for cooling is so great (on the megawatt scale) that these "Bitter magnets" are constructed using copper plates with holes drilled in them for water cooling connected in a helical structure. Actually, the highest field magnets are hybrid structures (Fig. 3.36b); a superconducting solenoid around the outside produces a field ~ 10–20 T, and the copper magnet inside it adds another 10–30 T to the small central region. A record 45-T hybrid magnet is under construction at the National High Magnetic Field Laboratory in Florida. Finally, even larger magnetic fields (up to >60 T) are available as pulsed magnetic fields with durations on the millisecond scale. These are also generally constructed from copper, and the limit is provided by the maximum transient heat load obtainable without melting the copper as well as the maximum stress without exploding the magnet.

Magnetic Resonance

The largest commercial application of superconducting magnets is currently for magnetic resonance imaging (MRI), an important method of noninvasive three-dimensional medical imaging. But before discussing the imaging application, let us briefly describe the physical phenomenon known as nuclear magnetic resonance (NMR). Many atomic nuclei have net magnetic moments μ_n, notably the hydrogen nucleus H^1, that is, the proton. In the presence of an applied dc magnetic field, such a magnetic moment will tend to align parallel to the field. The magnetic energy takes the form $U = -\mu_n \cdot B$, where $\mu_n = 1.4 \times 10^{-26}$ J/T for the proton. A classical magnet will rotate to achieve this optimum configuration. For a quantum magnet, however, the energies are quantized; for a spin-$\frac{1}{2}$ nucleus like the proton, only two energy levels are allowed, corresponding to parallel and antiparallel to the applied field (Fig. 3.37). A transition between these two states can occur if it is accompanied by a photon with energy

$$\mathscr{E} = hf = \Delta U = 2\mu_n B \qquad (3.72)$$

For $B = 1$ T, the photon frequency is 42.6 MHz, in the rf range. This is also a very small energy, 3×10^{-26} J, orders of magnitude smaller than the thermal energy kT at room temperature (4×10^{-21} J). So in thermal equilibrium close to half of the protons should be in each of the energy levels, with marginally more

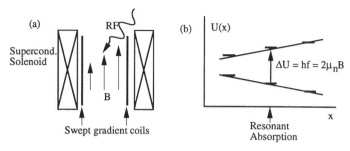

Figure 3.37. Magnetic resonance imaging. (*a*) Schematic cross section of magnetic field produced by main superconducting solenoid and swept gradient coils. The field gradient is exaggerated for clarity. An rf field is also coupled into the high-field region. (*b*) Proton energy level versus position, with resonant absorption where $\Delta U = hf = 2\mu_n B$.

in the lower level. However, if we irradiate a population of protons with photons at the proper resonant frequency, we can obtain resonant absorption of the photons and an increase in the fraction in the upper state. After the radiation is turned off, the distribution will return to thermal equilibrium, together with reemission of photons at the same resonant frequency. This thermalization process is normally rather slow (seconds) and depends on the atomic environment of the atoms. The basic measurement technique is that a sample in a fixed magnetic field is subjected to a pulse of rf radiation at the resonant frequency; this is followed by detecting the delayed emission at the same frequency. Information on the concentration of protons and on their chemical environment can thus be obtained.

The resonant absorption associated with NMR is typically extremely narrow; for a fixed rf frequency, the field must be within ~ 10 μT of the resonant field. This provides the basis for imaging of a large volume. The main magnetic field is designed to be extremely uniform, but there is a second coil (or combination of coils) that produces a very small field gradient that can be swept in three dimensions. Then, for a fixed frequency, only a small volume will be properly resonant at a given time. The resulting signal will thus be characteristic of that volume alone; the rest of the volume will be off resonance. By sweeping this gradient field to achieve resonance sequentially in the entire volume, we can obtain a three-dimensional map of the concentration and environment of the protons within the object under examination. This forms the basis of MRI. (The word "nuclear" was dropped to avoid any implication of nuclear radiation.)

We have been focusing on protons (H^1 nuclei), and indeed, that is what MRI normally uses. This is because the human body is mostly H_2O, and H is indeed the most common atom in the body. Furthermore, some of the other common atomic nuclei, such as C^{12} and O^{16}, have $\mu_n = 0$ and hence no magnetic resonance. So only the hydrogen signal is sufficiently strong to give high-resolution images. This signal is related to the density of water and to its chemical environment, which are sufficient for strong contrast in soft tissues. For full-body MRI systems, a superconducting solenoid with B equal to 1.5 T or 2 T operating in the persistent-current mode is typically used for the dc magnetic field. These solenoids have been constructed using LTS wire, since there is not yet commercial production of HTS wire on the 100-km scale. These fields are just inside the range that could in principle be achieved using conventional magnets, but the superconducting magnet is more practical. An iron core electromagnet on the necessary scale would be extremely heavy, require enormous electrical power, and be very noisy electrically. Recently MRI systems with $B \sim 0.5$ T have been developed that use permanent magnets, but these exhibit somewhat inferior spatial resolution as compared to the higher field superconducting systems.

In addition to MRI, NMR spectroscopy is widely used as a tool in chemical and biological research and diagnostics. Smaller superconducting magnets with fields up to 10 T have recently been widely developed for this application. Superconductors are also being applied to other parts of magnetic resonance systems; recent developments include the use of a superconducting rf resonator to reduce the noise of the rf detection system (Magin, Webb, and Peck, 1997).

Magnets for High-Energy Physics Research

Another class of applications of superconducting magnets is to bend the paths of high-energy charged particle beams in high-energy physics research. If a proton with charge e is immersed in a magnetic field perpendicular to its motion, it will be subject to the transverse Lorentz force $\mathbf{F} = e\mathbf{v} \times \mathbf{B}$, causing it to rotate in a circle (Fig. 3.38). Since the centripetal force for a circular path is mv^2/r, the

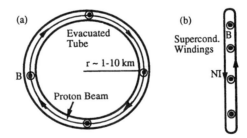

Figure 3.38. Superconducting magnets for high-energy accelerator. (*a*) Accelerator ring for high-energy proton beam. Inside the evacuated tube, a vertical magnetic field (which may be produced using superconducting magnets) causes the beam to bend clockwise around the ring. An antiproton beam of the same energy can also traverse the same ring in the opposite direction. (*b*) Configuration of bending magnet for proton beam. Many such magnets are placed around the circumference of the ring. The current must be ramped up as the beam energy is accelerated.

radius of the circular motion is

$$r = \frac{p}{eB} \tag{3.73}$$

where $p = mv$ is the momentum, and the angular velocity of rotation (also known as the "cyclotron frequency") is

$$\omega_c = \frac{v}{r} = \frac{eB}{m} \tag{3.74}$$

These formulas were derived here using nonrelativistic mechanics, but they turn out to remain valid for highly relativistic particles traveling near the speed of light. In this case, $p \approx \mathcal{E}/c$, so that Eq. (3.73) can be written

$$r \approx \frac{\mathcal{E}}{ecB} = \frac{\mathcal{E}}{cB} \tag{3.75}$$

where the latter expression is for the energy in units of electron-volts. For a proton, this should be valid for $\mathcal{E} \gg 1$ GeV. For example, the Tevatron at the Fermi National Accelerator Laboratory (Fermilab, outside Chicago) accelerates protons up to 900 GeV using superconducting bending magnets (called "dipole magnets") with $B = 4$ T. This yields $r = 750$ m, or a circular ring with a circumference of 4.7 km. The actual ring circumference is somewhat larger (6.3 km), but this includes some straight sections as well. As with all of these accelerator applications, this uses LTS coils cooled with circulating liquid helium.

Note that these dipole magnets cannot be wound as simple solenoids, since they must produce a vertical field inside a narrow horizontal tube. Also, for a ring of a fixed radius, the protons can be increased in energy only if the magnetic field is also increased proportionately. This accelerator works by injecting protons into the ring at a lower energy (~ 150 GeV), accelerating them up to 900 GeV using multiple passes through a special accelerator section and then bombarding them into either a fixed target or a beam of antiprotons (with the same energy but opposite charge) going in the other direction. This is designed to occur at a repetition rate of more than one injection per minute. This means that these magnets cannot be operated in the persistent-current mode but must instead ramp repeatedly from a low field to a high field and back again. This rapid cycling brings about some ac losses in the superconductor, and in addition careful design is necessary to avoid quenching due to flux jumps. These magnets have now been in operation for some years, working as predicted.

Other superconducting proton accelerator systems have been designed and at least partly developed. The Superconducting SuperCollider (SSC) in Texas was canceled a few years back, but it had been designed for 20-TeV protons with

6.5-T bending magnets and a ring circumference of 87 km. The current next-generation accelerator is the Large Hadron Collider (LHC), under design for construction at the European Center for Nuclear Research (CERN) in Switzerland. This will incorporate a ring that is 27 km in circumference with 8.6-T bending magnets and a maximum energy of 7 TeV and is projected for completion in the year 2005.

For all of these accelerators, an additional set of superconducting magnets, called "quadrupole magnets," produce gradients in magnetic fields that are used for maintaining a narrow focused beam (which would otherwise spread out since the protons repel each other) as it moves around the ring. Superconducting bending magnets are also used in detectors and spectrometers for high-energy charged particles produced as a result of collisions. From an observation of the bending radius in a given magnetic field, one can infer the momentum of the charged particle.

Power Applications and SMES

The promise of storing and transporting large currents with little or no loss lends itself naturally to applications in the field of electric power. A number of prototype systems have been developed, designed, or proposed, although they generally have not yet achieved acceptance in the industry. In addition, some future power generation technologies may require very large magnetic fields, with a key role for superconducting magnets.

One of the conceptually simplest power applications is known as *superconducting magnetic energy storage* (SMES). It uses a large superconducting solenoid for lossless storage of electric power for load leveling, that is, to better match power production to demand on an hourly or daily basis. This requires truly massive solenoids, with energies on the order of 10 GW-h, or ~ 40 TJ. If we assume a field of 10 T, the energy density is $B^2/2\mu_0 = 40$ MJ/m^3, so this corresponds to a volume of order 10^6 m^3, corresponding to a cube 100 m on a side. The stress on these coils would also be tremendous, so that design studies have suggested embedding the entire solenoid underground in bedrock. This would, of course, have to be cryogenically cooled, but the surface-to-volume ratio, which determines heat loss, becomes more favorable as the scale becomes larger. It is also worth noting that the energy density in a superconducting magnet is comparable to other energy storage mechanisms. For example, a typical lead-acid battery has an electrochemical energy density of order 300 MJ/m^3, and natural gas has a chemical energy density of 35 MJ/m^3 at standard pressure. A competitive storage technology is hydroelectric storage, but the requisite bodies of water at differing elevations are not always available.

In the near term, much smaller SMES systems (sometimes called micro-SMES) may be more practical. These are superconducting solenoids with 1–10 MJ storage that could be used for energy storage and power conditioning on a kilowatt scale for times of order 1 min or less. This would permit a dedicated piece of manufacturing equipment, for example, to continue operating despite occasional brief power dips or interruptions. Such a system would be combined with a rectifier and inverter for interfacing the dc magnet with the ac power network and a sensor that detects power dips and enables the system to respond in well under a second. Such a superconducting system may be competitive with other short-term storage devices such as batteries and flywheels, and several micro-SMES systems (again using LTS) have been installed at industrial sites.

AC Power Generation and Transmission

Another important class of potential applications involves the generation and transmission of ac power. As we discussed earlier, superconductors do exhibit some ac losses, but if properly designed, these can be much less than for resistive systems. Development efforts have focused on several key components, including generators and motors, transformers, and power transmission cables. While there are distinct issues for each of these, a common theme is that superconducting components operate at higher current densities and magnetic fields and without iron cores. This permits the superconducting version to be much smaller and lighter than its conventional equivalent for the same power rating. On the other hand, the superconducting system requires a cryogenic environment, which tends to limit potential applications to very large systems (otherwise, refrigeration losses become prohibitive). Furthermore, the conventional components have been developed and refined over many decades and are really extremely efficient. For example, large motors are typically 97% efficient in converting electrical to mechanical energy. Superconducting motors, generators, and transformers have all been successfully demonstrated on an intermediate scale (megawatts and above), both in LTS and HTS materials, but they are not yet currently in use in utility systems.

Superconducting power transmission cables may offer some important advantages in utility systems. These would be underground coaxial lines, which would operate at lower voltages and higher currents than conventional underground lines. This would permit somewhat reduced insulation thickness corresponding to substantially reduced cross section, even considering the cryogenic requirements. Thus, one could project conducting greater ac power in an existing underground conduit. Furthermore, dc superconducting power transmission cables may offer some clear advantages in terms of reduced ac losses and simpler superconducting cable design. Despite the losses associated with ac–dc conversion, such a dc line might be preferable in systems with very long lengths (many hundreds of kilometers). As with the ac superconducting cables, this could operate at lower voltages and higher currents than equivalent conventional high-voltage dc lines, permitting a more compact cable, and perhaps bundling of multiple cables in a single cryogenic conduit. Both LTS and HTS power transmission cables have been demonstrated in short lengths but remain to be incorporated into power networks.

Fusion Magnets

Several proposals for future power generation technologies require very large magnetic fields. For these to be practical, superconducting magnets are likely to be a necessity. The scheme that has drawn the most attention is nuclear fusion, where a hot thermonuclear plasma is confined in a strong magnetic field. Although there are other general approaches to fusion reactors that do not require magnetic confinement (in particular, the inertial confinement associated with laser bombardment), the "tokamak" and related reactors are the most advanced worldwide. The basis for this confinement is closely related to the use of magnetic fields for bending particle beams but is substantially more complicated because there are both ions and electrons moving in random directions. A typical reactor has a toroidal vacuum vessel with a magnetic field that is also toroidal (Fig. 3.39). The motion of the ions can be approximated as helical orbits around the field lines, with radius determined by Eq. (3.73). For example, for an H^2 nucleus (deuteron) with kinetic energy $\mathscr{E} = 10$ keV, taking $B = 10$ T, this gives $r \sim 2$ mm. The larger the field, the tighter the spirals, and the more effectively the particles are confined. The major project of this type under development is the International Thermonuclear Experimental Reactor (ITER), planned for completion in 2008. The design calls for a 13-T toroidal field (using Nb_3Sn) over a volume of more than 1600 m^3 for a total stored magnetic energy of more than 100 GJ.

Magnetic Separation

There are also applications of strong magnetic fields that depend not only on the field itself but also on its gradient. In particular, a particle with magnetic dipole moment μ will rotate to align with a uniform magnetic field, but a gradient in the field is required to provide net motion of the particle:

$$F = \mu \cdot \nabla B \tag{3.76}$$

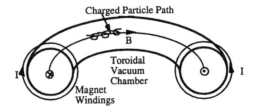

Figure 3.39. Magnetic confinement of plasma in a tokamak fusion reactor with toroidal magnetic field produced by superconducting windings. The ions follow helical paths rotating around the magnetic field lines and are thus largely confined.

The particle will move in the direction of increasing field. This is why, for example, a permanent magnet will attract iron filings to both its north and south poles; the magnitude of the field gradient is greatest there. This principle can be used to separate magnetic particles from nonmagnetic particles in a mixture and is the basis for magnetic separation.

We can estimate the size of the effect on a strongly magnetic particle by considering the magnetic force density per unit volume $f_m = M\nabla B$, where M is the magnetization of the particle, typically $\sim 10^6$ A/m. For a moderate $\nabla B \sim 1$ T/m, the force density $f_m \sim 10^6$ N/m^3. This can be compared to the gravitational force density $f_g = \rho g \sim 10^4$ N/m^3, where ρ is the mass density and $g \approx 10$ m/s^2 is the gravitational acceleration. As long as the magnetic force is comparable or larger than the gravitational force, separation can be achieved. (This also suggests the possibility of magnetic levitation, which will be discussed in more detail below.) So a superconducting magnet is not needed to separate highly magnetic particles.

However, this principle can also be used to separate weakly paramagnetic materials, where the presence of the field induces a weak magnetic moment, which then is attracted up the field gradient. In this case, the induced magnetization is

$$M = \chi H = \frac{\chi B}{\mu_0} \qquad (3.77)$$

where χ is the paramagnetic susceptibility. Then, the force density is

$$f_m = \frac{\chi B \nabla B}{\mu_0} = \chi \nabla \frac{B^2}{2\mu_0} \qquad (3.78)$$

Many materials that are not ferromagnetic are weakly paramagnetic, with $\chi \sim 10^{-4}$–10^{-5}. For $B \sim 10$ T, with a characteristic decay length ~ 0.1 m, this gives $f_m \sim 10^4$ N/m^3, comparable to the gravitational term. These are fields and field gradients that cannot be efficiently achieved without superconducting magnets, and indeed, separation and purification of certain clays using superconducting magnets have already been commercialized.

Furthermore, most nonmagnetic materials have a background diamagnetic susceptibility on the atomic level, with $\chi \sim -10^{-6}$. This would cause weak repulsion from a magnetic source (rather than attraction as for a paramagnetic material), which in a very strong field gradient can even lead to magnetic levitation. The diamagnetic levitation of small nonmagnetic objects on the centimeter scale has in fact been recently demonstrated using a superconducting solenoid with a room temperature bore, including levitation of a droplet of water and even a small, live frog! It has also been suggested that this effect might be used to simulate weightlessness in the laboratory.

Magnetic Levitation

Let us now discuss magnetic levitation as it applies to superconductors. Consider first a superconducting plate in the Meissner state, with $B = 0$ inside the superconductor. Let us assume also a permanent magnet in the form of a long cylinder of radius a and height $h \gg a$ magnetized in the axial direction with uniform magnetization M. The permanent magnet can also be viewed in the "magnetic pole" picture, where the effective "magnetic charge" Q_m on the ends of the cylinder is given by the relationship

$$M = \frac{Q_m h}{\pi a^2 h} = \frac{Q_m}{\pi a^2} \tag{3.79}$$

since M is the density of magnetic dipoles. If the end of the magnet is brought a distance $d < a$ from the superconducting surface, it will "see" its reflection, the same distribution of magnetic charge a distance d below the surface (Fig. 3.40). This image magnet is due to the screening currents at the surface of the superconductor and assures that $B = 0$ inside. The force between the magnet and its image is repulsive, with a form that is analogous to the force between two

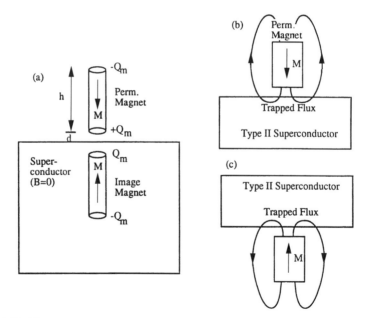

Figure 3.40. Magnetic levitation using type II superconductor. (*a*) Levitation of permanent magnet above superconductor in Meissner state using method of images to calculate force. (*b*) Levitation of magnet above superconductor with partial flux penetration giving flux trapping near surface of superconductor. This increases the transverse stability of the levitation. (*c*) Attractive magnetic suspension of magnet below superconductor with trapped flux.

plates of a parallel-plate capacitor.

$$F = Q_m B = (\pi a^2 M)(\mu_0 M) = \mu_0 \pi a^2 M^2 \tag{3.80}$$

In other words, the force per unit area is $\mu_0 M^2$, which can be very large indeed. For example, a typical high-performance permanent magnet may have $M \approx 1$ MA/m, which yields $F \sim 1$ MN/m² (10 atm of pressure!), corresponding to the ability to levitate a mass of ~ 100 tons/m². This is clearly an upper estimate [a more accurate calculation falls off with increasing separation; see Orlando and Delin (1991, p. 199)] but it does suggest that rather large levitation forces are possible. Alternatively, the same levitation force could be exhibited by a superconducting solenoid that produced the same magnetic field $H = 1$ MA/m. Superconducting solenoids producing fields 10 times as great are commercially available, producing a maximum levitation force $\mu_0 H^2$ up to 10,000 tons/m². For a large-area magnet, the superconductor could be much lighter, since the core could be hollow, and it could also be operated in the persistent-current mode, consuming no power.

There is an important problem with the analysis above, namely that the assumption of the Meissner effect does not remain valid for arbitrarily large magnetic fields. However, even for fields above H_{c1}, a magnetic field will still be excluded by near-surface screening currents associated with the critical state. For example, if we have $H = 1$ MA/m and $J_c = 10^9$ A/m², then the screening currents will penetrate a distance $d = H/J_c = 1$ mm. So the above analysis may continue to be approximately valid up to fields much larger than H_{c1}.

However, there are some important ramifications of flux penetration, namely that some flux is "trapped" in the superconductor underneath the magnet, even if the external field is removed. This means that the screening currents in the superconductor will try to maintain the flux at this position, and implicitly, restoring forces will act to maintain the magnet at its current position. This will not restrict the rotation of a (symmetric) magnet about an axis perpendicular to the superconductor. In fact, the stability of this rotation will be increased, since it will be prevented from moving to the side, or even moving up. It is even possible in some cases to take advantage of flux trapping to suspend a magnet *beneath* the superconductor. It is ironic that a normally diamagnetic superconductor can be used for attractive levitation, but this is a natural consequence of flux trapping.

The high-field magnetic levitation properties of a superconductor lend themselves naturally to applications in low-loss rotating magnetic bearings. Magnetic bearings based purely on permanent magnets are possible, but the inclusion of a superconductor (with its intrinsically repulsive interactions) enhances the overall stability of the system. Such a bearing is practically frictionless, particularly if it operates in a vacuum. One prototype system incorporating such bearings is a rapidly rotating flywheel, which may be used for energy storage in utility systems.

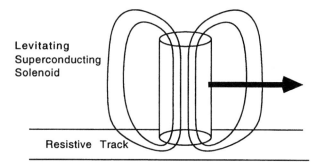

Figure 3.41. Schematic of superconducting solenoid levitating above a resistive metallic track that forms the basis for a magnetically levitated train. Provided the solenoid is moving fast enough, the flux is excluded from the track, giving a repulsive levitation force. Propulsion can also be achieved electromagnetically.

Finally, there has been continuing interest in the use of magnetic levitation in the context of transportation, more specifically, magnetically levitated trains. Let us consider a configuration with a superconducting solenoid located above a normal resistive plate, rather than a superconducting plate (Fig. 3.41). When the solenoid is turned on, eddy currents in the conductor will initially act to levitate the magnet in the same ways as for the superconducting plate. However, in the case of the resistive plate, these will die out in a short L/R time. For solenoid of radius $a \sim 1$ m and a plate with resistance $\rho \sim 10^{-8}$ Ω-m (typical of copper) and thickness $d \sim 1$ cm, we can crudely estimate

$$t \sim \frac{\mu_0 a}{\rho/d} \sim 1 \text{ s} \qquad (3.81)$$

Now let us assume that the solenoid is moving along the track in the transverse direction. Since the track is resistive, there is no flux trapping and hence no restriction on transverse motion. If the transverse speed of the solenoid is much greater than $2a/t \sim 2$ m/s $= 7$ km/h, the eddy currents will not have time to relax, and the solenoid will remain levitated. This provides the basis for magnetically levitated trains, and indeed working prototypes of such trains (using superconducting solenoids and copper plates in the track) have been developed and extensively tested, particularly in Japan. Such a train must have wheels to start moving, but once it is above a threshold speed, it lifts off the track and is then subject only to air resistance and drag due to eddy current losses. Furthermore, the eddy current losses decrease at high speeds, so that extremely high speeds can be achieved. We have not explained how such a train can be propelled, but this too can be accomplished electromagnetically (Orlando and Delin, 1991, p. 203). Two key approaches are a passive track with a linear induction motor on the train and an active track incorporating a linear synchronous motor; the latter has recently received the most attention. "Maglev" systems with speeds up to 500 km/h have proven to be technically feasible, but the commercial viability of such an endeavor remains to be established.

Impact of HTS

There have been some 30 years of development of LTS magnet technology, and much of this technology is quite mature. In contrast, very long lengths of high-quality HTS magnet wire are not yet available. So although there have been design studies and small-scale prototypes of HTS systems, very few full-scale HTS systems have yet been demonstrated. This may be partly a matter of time; as we discuss in the next chapter, the materials and manufacturing technology for HTS wire and tape continues to improve. However, in the near term, most large-scale applications are likely to continue to rely on LTS systems operating at liquid helium temperatures.

On the other hand, there are several applications that do not require kilometers of wire and where HTS devices may be expected to have a near-term impact. One that is already a commercial product is HTS current leads for an LTS magnet system. Conventional technology requires the use of copper leads going from a room-temperature power supply to the LTS magnet at 4 K. Given the large thermal conductivity of copper, this contributes a rather large heat leak from room temperature into the cryogenic environment as well as that due to Joule heating in the copper. In contrast to copper, HTS materials are generally relatively poor thermal conductors. An HTS lead that goes from liquid nitrogen temperature (77 K) to 4 K can carry the needed current while blocking much of the heat.

Another near-term HTS application may include low-loss bearings operating at 77 K. These are based on the bulk magnetic properties of HTS ceramics, which are much more forgiving than the transport properties. For example, a microcrack at a grain boundary will stop a transport current but has little effect on the diamagnetism of a bulk ceramic. Such HTS bearings made from YBCO with needed properties are already available and are being developed into flywheel systems, for example.

Finally, one possible power application that we have not discussed is a fault current limiter, essentially a very fast circuit breaker in case of a short in an electrical load. Given the highly nonlinear properties of superconductors in large currents and fields, this may be a natural application. Both resistive and inductive designs are being designed and tested using HTS materials.

In the longer term, as HTS wires and tapes become a more mature technology, we can expect that HTS magnets, power lines, rotating machines, and the like will be successfully developed. However, it is likely that they will not fully supplant LTS applications but instead supplement them. HTS magnets are likely to be more expensive than LTS magnets for some time; for many systems, the cryogen is *not* the major expense. We have already discussed a hybrid magnet consisting of an LTS magnet on the outside and a resistive solenoid inside. One can envision replacing the resistive insert with an HTS insert to produce the very large fields without the massive power requirement. Furthermore, as we will discuss in the next chapter, many HTS materials exhibit their optimum high-field performance at quite low temperatures, below 20 K, so that it may be preferable to operate the device in the low-temperature regime. Finally, as the technology of closed-cycle cryogenic refrigerators continues to be developed, both LTS and HTS systems may be operated without liquid cryogens, and the distinction between them may eventually become less significant.

SUMMARY

- In the Meissner effect, a superconductor expels the internal magnetic flux upon cooling below T_c in a small magnetic field, except within $\lambda(T)$ of the surface.
- A bulk Meissner superconductor acts like a diamagnetic material with permeability $\mu = 0$. This has applications to magnetic shielding.
- Magnetic flux Φ (or more accurately, fluxoid Φ') is quantized in units of Φ_0 in a superconducting loop, corresponding to a phase change $\Delta\theta = 2\pi n$ around the loop.
- The large-field behavior of a superconductor depends on the GL parameter $\kappa = \lambda/\xi$. In a type I superconductor, with $\kappa < 0.7$, the Meissner state breaks down for $H > H_c \approx \Phi_0/10\mu_0\xi\lambda$.
- Practical superconductors are type II, with $\kappa > 0.7$, and magnetic flux in discrete fluxons, each with Φ_0, for $H > H_{c1} \approx (\Phi_0/4\pi\mu_0\lambda^2)\ln\kappa$ up to the upper critical field $H_{c2} = \Phi_0/2\pi\xi^2$.
- Each vortex consists of a "normal" core $\sim \xi$ surrounded by screening currents out to a distance $\sim \lambda$. The vortices repel each other and form a vortex lattice.
- Vortices are subject to a transverse Lorentz force $\mathbf{J} \times \mathbf{\Phi}_0$, analogous (dual) to the electrical force on carriers in a semiconductor. This provides the basis for the flux flow transistor.
- $R = 0$ only if all vortices are locked in place by pinning sites and the forces of other vortices. This leads to a maximum depinning current density $J_p \approx \Phi_0/100\mu_0\xi\lambda^2$.
- In the critical state the distribution of magnetic flux is given by $dH/dz = \pm J_p$. This leads to hysteretic behavior and trapped flux as magnetic field or current is changed.
- A composite of small superconducting filaments in a normal metal matrix (Cu or Ag) serves to maintain electrothermal stability against runaway heating due to flux jumps.
- High-field applications include magnetic resonance imaging, high-energy accelerators, and superconducting magnetic energy storage (SMES).

- Diamagnetic repulsion may lead to frictionless superconducting bearings. Magnetic levitation of a superconducting magnet on a normal-metal track may provide the basis for high-speed trains.

REFERENCES

M. R. Beasley, J. E. Mooij, and T. P. Orlando, "Possibility of Vortex–Antivortex Pair Dissociation in Two-Dimensional Superconductors," *Phys. Rev. Lett.* **42**, 1165 (1979).

A. Davidson and M. R. Beasley, "Duality between Superconducting and Semiconducting Electronics," *IEEE J. Solid State Circuits* **14**, 758 (1979).

P. M. Grant, "Superconducting and Electric Power," *IEEE Trans. Appl. Supercond.* **7**, 112 (1997).

A. M. Kadin, "Duality and Fluxonics in Superconducting Devices," *J. Appl. Phys.* **39**, 5741 (1990).

R. L. Magin, A. G. Webb, and T. L. Peck, "Miniature Magnetic Resonance Machines," *IEEE Spectrum* **34**, 51 (Oct. 1997).

D. B. Montgomery, "Future Prospects for Large Scale Applications of Superconductivity," *IEEE Trans. Appl. Supercond.* **7**, 134 (1997).

J. Nordman, "Superconductive Amplifying Devices Using Fluxon Dynamics," *Supercond. Sci. Technol.* **8**, 681 (1995).

T. P. Orlando and K. A. Delin, *Foundations of Applied Superconductivity*, Chap. 5–7 (Addison-Wesley, Reading, MA, 1991).

T. P. Sheahen, *Introduction to High-Temperature Superconductivity*, Parts III and IV (Plenum, New York, 1994).

M. Tinkham, *Introduction to Superconductivity*, 2nd ed., Chaps. 4 and 5 (McGraw-Hill, New York, 1996).

PROBLEMS

3.1. Persistent current. A circular superconducting loop 10 cm in diameter constructed of wire 1 mm in diameter is placed perpendicular to an external magnetic field of 10 mT for $T > T_c$ and then cooled to $T < T_c$. Then the external magnetic field is removed.

(a) What is the trapped flux in the loop, and how many flux quanta is this? What is the phase change $\Delta\theta$ in going around the loop?

(b) The inductance of a circular loop of outer radius b and wire radius a is given by $L = \mu_0 b \ln(2.8b/a)$. Determine the inductance of the loop and the circulating current.

(c) This loop is kept cool for one year, and the magnetic field measured. No change is observed to better than 0.1%. Assume that the current flows at the surface over a distance $\lambda \sim 100$ nm. Estimate the maximum resistivity of the superconductor.

(d) Now the superconductor is suddenly heated above T_c, where it has a resistivity of 10 μΩ-cm. How quickly does the trapped flux leak out of the loop?

3.2. Meissner effect and diamagnetic susceptibility. A superconducting film of thickness d is in the Meissner state with a magnetic field H_0 parallel to both its surfaces.

(a) In the M-picture, determine the average magnetization $\langle M \rangle$ in the film as a function of d/λ.

(b) Determine the diamagnetic susceptibility $\chi = \langle M \rangle / H_0$.

(c) Determine χ in this thin film with $d \ll \lambda$.

(d) Consider the temperature dependence of χ using the conventional two-fluid temperature dependence of $\lambda(T)$. Plot χ versus T for $d/\lambda(0) = 0.3, 1, 3$.

3.3. Superconducting magnetic shielding. A long, hollow superconducting cylinder is placed in a magnetic field H_0 perpendicular to the axis. Assume that the superconductor is in the Meissner state and that the shell prevents penetration of magnetic flux inside the cylinder. Assume also that the shell thickness $d < \lambda$ and that the cylinder radius $r \gg d$ and $r \ll L$, the length.

(a) What is the demagnetizing coefficient for this geometry? What is H_{int} in the M-picture?

(b) What is the maximum magnetic field H_{max} at the outer surface of the cylindrical shell? What are the corresponding surface current and current density J_{max}?

(c) If $B_0 = 50$ μT (the ambient field of the earth), $\lambda = 200$ nm, and $d = 100$ nm, estimate H_{max} and J_{max}. Will the superconductor remain in the Meissner state? (Estimate H_{c1} and compare to typical values of J_c.)

(d) Now an external field of 1 T is applied. What values of J_{max} would be required to screen the field? Is this achievable?

3.4. Solenoid design. Design a laboratory-scale superconducting solenoid that will produce a magnetic field of 10 T in a cylindrical region 10 cm in diameter by 30 cm long. The available multifilamentary superconducting wire has an outer diameter of 0.6 mm and can carry 200 A in 10 T.

(a) How many turns of wire are needed? How many wire layers is this?

(b) Estimate the total length of wire needed for this.

(c) What is the total inductance of the magnet? How long would it take to charge it with a 1-V power supply?

(d) Determine the effective outward pressure and total force on the framework of the magnet when it is fully charged.

(e) If one were to make the same solenoid using Cu at room temperature (with resistivity $\rho = 1.7$ μΩ-cm), how much power would be needed for such a magnet? Does this seem feasible?

PROBLEMS

3.5. Diffusion of magnetic flux and heat. At the end of Section 3.5, it is pointed out that the motion of magnetic flux in a normal resistive metal may be modeled in terms of a diffusion constant similar to that for heat flow.

(a) Show how this magnetic diffusion follows from Maxwell's equations and obtain the diffusion constant D_{em}.

(b) Use the transmission line picture of wave propagation in a metal to obtain the dual circuit that represents the flow of flux in the metal.

(c) Sketch a transmission line picture to represent the flow of heat and compare to that in (b). Does this suggest why a good conductor of heat is a bad conductor of flux, and vice versa?

3.6. Type II superconductor. A given type II superconducting material exhibits $H_{c1} = 10^5$ A/m and $H_{c2} = 10^7$ A/m.

(a) Determine λ, ξ, and $\kappa = \lambda/\xi$.

(b) Estimate the maximum depinning current density J_p, and determine the maximum total current that could be carried in the critical state by a wire 1 mm diameter.

(c) Using the critical state model, assuming that J_c is this maximum J_p, estimate the penetration field H_p for a wire 1 mm in diameter. (This would be a great overestimate for a real sample, since J_c is always much less than the maximum.) How does this compare to H_{c2}?

3.7. Vortex and vortex lattice.

(a) Given the circulating currents surrounding a vortex, determine the radius at which the current will exceed the ideal critical current I_c, thereby destroying superconductivity. Compare this to the coherence length ξ. This is why the vortex core is effectively a normal metal.

(b) At the upper critical field H_{c2}, what is the distance between neighboring vortices in the vortex lattice? Is it surprising that the superconductor is now fully normal?

(c) Show that the force between two adjacent vortices is repulsive, and estimate its magnitude when the intervortex distance is several ξ. How does this compare to the maximum pinning force? What does this suggest about the relative importance of pinning and the vortex lattice for immobilization of vortices?

3.8. Superconducting permanent magnet. Permanent magnets are typically made by subjecting a magnetic material to a large pulsed magnetic field in order to align the magnetic domains. Something similar may be possible for superconductors using trapped magnetic flux in the critical state. Consider a hollow superconducting cylinder of radius r, thickness $d \ll r$, and height $h \gg r$. A large magnetic field H_0 is applied parallel to the axis of the cylinder and then removed. Assume that the critical current density of the cylinder is a constant J_c within the critical state model and that $H_0 > J_c d$.

140   MAGNETIC PROPERTIES OF SUPERCONDUCTORS

(a) Sketch the spatial dependence of the residual current distribution in the cylinder wall and the field inside the cylinder, in the cylinder wall, and outside.

(b) How does this field distribution compare to that for a solenoid wound using many turns of superconducting wire having the same value of J_c and operating at its maximum field?

(c) Now consider a uniformly magnetized permanent magnet with internal magnetization M. Determine the value of M that will give a similar external magnetic field to that for the superconductor.

(d) A small high-performance permanent magnet can be made that produces a field $B \sim 1$ T at its poles. If we wanted to design a superconducting magnet 1 cm in diameter to do the same thing, estimate the necessary critical current.

(e) Comment on the feasibility and stability of such a small superconducting permanent magnet.

3.9. Flux flow transistor. A flux flow transistor consists of a control line inductively coupled to a vortex flow channel (see Fig. 3.23b). Assume that the control line is a film 2 μm wide and 0.5 μm thick and is 2 μm away from the vortex channel, which is also 2 μm wide but may be thinner. Assume that the vortex channel has a normal state resistance $\sim 1\,\Omega$ and an effective upper critical field $B_{c2} \approx 100$ T.

(a) Estimate the maximum control field that may be produced by this control line on the vortex flow channel if it is to operate below its critical density $J_c = 10^6$ A/cm^2.

(b) What is the effective areal vortex density n_v for this control field?

(c) Estimate the flux flow resistance with the control current turned on.

(d) Assume that the vortex flow channel operates at a current of 1 mA, so that the self-field remains small compared to the control field. Estimate the vortex velocity and the vortex traversal time.

3.10. Magnetic levitating train. Consider a large superconducting solenoid 1 m in diameter that produces an internal magnetic field $B = 5$ T. Two such solenoids are mounted vertically at the two ends of a section of a levitating train and are designed to support it. The train moves above the surface of a copper track (resistivity $\rho = 2$ μΩ-cm) 1 cm in thickness.

(a) Assume first that the train is moving fast enough so that eddy currents in the copper can be ignored and it is levitating. Estimate the maximum diamagnetic levitation force and the equivalent maximum mass that may be supported.

(b) One way to examine the effect of the velocity on eddy current losses in the track is through the effective flux diffusion constant $D_{em} = \rho/\mu_0$. If the velocity is large, then the flux will not have time to penetrate the track, and the track acts more like a perfect conductor. Estimate the velocity needed to achieve this.

4
SUPERCONDUCTING MATERIALS AND THIN-FILM TECHNOLOGY

Superconductivity can be found in a wide range of materials. Unfortunately, in most of them it can only be observed below the boiling temperature of liquid helium (4 K), which makes it of mostly academic interest. This chapter will focus on materials aspects of more practical superconductors, including sample fabrication. For conventional superconductors (now called low-temperature superconductors, or LTSs), that generally means materials based on the element niobium (Nb). The newer high-temperature superconductors (HTSs) are all based on a family of layered copper oxides, the discovery of which came as a complete surprise to much of the scientific community in 1986. For both classes, we will describe in more detail the current technologies for making devices and circuits using both HTS and LTS thin films. In addition, we will discuss more briefly some of the other classes of superconducting materials and microstructures that may appear promising for future development.

4.1 LOW-TEMPERATURE METALLIC SUPERCONDUCTORS

Elements and Alloys

In order to become a superconductor, a material must first be a conductor; semiconductors generally do not have a sufficient density of carriers. In fact, all of the early superconductors were simple metallic elements, alloys, and compounds. Indeed, most metals become superconductors at sufficiently low temperatures. A Periodic Table with the critical temperatures of the elements is shown in Fig. 4.1. It is evident that there are some regularities in the distribution of T_c's, but it is not simple. First of all, superconductivity does not occur in

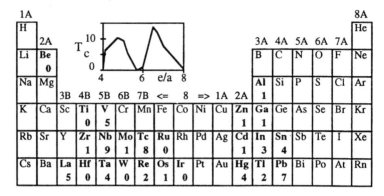

Figure 4.1. Variation in superconducting critical temperatures among elements. Periodic Table of Elements, with superconducting elements marked in bold and values of T_c (to nearest degree kelvin) listed. The inset shows the maximum T_c for transition elements and metallic alloys (for the row containing Nb) plotted versus number of valence electrons per atom (e/a). Note the double-peaked structure.

metals with magnetic ions, which break Cooper pairs. These include Cr through Ni in the top row of transition metals as well as the whole series of rare-earth metals and most of the actinides with partially filled f-bands (not shown in Fig. 4.1). Furthermore, it seems that more than two valence electrons per atom are generally necessary to exhibit superconductivity. The univalent noble metals (Cu, Ag, Au) and the alkali metals, all of which are excellent normal conductors, do not become superconducting even for $T \to 0$.

The highest T_c, 9.2 K, among the pure elements is the transition metal Nb, which has five valence electrons. The second highest (8 K) is technetium, with seven valence electrons, whose symbol just happens to be Tc. Unfortunately, all isotopes of Tc are radioactive, so the element exists only in trace amounts. The more practical second highest is lead (Pb), a p-band metal with a valence of 4 and $T_c = 7.2$ K. It is noteworthy that the columns in the Periodic Table with these elements are local maxima in the other rows as well. This suggests that the number of electrons/atom (e/a) might be a good predictor of T_c in metallic alloys and compounds as well. This observation, known as Matthias's rule, is empirically found to be obeyed quite broadly for many materials (White and Geballe, 1979; Roberts, 1976). It does not have a firm theoretical basis but appears to be related to peaks in the density of states $D(\mathscr{E}_f)$ at the Fermi level. There appears to be a double maximum (at least for transition metals) at around 4.7 and 6.5 (see Fig. 4.1). Therefore, an alloy of Nb with a small amount of the adjacent element Zr can have T_c raised up to 11 K, whereas a similar alloy of Nb with a small amount of Mo reduces T_c. And an alloy of Mo with (radioactive) Tc and $e/a \sim 6.5$ has exhibited T_c up to 14 K!

When analyzing alloys in this way, one must be careful to confirm that the elements are actually mixing on the atomic scale rather than separating into different phases. One technologically very important superconducting alloy is Nb–Ti, typically with about 50% Ti by atomic fraction. One might expect a significant increase in T_c (since e/a should move to the peak value), but in fact there is only a slight increase to 9.5 K. This is because most of the Ti actually goes into small grains of a separate Ti-rich phase, which is not even superconducting (at 4 K). While this might initially seem to be a disadvantage, these normal Ti grains can act as pinning sites for vortices, while the currents flow in the Nb-rich phase that remains superconducting. We will continue our discussion of vortex pinning later.

Intermetallic Compounds

Even higher values of T_c can be achieved using binary intermetallic compounds, where there is an ordered crystal structure of the two types of atoms, rather than the random atomic mixture in an alloy. Analysis of such intermetallic compounds is rather complex, but even here this correlation with the number of electrons per atom may continue to apply. For optimizing T_c, Nb is the main constituent here as well. Table 4.1 lists several of these compounds, together with several of their typical properties. There are two cubic crystal structures which are important for the highest values of T_c among conventional superconductors. These are the A-15 structure (sometimes called the beta-tungsten structure) and the B-1 (or NaCl) structure, which are illustrated in Fig. 4.2. The record high-T_c for 14 years prior to 1986 was Nb_3Ge in the A-15 structure, with $T_c = 23$ K.

These intermetallic Nb compounds are similar in several respects to conventional metallic alloys. In particular, they look metallic and have a comparably high carrier density ($\sim 10^{28}$ m^{-3}) and exhibit an electrical resistivity (above T_c) that is in the metallic range (~ 1 μΩ-m) although much higher than a pure metallic element such as Cu. On the other hand, their mechanical properties are somewhat closer to those of ceramics. Partly because of directional covalent bonds, they tend to be very hard, more brittle, and much less ductile than conventional metals.

Table 4.1. Some Practical Nb-Based Superconducting Alloys and Compounds

	T_c (K)	e/a	λ (nm)	ξ (nm)	2Δ (meV)	B_{c2} (T)	B_{c1} (T)
Nb	9	5	50	30	3	0.4	0.1
Nb–Ti	9	4.5	150	5	3	13	0.025
NbN	16	5	100	5	5	15	0.05
Nb_3Sn	18	4.7	30	4	7	23	0.04
Nb_3Ge	23	4.7	45	3	8	38	0.025

Note: Parameters are typical values at $T = 4$ K.

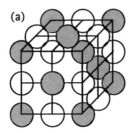 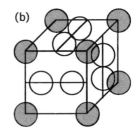

Figure 4.2. Cubic unit cells for practical LTS crystals. (*a*) B-1 structure for NbN (open, Nb; shaded N). (*b*) A-15 structure for Nb_3Sn, showing Nb chains in three directions (open, Nb; shaded, Sn)

All of the materials in Table 4.1 are type II superconductors. In fact, apart from the pure element Nb, all of them are extreme type II superconductors, with $\lambda \gg \xi$. This is in part because they are "dirty metals" with a short electronic mean free path ($l \ll \xi_0$). Even the crystalline compounds have a high density of atomic vacancies and other defects. As discussed earlier, this has the effect of increasing λ and decreasing ξ. This, in turn, increases H_{c2}, thus making these materials more suitable for high-field applications. On the other hand, H_{c1} is correspondingly reduced, limiting applications that require a true Meissner state without flux penetration. For these reasons, pure Nb tends to be used for high-frequency and low-field electronic applications; Nb–Ti and Nb_3Sn are used for high-field magnet wire (Fig. 4.3). Unfortunately, Nb_3Ge turns out to be metastable and rather difficult to fabricate in the proper crystalline structure, and so has not had wide application.

Note that despite our discussion of crystal structures, superconductivity in metals does *not* require a single crystal or anything close to it. The superconductivity normally couples right across grain boundaries, and the materials are essentially isotropic. In fact, amorphous metals (having no structure beyond the atomic level) can sometimes be superconducting, and as we have discussed, a certain amount of disorder on several scales is necessary to obtain high values of H_{c2} and J_c. On the other hand, it is necessary that there be a continuous superconducting path within the material. Niobium does tend to form an insulating oxide (Nb_2O_5), and it is possible to form a "granular superconductor" with weak coupling (essentially, tunnel junctions) between superconducting grains. Depending on the strength of the intergrain coupling, such a granular superconductor might still exhibit long-range supercurrents, but the critical current density J_c could be greatly reduced. This is a more serious problem for HTS materials, as we will discuss later.

Most LTS systems are designed to operate immersed in a bath of boiling liquid helium at 4.2 K. This is less than half the T_c of Nb or Nb–Ti, and so the superconducting properties (H_{c2}, J_c, and Δ) for all of the materials in Table 4.1 have reached close to their $T = 0$ values. In these systems, Nb or Nb–Ti may be perfectly satisfactory; there may be no need to go to a material with a higher

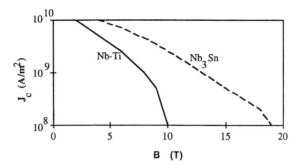

Figure 4.3. Typical magnetic field dependence of critical current density for practical conventional superconducting wires (Nb–Ti and Nb$_3$Sn) at 4 K. A value of $J_c \sim 10^9$ A/m^2 or greater is generally necessary to make a practical superconducting magnet.

value of T_c. Of course, this does require that the liquid helium bath be replenished on a regular basis (on the scale of days). On the other hand, in recent years there have been significant improvements in the reliability of closed-cycle cryocoolers (see Appendix C) that can attain around 10 K without the presence of liquid helium. A superconducting system could then use NbN or Nb$_3$Sn as the superconductor, with only a continuous supply of electric power to run the cryocooler.

Bulk Wire Fabrication

Let us first describe the major steps in the fabrication process of multifilamentary Nb–Ti superconducting wire. As we briefly discussed earlier, this requires that microscopic (~ 10 μm) Nb–Ti filaments must be embedded in a matrix of copper in order to maintain electrothermal stability in the wire under high-current, high-field conditions. The conventional way to manufacture such a composite (Fig. 4.4a) takes advantage of the ductile nature of both Cu and Nb–Ti. We can start with a block of Nb–Ti made by melting the elements together. A macroscopic composite with just a few large blocks of Nb–Ti can then be placed in holes drilled in a larger Cu block, or blocks of both materials can be assembled together. Then, this structure can be drawn down to successively smaller cross sections, with more sections being grouped together at several times during the process to result in a composite with hundreds of filaments. The resulting wire (which can be many kilometers long) can then be wound into whatever form is necessary.

This is now a standard commercial technology, and indeed, Nb–Ti is the primary material for superconducting magnets in applications as diverse as magnetic resonance imaging and high-energy accelerators. Magnetic fields as high as 9 T can be reliably produced by such magnets. This compares to a maximum of about 2 T for a conventional iron-core electromagnet.

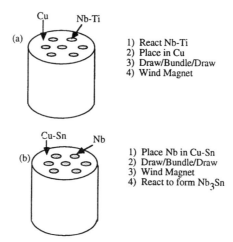

Figure 4.4. Fabrication procedures for multifilamentary LTS wire. (*a*) Initial ductile composite of reacted Nb–Ti in Cu. (*b*) Bronze process for Nb$_3$Sn, with initial ductile composite of Nb in Cu–Sn (bronze) and reaction to form brittle compound at end.

The critical current density J_c in such a wire is limited by the effectiveness of pinning sites in the Nb–Ti filaments. As mentioned above, a standard metallurgical process forming this alloy from the melt leads to the precipitation of small Ti-rich crystallites that are not superconducting and can act as pinning sites. A great deal of research and development has gone into optimizing the scale and density of these "natural" pinning sites in order to enhance J_c in large magnetic fields. In this way, values of $J_c \sim 10^9$ A/m^2 at $B \sim 9$ T are now standard (see Fig. 4.3). In addition, in recent years there has been further research into "artificial pinning sites," which may include small particles or layers of a different nonsuperconducting material. Even magnetic inclusions have been investigated.

This general process for making multifilamentary wire must be modified for Nb$_3$Sn and other intermetallic compounds (Fig. 4.4*b*). If the same process of drawing down the wire were attempted, the brittleness of the superconductor would lead to broken filaments. On the other hand, we can make a multifilamentary wire with Nb filaments in a Cu–Sn alloy ("bronze") matrix by the same process; both of these materials are suitably ductile. Then, once the wire has been made and wound into its final form, the whole structure can be heated to a high temperature (600–750°C) to react the Sn in the Cu matrix with the Nb filaments, creating the desired compound Nb$_3$Sn. (The Cu atoms generally do not enter this reaction.) An alternative to this "bronze process" is the "internal tin process," which uses separate layers of Sn in the initial composite before drawing down, winding the wire, and then reacting at high temperatures. Superconducting magnets using Nb$_3$Sn wire can produce fields of up to about 15 T at 4 K, considerably higher than is possible using Nb–Ti wire. They can also operate at 10 K (for a somewhat smaller field), cooled by a closed-cycle refrigerator. However, because of the additional complexity of the fabrication process, Nb_3Sn superconducting magnets tend to be much more expensive than those made of Nb–Ti. Critical depinning currents tend to be rather large (again, $\sim 10^9$ A/m^2 at the operating field), probably due to defects and impurities in the material.

Finally, it is worth noting that a drawn wire form is not the only possibility for LTS magnets. An alternative that has also been used in some cases consists of a thin tape made of a layer of superconductor clad with a layer of copper. This may not be quite as stable as a multifilamentary geometry (and hence more prone to quenches), but for certain materials that are difficult to fabricate, this may be acceptable.

Bulk NbN may be formed from Nb by reacting with pure N_2 gas at very high temperatures; the diffusion of the N atoms through the solid is otherwise extremely slow. If there is insufficient nitrogen, the structure may still form, but with vacancies at some of the N sites and a reduced value of T_c. Bulk NbN has not been used for making wires, but the surface layers of bulk Nb can be nitrided for some high-frequency applications.

Thin-Film Fabrication

For many applications of superconductors, particularly at high frequencies, it is important that vortices not enter the superconductor. In this case, the currents flow only within λ of the surface, and we need only deal with thin films on the scale of less than 1 μm thick. Thin films are also used throughout small-scale electronic applications of superconductors. The current technology of LTS thin films is dominated by Nb, with NbN in second place, and we will focus on these two materials.

However, an earlier superconducting electronics technology was built around the use of Pb alloys, with T_c around 7 K. This was the basis of the Josephson computer project that IBM pursued through the 1970s. Lead films are particularly easy to produce, given the low melting point of Pb (350°C, less for alloys). One simply melts a sample of Pb in a vacuum (usually in an electrical heater), and Pb atoms evaporate from the liquid surface, condensing on a substrate (Fig. 4.5a). It is also straightforward to produce a thin (several-nanometer) PbO_x layer between two superconducting layers that acts as a tunneling barrier for a Josephson junction. On the other hand, Pb corrodes easily, and the atoms are somewhat mobile at room temperature. Even more problematic, stresses caused by thermal cycling of Pb junctions can be relieved by formation of surface "hillocks" that short out the Josephson junctions. These problems were addressed to some degree in the course of the IBM project, but related problems contributed in part to the cancellation of that project in 1983.

In contrast to Pb, Nb is a "refractory material" with a melting temperature of 2500°C, and simple thermal evaporation cannot be used to prepare Nb films. But once they are formed, these are extremely stable. The same thing is true of

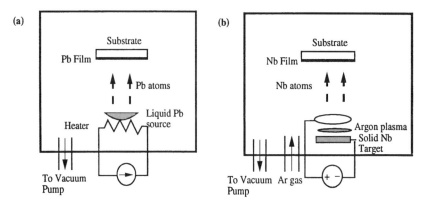

Figure 4.5. Deposition methods for low-temperature superconducting thin films. (*a*) Thermal evaporation of a soft metal such as Pb. (*b*) Sputtering of a hard metal such as Nb.

Nb-based Josephson junctions; they can be thermally cycled repeatedly without any degradation. This has contributed to making Nb-based superconducting integrated circuits a reality.

The major method to produce Nb films is known as "sputtering." Sputtering involves ejection of atoms from the surface of a solid target under ionic bombardment in a plasma (Fig. 4.5*b*). In the simplest form, a high voltage (~ 1000 V) is applied across a gap in a low pressure of an inert gas (usually Ar, at a pressure of order $1 \text{ Pa} = 1 \text{ N/m}^2 \sim 10^{-5}$ atm ~ 7 m Torr). An electric discharge leads to the formation of a plasma, a mixture of electrons and positive Ar ions. A dc voltage drives the Ar^+ ions toward the "target," which is the source of material to be deposited. The resulting ion bombardment knocks atoms, ions, and electrons off the surface of the target, and some of these atoms move across to the substrate, forming the thin film. It is important that the target be cooled, since most of the bombarding energy gets dissipated as heat. In this way, the target remains solid and stable and will erode at a steady rate. This method is not extremely fast, but it has the key advantage of being highly reproducible and controllable electrically. If the gas pressure and the bias voltage are kept constant, then the rate of thin film deposition will be constant. A common variant of sputtering, known as "magnetron sputtering," employs permanent magnets to help confine the plasma near the target, thus increasing the deposition rate. A typical deposition rate is as high as 10 Å/s, so that a typical film ~ 0.3 μm thick would take as little as 5 min.

Niobium thin films are typically prepared by sputtering from a pure Nb target in a pure Ar plasma. Since Ar is an inert gas, very little of it is incorporated in the film. A variety of substrates are used, such as oxidized silicon, and are typically not heated during deposition, resulting in polycrystalline Nb films. (This is in sharp contrast to the typical situation with HTS films, as discussed in the next section.) It is important to avoid contamination of the gas with oxygen or water vapor, since that will lead to Nb films with sharply depressed values of T_c.

It is possible to prepare NbN films by a similar sputtering process but with some nitrogen gas added to the Ar plasma. Here, the target material is still Nb, but the nitrogen is incorporated at the surface of the growing film. This is referred to as "reactive sputtering." Sometimes, some C is deliberately incorporated by also including some methane in the gas, but again, oxygen contamination must be avoided. It is difficult to obtain films with the optimum $T_c \sim 17$ K unless the substrate is heated (to $\sim 300°$C). This is because the surface mobility of the atoms at ambient temperatures may not be enough to form the fully ordered crystal structure. This reflects a general trend that more complex crystal structures (with larger unit cells) tend to require higher growth temperatures. Still, values of $T_c \sim 15$ K have been reported for NbN on substrates that do not exceed 200°C. Also, the GL parameters (λ and ξ) are often significantly different from ideal values, due in part to the increased local disorder; $\lambda \sim 300$ nm is more typical of these thin films.

In certain cases, it is possible to use thin-film techniques to fabricate a superconducting compound that cannot be obtained using standard thermal processing of bulk materials. An important example in Nb_3Ge in the A-15 structure, which is metastable in this stoichiometry. Indeed, the record high T_c, 23 K (prior to HTS), was obtained using a thin film of Nb_3Ge deposited on a substrate heated to $> 700°$C by sputtering from a single composite target. Although this could be deposited on a long, continuous tape, such a method would probably be much too slow to be practical for bulk-type applications. For these and other reasons, Nb_3Ge is not currently being developed for superconducting applications.

Josephson Junction Integrated Circuit Technology

For most superconducting electronic applications, the active elements are based on Josephson junctions. A classic Josephson junction consists of two layers of superconductor with a very thin insulating layer between them. For Nb films, one might initially expect that the saturated oxide of Nb, Nb_2O_5, would be a good insulating layer for a tunnel junction, since it is an excellent insulator. However, it turns out that there are other compounds of Nb and O (such as NbO) that are conducting but not superconducting at the operating temperature, and tend to form at the interface between Nb and Nb_2O_5. This leads to a depressed value of the energy gap near the junction and consequently weakens the supercurrent across the junction.

In fact, the standard insulating layer in modern Nb circuit technology is now based on aluminium oxide, Al_2O_3. The process, as illustrated in Fig. 4.6a, involves depositing an ultrathin layer of metallic Al (~ 1 nm) on top of fresh Nb in a vacuum. Then a controlled amount of oxygen is added to the system, permitting most of the Al to oxidize to form the electrical insulator Al_2O_3 but not oxidizing the underlying Nb. Finally, another layer of Nb is deposited on top of the oxidized Al. This Nb/AlO$_x$/Nb trilayer turns out to be extremely reliable and remarkably free of shorts. Since metallic Al is not superconducting

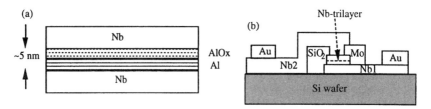

Figure 4.6. Process for Nb-based Josephson junctions. (*a*) Nb/AlO$_x$/Nb trilayer, consisting of Nb base electrode, oxidized Al insulating barrier, and Nb counterelectrode. (*b*) Typical multilayer process for Nb junction technology, including Nb trilayer, two Nb wiring layers (Nb1 and Nb2), SiO$_2$ insulating layer, a Mo resistive layer, and a Au contact layer, on a Si substrate.

at 4 K, one might expect that a residual layer of Al in the bottom electrode might depress the energy gap at the lower surface. However, this is generally not a problem, perhaps because the layer is so thin and the interface so clean. These Nb trilayer junctions are now standard in the superconducting electronics community. Typical junction critical current densities are $J_c \sim 10^7$ A/m^2, several orders of magnitude smaller than that for the films themselves.

In recent years, a true integrated circuit technology has developed based on Nb films and Nb-trilayer Josephson junctions (Abelson et al., 1993). The entire process typically has ten or more distinct layers. This includes several Nb wiring layers, layers of a thick insulator (such as SiO$_2$) to separate wiring layers, the Nb trilayer for junctions, a layer for non-superconducting resistors (such as Mo), and a top Au contact for connecting non-superconducting wires leading to room temperature circuits. The layout of each layer can be designed using standard integrated circuit design tools, and standard microlithographic processes can be used to pattern and align each layer with the one below it. The substrate is frequently a Si wafer, although in most cases this is used simply as a flat insulating surface. This is illustrated by a cross-sectional view of a simple thin-film Josephson junction circuit in Fig. 4.6*b*. Here the junction consists of a Nb-trilayer connected to top and bottom Nb wiring layers, a resistive (Mo) shunt across the junction, an insulating layer (SiO$_2$), and Au contacts to the Nb wires. A large number of micron-scale Josephson junctions can be made on a chip less than 1 cm across. For a typical junction scale of order 3 μm, the junctions have critical current $I_c \sim 0.1$ mA and normal-state resistance $R_n \sim 20\,\Omega$. The thin films are polycrystalline, of order 200–500 nm thick, and are deposited at ambient (or slightly elevated) temperatures.

A similar Josephson junction circuit technology has also recently been developed based on NbN films (Thomasson et al., 1993). Here the thin insulating layer for the Josephson junction is typically ~ 1 nm of MgO, which happens to form with the same crystal structure (B-1) as NbN.

In order to fabricate complex circuits that perform as expected, it is necessary that the quantitative properties of the individual Josephson junctions and related elements be controllable, uniform, and reproducible to within a margin of better than 10%. This has recently been achieved with the standard Nb processes, and NbN is not far behind. This reflects a development that goes back more than 30 years. The same thing has not yet been achieved with HTS films and junctions (which will be discussed in the next section), which is why complex microelectronic circuits are still largely restricted to LTS materials.

4.2 HIGH-TEMPERATURE COPPER–OXIDE SUPERCONDUCTORS

The Breakthrough

By 1986, most researchers had given up searching for higher temperature superconductors. The maximum T_c had been fixed at 23 K for the previous 14 years. There had been a number of preliminary observations suggesting much higher values of T_c, but none of these had been reproduced. Indeed, a number of theoreticians had presented convincing arguments that T_c above about 25 or 30 K was probably impossible. So an obscure report in 1986 suggesting superconductivity up to 30 K in a compound of La–Ba–Cu–O did not get much attention, particularly since the resistance did not go to zero until 13 K (Bednorz and Müller, 1988). However, by early 1987, the situation had changed dramatically, with wide verification of a closely related superconducting compound $YBa_2Cu_3O_7$ (also called YBCO or 1–2–3) with $T_c = 92$ K. This is well above the boiling temperature of liquid nitrogen of 77 K.

Superconductivity of YBCO in liquid nitrogen can be demonstrated very easily and dramatically (Fig. 4.7). The compound can be made simply by heating the constituent elemental oxides (Y_2O_3, BaO or $BaCO_3$, and CuO) together in a high-temperature furnace (at ~950°C) in oxygen, and the resulting black powder can be formed into a ceramic pellet (Chen et al., 1987). If this YBCO pellet is placed in a dish with a small permanent magnet on top, and liquid nitrogen is added to the dish, then the magnet will suddenly rise up and levitate (a few millimeters) above the pellet when the temperature cools below 92 K and will remain stably levitated for as long as the liquid nitrogen remains. This is clear demonstration of diamagnetism requiring at least a partial Meissner effect. Such spontaneous levitation can only be caused by superconductivity.

(a) RECIPE FOR YBCO PELLET

Mix and grind oxide powders with Y:Ba:Cu = 1:2:3
Heat to 900°C in oxygen ~ 10 hrs.
Regrind/Reheat/Regrind
Press Pellet, Heat to 900°C ~ 10 hrs.
Cool in oxygen at 500°C ~ 10 hrs.

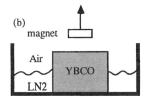

(b)

Figure 4.7. Fabrication of YBCO ceramic pellet and demonstration of Meissner effect. (a) Recipe for solid-state reaction for YBCO pellet with $T_c \sim 92$ K. (b) Levitation of small high-field permanent magnet (such as Nd–Fe–B) above YBCO pellet sitting in liquid nitrogen.

Table 4.2. Some Representative High-Temperature Superconducting Compounds with Critical Temperatures and Number of Closely Spaced CuO₂ Planes

Compound	T_c (K)	Number of Layers	Comments
$La_{1.7}Sr_{0.3}CuO_4$ (LSCO)	38	1	First family discovered in 1986
$YBa_2Cu_3O_7$ (YBCO or Y-123)	92	2	Thin films for electronics
$Bi_2Sr_2CaCu_2O_8$ (BSCCO-2212)	85	2	
$Bi_2Sr_2Ca_2Cu_3O_{10}$ (BSCCO-2223)	110	3	Being developed for wires
$Tl_2Ba_2CaCu_2O_8$ (TBCCO-2212)	105	2	
$Tl_2Ba_2Ca_2Cu_3O_{10}$ (TBCCO-2223)	125	3	
$HgBa_2Ca_2Cu_3O_8$ (HBCCO-1223)	135	3	Record $T_c = 160$ K under pressure
$Nd_{1.85}Ce_{0.15}CuO_4$ (NCCO)	30	1	E-doped superconductor

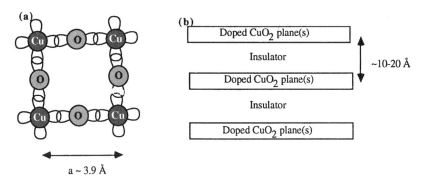

Figure 4.8. Structure of cuprate superconductors. (*a*) Unit cell of CuO₂ plane, showing Cu–O bonds along which carriers can flow. (*b*) Alternating layers of conducting planes and insulators.

During and since this early period, a large number of high-temperature superconducting compounds have been identified, with T_c up to 135 K, some of which are listed in Table 4.2. But all of them are generically the same material: They all consist of conducting CuO₂ planes, as indicated in Fig. 4.8, separated by insulating layers. If these planes are optimally doped with carriers (generally holes, but in a few cases electrons), then they can both conduct and superconduct within the planes, along alternating Cu and O in two directions (the adjacent Cu $3d$ orbitals and the O $2p$ orbitals line up in energy). In addition, since there are multiple parallel conducting layers that are ~1 nm apart, they are weakly coupled by tunneling currents. So the intervening layers serve both as charge reservoirs to obtain proper doping in the CuO₂ layers and as spacers that couple the CuO₂ layers together. The entire material is a superconductor, but one that is very anisotropic.

It is remarkable how different these cuprate superconductors are from conventional Nb-based superconductors. For example, none of the constituent atoms of Y–Ba–Cu–O are superconducting in elemental form. Neither the empirical work on metallic superconductors nor the BCS theory would have suggested focusing on anisotropic conducting oxides. Most oxides are insulators, and earlier research on conducting oxides had focused mostly on the metal–insulator transition rather than on any search for superconductivity. Indeed, there are reports that conducting cuprates may have been synthesized as far back as the 1950s, but no one attempted to cool them down to cryogenic temperatures. It is intriguing to consider how the development of superconductivity might have been different if YBCO had been discovered in 1957 rather than in 1987.

Crystal Structures

All of the copper oxide superconductors are based on the perovskite crystal structure, although with some significant variants. The name "perovskite" refers to the compound $CaTiO_3$, with a cubic (or almost cubic) unit cell with Ca in the center of the cube body, Ti on each of the eight vertices, and O in the middle of each edge (Fig. 4.9a). This is an electrical insulator, since all of the valences are saturated: Ca is $+2$, Ti is $+4$, and O is -2, for a net total of 0 for each unit cell. This is a fairly common crystal structure, but it was perhaps better known for ferroelectrics and piezoelectrics (such as lead zirconate–titanate) than for superconductors.

Compare this to the crystal structure of YBCO (Fig. 4.9b), with a unit cell of dimensions $c = 11.6$ Å, $a = 3.84$ Å, and $b = 3.88$ Å. This unit cell can be thought of as a stack of three basic perovskite cells, with Y and Ba in the Ca sites, Cu in the Ti sites, and some of the O sites empty. If all of the oxygen sites were filled, then the formula unit for this triple perovskite cell would contain nine oxygen atoms instead of seven. The locations of the oxygens are not random and in fact are critical to the operation of the superconductor. First, the oxygens at the level of the Y atom are missing, but above and below the Y are complete CuO_2 planes. Second, on the top (and bottom) of the triple cell, the oxygens in the front and back are missing, so that at this level there are straight Cu–O chains but no complete planes. So if we assume that the superconductivity occurs in the CuO_2 planes, then we have a set of two closely coupled planes, with about 10 Å to the next set of two coupled planes.

It is also useful to consider the valences of the atoms in YBCO. Since Y is $+3$, Ba is $+2$, Cu is $+2$ and O is -2, we have $3 + 4 + 6 - 12 = +1$, indicating the presence of one excess hole per unit cell. This suggests an effective carrier density (assuming that all such holes are mobile) of $1/abc \sim 6 \times 10^{27}$ holes/m³, which is somewhat lower than in conventional metals. Further, the oxygen content x per formula unit in $YBa_2Cu_3O_x$ can also be reduced below 7, in which case the carrier concentration decreases. This occurs by removing some of the oxygen atoms in the Cu–O chains; those in the CuO_2 planes remain. As x

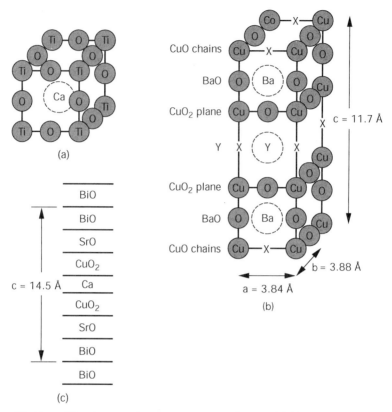

Figure 4.9. Perovskite structures in cuprate superconductors. (*a*) Basic perovskite unit cell. (*b*) Triple perovskite unit cell for YBCO. (*c*) Schematic of layers in BSCCO-2212.

approaches 6.5, this argument suggests that the carrier concentration goes to zero, and the material should become an insulator. And indeed, this does occur (at $x \sim 6.4$), together with the reduction of T_c. However, this is somewhat oversimplified, for it also suggests that reducing x below 6.5 should produce an electron-doped conductor. In fact, for $x \approx 6$, there are ions with uncompensated valence, but rather than leading to mobile carriers, these produce an antiferromagnetic insulator, with uncompensated electron spins on the Cu ions pointing in alternating directions on consecutive ions. A rough phase diagram of YBCO as a function of oxygen content is given in Fig. 4.10. Also, most of the trivalent rare-earth atoms (Nd, Gd, Dy, etc.) may be substituted in for Y without significantly changing the superconductivity, even though the atoms may be magnetic.

Other copper oxide compounds (cuprates) are similar in several respects. All have conducting CuO_2 planes separated by insulating layers. Apart from YBCO, they do not have Cu–O chains, but these do not appear to be essential to

4.2 HIGH-TEMPERATURE COPPER–OXIDE SUPERCONDUCTORS

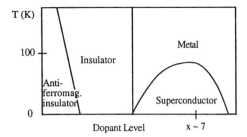

Figure 4.10. Rough phase diagram of YBCO and other cuprates on plane of T versus x, where x is oxygen content for YBCO (equivalent to dopant concentration).

the superconductivity. The unit cell of La_2CuO_4 has a single CuO_2 plane with a La–O spacer. This is an antiferromagnetic insulator, but when it is doped with holes, by partially substituting divalent Sr (or Ca, or Ba) for the trivalent La, it can become superconducting, with T_c up to 38 K (for $La_{1.7}Sr_{0.3}CuO_4$). Another cuprate incorporates Bi, Sr, and Ca to form BSCCO with the formula $Bi_2Sr_2CaCu_2O_8$ (sometimes called Bi-2212). This has two closely spaced CuO_2 planes (with a Ca in between) with a spacer of Bi and Sr oxides (Fig. 4.9c) and exhibits superconductivity with $T_c = 85$ K. If there are three closely coupled CuO_2 planes, this adds $CaCuO_2$ to the formula to yield $Bi_2Sr_2Ca_2Cu_3O_{10}$ (sometimes called Bi-2223). This has an even higher T_c, 110 K, following a general trend that T_c increases with the number of close cuprate planes. The nature of the doping is not quite as clear in BSCCO; sometimes Pb is partially substituted for Bi, which would be expected to contribute holes, but the superconducting phase forms even without this. These same two BSCCO structures can also be made with Tl and Ba substituting for Bi and Sr, respectively, to form TBCCO. The Tl-2212 structure has $T_c = 105$ K, and Tl-2223 has $T_c = 125$ K. The current record has been found for a compound containing mercury, with the formula $HgBa_2Ca_2Cu_3O_8$, with three closely coupled CuO_2 layers and $T_c = 135$ K. Under a very large pressure (3×10^{10} N/m^2 = 300,000 atm), this increases to more than 160 K. In all of these materials, the carriers are believed to be holes. For completeness, let us also mention the compound $Nd_{2-x}Ce_xCuO_4$, for which tetravalent Ce^{4+} substitutes for trivalent Nd^{3+} to yield an electron-doped superconductor with T_c up to about 30 K.

Anisotropy and Grain Boundaries

Above T_c, these copper oxides behave in many respects like metals, if somewhat unusual ones with low carrier densities. Their resistance parallel to the CuO_2 planes (in the a–b plane) behaves similar to that in a metal, decreasing as the temperature is decreased; for a clean sample, one typically has $\rho_{ab} \sim T$. Further, the magnitude of ρ_{ab} just above T_c is $\sim 10^{-6}$ Ω-m, certainly in the metallic range. These materials certainly do not appear metallic in visible light; all of them are black. However, beyond a wavelength of several micrometers in the infrared, they do indeed become reflecting, more like conventional metals.

However, all of the copper oxides are highly anisotropic, with normal-state electrical resistivity in the *a–b* plane being orders of magnitude larger than that along the *c*-axis perpendicular to the conducting planes. The degree of anisotropy is always large but differs among the various copper oxides. For YBCO, where the Cu–O chains may assist interplane coupling, this resistivity anisotropy ratio is of order 100, while for BSCCO and TBCCO, it can be as high as 10^5. So to a first approximation, we can regard these materials as consisting of parallel isolated thin metallic films of thickness $d \sim 10$ Å (depending on the number of closely coupled layers), each with sheet resistance $R_s = \rho_{ab}/d \sim 1000\ \Omega$.

The same anisotropy also dominates superconducting properties. Although the entire sample does become superconducting below T_c, a crystal can carry much larger supercurrents in the *ab* plane than along the *c*-axis. This is reflected in a superconducting energy gap that is also believed to be anisotropic, being much larger for Cooper pairs moving in the *a–b* plane than for those moving along the *c*-axis. That being said, YBCO and related materials act similar, in certain respects, to a classic BCS superconductor with the energy gap $2\Delta \sim 30\text{--}50$ meV, 10–20 times that of Nb. This probably reflects an average energy gap in the *a–b* plane.

The situation may even be more complicated than this; there is strong evidence suggesting anisotropy within the *ab* plane as well (Tinkham, 1996). The "*d*-wave model" of the superconducting energy gap for cuprate superconductors (strictly called $d_{x^2-y^2}$) predicts that the gap should be a maximum along the *a* and *b* directions and go to zero between the two (see Fig. 4.11*a*). Furthermore, this model predicts that the superconducting wave function should actually change sign between the two directions (i.e., the phase shifts by 180°), in a way that is similar to *d* orbitals of electrons in atomic physics, such as the electronic states of the copper atoms in the CuO$_2$ planes (see Fig. 4.8*a*). Consider an HTS single crystal with flat surfaces perpendicular to the *a*, *b*, and *c* directions. If we take a wire of a conventional isotropic superconductor (such as Pb) and make

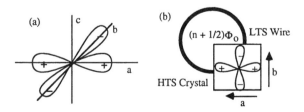

Figure 4.11. Proposed *d*-wave symmetry of superconducing energy gap in cuprates. (*a*) The lobes point along the *a* and *b* directions in the CuO$_2$ planes. Note that the sign changes between adjacent lobes; the magnitude of the energy gap goes to zero between them. (*b*) Half-integral flux quantization in loop with isotropic superconductor contacting two sides of cuprate single crystal.

contact with the *a* direction on one side of the crystal and the *b* direction on the other, we have a closed superconducting loop that should permit only quantized values of magnetic flux (Fig. 4.11*b*). By the argument we used earlier, each 360° rotation of the quantum mechanical phase corresponds to one flux quantum Φ_0. However, there is an extra sign change in going around the loop, which corresponds to an extra 180° and an extra $\frac{1}{2}\Phi_0$, so that the flux should be quantized in the form $\Phi = (n + \frac{1}{2})\Phi_0$. This has indeed consistently been observed in experiments. This may have subtle effects on other properties as well, but we will not address them here.

We can make use of much of what we discussed in Chapters 2 and 3 if we use anisotropic values of penetration depth λ and coherence length ξ (see Table 4.3). For YBCO, for current flowing within the planes, we have $\lambda_{ab} \sim 150$ nm and $\xi_{ab} \sim 1.5$ nm, making this an extreme type II superconductor with $\kappa = \lambda/\xi \sim 100$. [These are the values at low temperature; they diverge as $(1 - T/T_c)^{-1/2}$ close to T_c.] These values reflect the low carrier density and large energy gap, rather than a short mean-free path as for extreme type II LTS materials. For currents flowing in the *c* direction, $\lambda_c \sim 600$ nm and $\xi_c \sim 0.4$ nm. Note that ξ_c is less than the distance ~ 1 nm across the spacer layer, so that this can be regarded (except close to T_c, when ξ diverges) as a collection of discrete superconducting layers that are weakly coupled together. The same situation also holds for the other HTS copper oxides (see Table 4.3), with the understanding that the more weakly coupled the layers, the shorter is the coherence length ξ_c perpendicular to the layers. The extremely small values of ξ imply very large values of the upper critical magnetic field B_{c2}. For *B* applied parallel to the *c* axis, $B_{c2} \sim \Phi_0/2\pi\xi_{ab}^2 \sim 150$ T; for *B* applied in the *ab* plane, $B_{c2} \sim \Phi_0/2\pi\xi_c\xi_{ab} \sim 600$ T. They are somewhat smaller closer to T_c when ξ diverges, but they are still extremely large even at 77 K. (However, as we point out below, these materials may *not* exhibit zero resistance all of the way up to B_{c2}.)

Given this anisotropy, if we want to carry a large supercurrent in one of these materials, this can only be done in the direction parallel to the *ab* planes. One way to achieve this, at least in principle, is to make the entire superconducting wire in the form of a macroscopic single crystal. This may be practical on a small scale for microelectronic purposes but is clearly not practical for bulk applications. In the latter case, we need to consider a packed assembly of crystals with

Table 4.3. Anisotropic Characteristic Lengths λ and ξ and Interlayer Spacing c^a for High-Temperature Superconductors for $T \ll T_c$

	λ_{ab}	λ_c	ξ_{ab}	ξ_c	c
YBCO	150	600	1.5	0.4	1.2
BSCCO-2223	200	1000	1.3	0.2	1.5

[a] All in nanometers. Values from Cyrot and Pavuna, 1992.

different orientations. Here, the current does not go in a straight line but follows a percolative path along the "easy directions" of these crystallites. In fact, the situation turns out to be even worse than this suggests. If two YBCO crystallites come together at a large angle, the disordered material in the grain boundary near the interface is not a good superconductor, so that this forms a weak link. This problem is exacerbated by the short coherence length even in the optimum direction. Even if the planes are parallel, the connection across the grain boundary is weak unless the angle between the a-axes is very small. (It is possible that the complex nature of the energy gap may also be contributing to this.) This is a critical problem in HTS materials, which has still not fully been solved, but considerable progress has been made (see below).

Vortex Lattice Melting and Irreversibility

Let us focus on a single CuO_2 bilayer in YBCO with thickness ~ 5 Å and assume that this is completely decoupled from the next bilayer ~ 10 Å away. Then, the location of a vortex in one layer would be largely uncorrelated with that in another layer, and we can regard this as an isolated thin film. Given the very short coherence length, the thermal energy needed to excite a vortex–antivortex pair in such an ultrathin layer is very small. As we discussed in Chapter 3, the characteristic temperature above which such vortex excitations will contribute to resistance is given by $T_v = \Phi_0^2/25\mu_0\lambda_\perp = 1$ cm-K$/\lambda_\perp$, where $\lambda_\perp = \lambda^2/d$. If we take $\lambda \sim 200$ nm, this gives $T_v \sim 100$ K, suggesting that this will become important in broadening the resistive transition close to T_c.

Even more important, in a large magnetic field, vortex interactions will cause the formation of a vortex lattice. This vortex lattice is important in vortex pinning, since without a reasonably rigid lattice, one would have to pin every single vortex in order to have $R = 0$. But like a molecular crystal, a lattice of vortices can "melt" if the temperature is high enough and the vortex interactions weak enough. For an ultrathin film, an analysis similar to that for the vortex unbinding transition (Tinkham, 1996) gives rise to a melting temperature $T_m \sim 0.05$ cm-K$/\lambda_\perp$, which would suggest $T_m \sim 5$ K $\ll T_c$. The situation is not quite as bad as this, since this analysis has ignored the interplane coupling, which strengthens the vortex lattice, substantially raising the melting temperature. Still, all of the HTS materials exhibit a line in the H–T phase diagram, called the irreversibility line (see Fig. 4.12). [Sometimes this is used to define an irreversibility field $H^*(T)$.] To the right of this line, the system is essentially in the "vortex liquid" state, where pinning is ineffective at large fields, magnetic properties are reversible, and there is flux flow resistance. To the left of the line, the magnetic properties are irreversible, as expected for a type II superconductor in the critical state. For YBCO, the irreversibility line is within a few degrees of T_c, while for BSCCO and TBCCO, which are more weakly coupled, this line can be far below T_c (Larbalestier, 1997). This is likely to limit some high-field applications in the "high-temperature" range at 77 K and above, despite the extremely large values of H_{c2}.

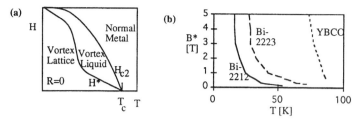

Figure 4.12. Vortex lattice melting and irreversibility line. (*a*) Generic HTS phase diagram in *H–T* plane. (*b*) Irreversibility field $H^*(T)$ for several cuprates (Larbalestier, 1997).

Bulk Processing and Wire Fabrication

It is generally rather easy to make ceramic pellets of the cuprates, by heating and compressing powder with the appropriate stoichiometry. Unfortunately these pellets are not very useful and certainly not for making high-field superconducting wire. The bulk intergranular critical current in such a polycrystalline assembly is small due to weak links between anisotropic superconducting grains and decreases further in only moderate magnetic fields. (We will discuss this later in Chapter 5 on Josephson junctions.)

As we discussed earlier in the context of LTS wire fabrication, a superconducting wire should consist of a multifilamentary composite of multiple small superconducting filaments embedded in a normal metallic matrix. Ideally, it should be fabricated from precursor materials that are reasonably ductile, so that they can be drawn out into fine wire. In addition, critical current densities of order 10^9 A/m² at the operating magnetic field and temperature are needed to make a practical high-field superconducting magnet. Substantial progress toward these goals has been achieved, but some additional years of development are needed before these HTS wires can become the basis for a commercial technology for magnets and other applications.

As indicated above, the key problem in making HTS wires is how to obtain parallel alignment of the crystalline orientations in adjacent crystal grains, so that high currents can flow between the grains as well as within them. The greatest development thus far has been pursued with BSCCO superconductors, both Bi-2212 and Bi-2223 (Larbalestier, 1997). These materials have mechanical properties that are called "micaceous," referring to the tendency to cleave easily to form flat platelets, somewhat like the mineral mica. This cleavage occurs mostly between the Bi–O double layers, parallel to the CuO_2 planes. This permits the material to be drawn out in a way that leads to parallel alignment of CuO_2 planes, with crystal plates sliding past one another, without breaking the conducting planes themselves. This is not so easily achieved in the other cuprates, such as YBCO. Unfortunately, as we pointed out above, BSCCO also has a relatively low irreversibility temperature T^* in large magnetic fields (20–30 K), limiting the high-field application of these materials to rather low temperatures.

The other key requirement is to form a composite with a metal having high thermal and electrical conductivity, in order to maintain electrothermal stability. It is critical here to have a low electrical resistivity between the metal and superconductor. Copper and most other conventional metals are not acceptable, since they tend to pull oxygen out of the cuprate, to form an insulating barrier at the contact. The only remaining candidates are silver and gold, which normally do not form insulating oxides. Since silver is much cheaper than gold, it is generally the preferred material. The most common path to fabricating BSCCO composite wire (Fig. 4.13) is called "powder in tube" (PIT). Reacted BSCCO powder (sometimes with the Bi partially substituted with Pb) is placed inside a silver tube, which is then drawn or rolled down in several stages, during which it is bundled with a number of other Ag/BSCCO filaments. At the end of the process, the wire or tape is annealed at a moderately high temperature to improve intergrain contacts. The best BSCCO wire made in this way shows fairly high critical currents at low T (< 20 K) up to magnetic fields as high as can be measured (20 T). Although critical currents must be improved further, this is very encouraging for application to a high-field magnet that operates at low T. In contrast, J_c for $T > 60$ K drops very sharply in large fields, presumably due to fundamental considerations (vortex lattice melting) in this very anisotropic material.

Several different techniques have been investigated for achieving proper grain alignment in YBCO bulk samples. Some of these approaches have been based on thin-film techniques, and we will discuss them in the context of the section below on HTS thin-film deposition. Another approach, called "melt texturing," involves melting a bulk YBCO sample at $T \sim 1200°C$ and slowly recrystallizing it in a temperature gradient. Since the crystal growth process is itself highly anisotropic (growth occurs much more quickly along the ab plane than perpendicular to it), YBCO melt texturing leads to highly aligned, fairly large

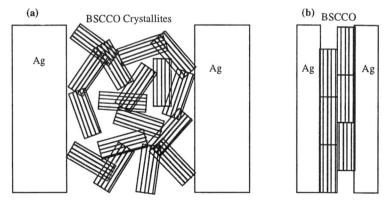

Figure 4.13. Fabrication process for BSCCO multifilamentary wires in silver matrix. (*a*) Initial BSCCO crystallites in Ag tube are randomly oriented. (*b*) After drawing down, platelets are oriented primarily with the CuO_2 planes in the same direction.

crystallites. On the other hand, this is an extremely slow process, producing typical lengths of less than 1 cm/h. For certain bulk applications, such as magnetic levitators and bearings to operate at 77 K, this approach produces reasonably sized samples with superior performance. However, it is difficult to see how this could be scaled up to manufacture a long wire.

There is one additional critical step that is necessary for the fabrication of superconducting YBCO samples by any method. In virtually all cases, the structure that is grown at high temperatures is $YBa_2Cu_3O_x$, with $x \sim 6$, which is essentially a semiconductor. This does not become a superconductor when it is cooled down. This exhibits a tetragonal crystal structure, with lattice constant $a = b \neq c$, since the oxygens that may be present at the level of the chains (see Fig. 4.9) are not ordered. However, if one cools slowly in oxygen through a temperature around 500°C, then additional oxygen enters the lattice and orders into chains, which brings about a transition into an orthorhombic crystal structure ($a \neq b \neq c$) with a and b slightly different and $x \sim 7$. This is the superconducting phase. This oxygen diffusion can be very slow in a dense bulk sample, sometimes requiring annealing for several days to optimize the oxygen content. A similar ordering transition does not seem to occur for the other cuprates, probably because they do not have Cu–O chains.

With the intergrain weak-link problem apparently solved in principle for both BSCCO and YBCO, attention has also been paid to improving the depinning critical currents within a superconducting grain. Given the short coherence lengths, the scale of pinning sites can be very small. A variety of different kinds of natural and artificial pinning sites have been explored, including second-phase precipitates, tracks from high-energy particles, atomic substitutions and vacancies, and other crystalline defects. In some cases, quite significant improvements in critical currents have been observed.

There has been much less development of Tl- or Hg-based cuprates for bulk applications, despite higher values of T_c and irreversibility temperatures. One difficulty is that at the high processing temperatures necessary to fabricate these compounds, the Tl and Hg atoms are volatile (and poisonous!), so that all heating must be done in sealed containers. Furthermore, it seems that the crystallites in these materials cannot be easily aligned by mechanical processing as in BSCCO. However, research into alternative processing methods is continuing.

Thin-Film Fabrication

For most applications of HTS thin films, we would like to carry a large supercurrent density parallel to the substrate. As with bulk samples, the anisotropy of the cuprates makes it essential that crystallites are highly aligned, with the c-axes perpendicular to the substrate, and preferably with in-plane alignment as well (Beasley, 1989). Most of the development in this area has focused on YBCO; the other cuprates exhibit similar considerations, but the problems are even more difficult to solve.

The key problem is that formation of the proper crystalline structure of YBCO requires heating to temperatures of ~ 700–$900°C$, at which both the films and most potential substrate materials tend to be very reactive. Most of the standard substrates, such as Si, glass, or even sapphire (Al_2O_3), are not acceptable. The best YBCO films are generally grown on single-crystal substrates (Phillips, 1996) that have the closest match to the YBCO crystal structure and lattice constant (see Table 4.4). The film grows almost epitaxially and can be almost a single crystal. Strontium titanate (STO) is a perovskite with lattice constant $a = 3.9$ Å, very close to that of YBCO; generally the (100) crystal surface is used. However, STO is only available in small sizes and has an enormous dielectric constant at low temperatures, which is undesirable for high-frequency applications. Some other perovskite substrates commonly used include $LaAlO_3$ and $NdGaO_3$. In addition, (100) cubic crystals of MgO and yttria-stabilized zirconia (YSZ) have also been used to prepare high-quality oriented YBCO films, even if the crystalline match is not as ideal.

One approach that was prominent early in the development of HTS films involved cold deposition of disordered YBCO onto an epitaxial perovskite substrate (such as STO) followed by high-temperature post-annealing in oxygen at $\sim 900°C$ in order to induce solid-state epitaxial growth of the proper crystalline structure. The most successful efforts in this direction actually deposited an Y–Ba–Cu–F film by coevaporating from separate Y, Cu, and BaF_2 sources (Fig. 4.14a). By post-annealing in oxygen with some water vapor, the excess F is removed as HF, and oxygen is incorporated in the film as it crystallizes.

More recently, most efforts at producing the highest quality YBCO films have focused on "in situ deposition" at temperatures of ~ 700–$800°C$, at which a "c-axis film" (i.e., one with the c-axis perpendicular to the substrate) will grow directly. As with bulk YBCO, the phase that is formed is actually insulating $YBa_2Cu_3O_{\approx 6}$; one must cool slowly through $\sim 500°C$ in oxygen in order to bring this up to O_7 and form the superconducting phase. Multisource evaporation techniques (including electron beam evaporation and molecular beam

Table 4.4. Crystalline Substrates Used for Deposition of YBCO Thin Films

Substrate	Crystal Type	Lattice Constant (Å)	Dielectric Constant
YBCO (a–b plane)	Perovskite	3.9	
$SrTiO_3$ (STO)	Perovskite	3.9	> 300
$LaAlO_3$ (LAO)	Perovskite	3.9	23
$NdGaO_3$	Perovskite	3.9	25
MgO	Cubic	4.2	10
$Y_2O_3:ZrO_2$ (YSZ)	Cubic	5.2	25

Note: YBCO is included for comparison.

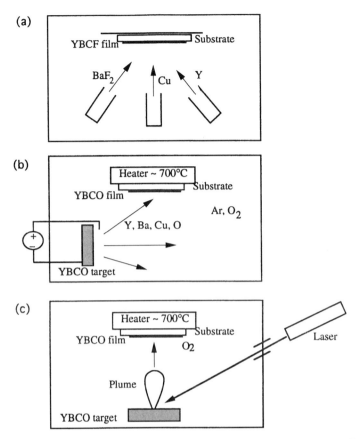

Figure 4.14. Deposition processes for YBCO thin films. (*a*) Coevaporation with BaF_2 followed by post-anneal. (*b*) In situ off-axis sputtering. (*c*) In situ pulsed laser deposition.

epitaxy) have been used by several groups but require special highly reactive oxygen sources (such as ozone) in order to grow the proper oxide crystal structure. The resulting films tend to be of very high quality but are also very slow and expensive to make. More common techniques use a single YBCO ceramic source together with either sputtering or pulsed laser deposition in a vacuum chamber containing a background pressure of oxygen (Fig. 4.14). Under optimum conditions, either method can transfer the proper composition from the source to the substrate, forming a high-quality *c*-axis film in situ. With sputtering, one must take care to avoid bombardment of the growing film by high-energy negative oxygen ions from the target, so that an off-axis configuration is often used. With pulsed-laser deposition, rapid high-energy ultraviolet pulses from an excimer laser (typically ArF, KrF, or XeCl) are scanned across a ceramic source in order to maintain even erosion and avoid particulate formation.

It is possible to broaden the range of available crystal substrates through the use of epitaxial buffer layers that prevent reaction between the YBCO film and the substrate. For example, an epitaxial buffer layer of YSZ permits one to deposit a YBCO film on sapphire (i.e., single-crystal Al_2O_3). One can also deposit such a buffer layer on Si, but an additional problem arises. Because of significantly different thermal expansion between Si and YBCO from room temperature to the growth temperature of $\sim 700°C$, a YBCO film deposited on a Si substrate has a tendency to be heavily strained or even develop cracks when cooled down to room temperature or below.

Optimum quasi-epitaxial YBCO films, typically 100–500 nm thick, exhibit $T_c = 90$ K (a bit lower than for bulk YBCO) and dc critical currents with $J_c \sim 10^{10}$ A/m^2 at 77 K and much more at lower T. For a typical narrow line ~ 2 μm wide by 0.5 μm thick, this corresponds to a current of 10 mA. It is notable that J_c is so large (even in large magnetic fields) without any special efforts to introduce pinning centers. These in situ deposited YBCO films are believed to contain a high density of microscopic defects that may be assisting in this regard. Furthermore, typical high-frequency properties include a surface resistance of ~ 1 mΩ at 77 K at 10 GHz, orders of magnitude below that of a normal metal such as Cu or Al. This typically scales with frequency as f^2, as expected from the two-fluid picture. Even lower surface resistance is available at lower temperatures, although the residual value of R_s tends to be somewhat higher than one would expect from a standard BCS superconductor.

The low values of R_s make it possible to fabricate a narrow-band low-loss passive microwave filter by patterning a single layer of YBCO (Zhang et al., 1997), as indicated in Fig. 4.15. The substrate is typically $LaAlO_3$, and the ground plane may be an unpatterned layer of YBCO deposited on the reverse side or on a separate $LaAlO_3$ wafer. A developing application for such filters lies in cellular telephone systems, where a bank of filters in an isolated base station can be cooled to ~ 60–70 K by a reliable, closed-cycle refrigerator.

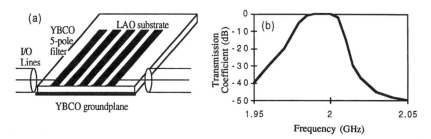

Figure 4.15. Multipole HTS microwave bandpass filter for use with cellular communications system (Zhang et al., 1997). (*a*) Layout of YBCO five-pole microstrip filter patterned on LAO substrate, with YBCO ground plane and coaxial input and output (I/O) lines. (*b*) Frequency dependence of transmission coefficient showing full transmission in a 20-MHz band near 2 GHz and a very sharp rolloff outside of the band.

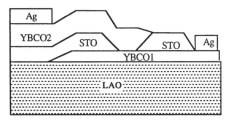

Figure 4.16. Multilayer process for YBCO superconducting interconnects or other complex circuits. Included are two superconducting layers (YBCO1 and YBCO2), an STO layer between them, and Ag contacts to both YBCO layers. Also shown is a patterned via between the two YBCO layers. All layers (except the top Ag contacts) are epitaxial.

Some other applications of HTS films may require a multilevel process with two or more HTS layers that are normally insulated from one another (Fig. 4.16). This would be the case, for example, if one wanted to apply HTS circuits to interconnects in a high-density integrated circuit; the ability to have crossovers is essential. Since the upper YBCO layer must also be an epitaxial c-axis film for optimum properties, the entire multilayer assembly must consist of lattice-matched epitaxial films. Fortunately, an insulating STO film grows on YBCO as well as YBCO on STO, so that a trilayer with the requisite properties can be fabricated. Other features of HTS thin-film technology include low-resistance Ag or Au contacts, and Ar ion beam etching and standard microlithography to define features down to submicron scales.

The previous discussion has focused on the growth of YBCO on single-crystal substrates, and indeed, this is the most straightforward way to obtain films with highly oriented and aligned crystallites, which are necessary for high critical currents. These YBCO films must be "biaxially textured": In addition to having their a–b planes parallel to the substrate, they must also have the a-axis strongly aligned as well. If such a biaxially textured film can be obtained on a polycrystalline substrate, this would open up a range of opportunities. For example, one could deposit a YBCO film on a long flexible metal tape and use this (at least in principle) to wind a superconducting magnet. One method that is being developed to achieve this is called "ion-beam-assisted deposition" (IBAD) (Larbalestier, 1997). This is based on the principle that crystallites of different orientations can be sputtered by an incoming ion beam at slightly differing rates. If a film is bombarded with a directional ion beam at the same time that it is being grown, then this differential effect can lead to preferential growth of a given orientation. This has been shown to produce biaxial texture in a YSZ film, which in turn can be used as the substrate for YBCO films. An alternative method was recently demonstrated [called "rolling assisted biaxial textured substrate" (RABiTS)] involving first producing grain alignment in a metal tape by mechanical processes and then depositing a biaxially textured buffer layer of YSZ and finally a layer of YBCO. The critical currents of these films are still not quite as good as those on single-crystal substrates, but they may be acceptable for some applications. However, any thin-film process is probably far too slow to be practical for producing the kilometers of length needed for large-scale applications.

It has also been possible to fabricate high-quality thin films of many of the other HTS materials, but it is generally somewhat more difficult than for YBCO. The BSCCO-2223 material has $T_c = 110$ K, but it is difficult to prepare without a significant mixture of the BSCCO-2212 phase (with $T_c = 84$). For the Tl and Hg cuprates, a major problem is that it is difficult or impossible to maintain sufficient Tl and Hg in the film during deposition temperatures. For this reason, one approach has been first to prepare precursor films without the volatile elements (i.e., Ba–Ca–Cu–O) using sputtering or pulsed laser deposition onto STO or LAO substrates. Then, this is post-annealed at high T ($\sim 800°C$) in a sealed container with an overpressure of Hg or Tl vapor. In this way, films with T_c well in excess of 100 K have been obtained, which have primarily c-axis orientation. However, these tend to be much rougher than typical YBCO films, and so are not yet ready for a multilayer process.

HTS Josephson Junctions

There is one final requirement for an HTS electronic technology: development of a process for making reproducible Josephson junctions (Gross et al., 1997). This has been the most challenging problem, and it has still not been fully solved. The problem is not that it is difficult to make HTS Josephson junctions; on the contrary, it is too easy, with almost every grain boundary a natural Josephson junction. Rather, the difficulty is in designing a junction or weak link that performs as predicted, so that we may produce hundreds of them with reproducible characteristics. We will start by reviewing some of the properties of natural HTS Josephson junctions and then go on to describe some of the efforts to design artificial YBCO junctions.

A classic Josephson junction consists of an insulating layer between two good superconductors, but in the context here, we mean simply a structure where there is weak coupling between the two superconductors, even if the fundamental basis for this weakening may not be totally understood. A junction is typically characterized by its critical current I_c (or sometimes J_c) and its normal-state resistance R_n. For an ideal classic Josephson junction, both of these parameters should be reproducible, and $I_c R_n \sim 2\Delta/e$, where 2Δ is the superconducting energy gap. For HTS junctions, we might therefore expect $I_c R_n \sim 40$ mV, but this is actually much smaller (typically $\lesssim 1$ mV), due perhaps to a depressed value of I_c and/or an internally shorted R_n.

There are two kinds of natural Josephson junctions that occur in high-temperature superconductors. The first is the "intrinsic Josephson junction" that occurs naturally between successive CuO_2 layers (or bi- or trilayers) even in a single crystal. Such intrinsic junctions have been observed in both single crystals and thin films of the highly anisotropic materials BSCCO and TBCCO. Ironically, the planes are too strongly coupled in YBCO; this must be weakened

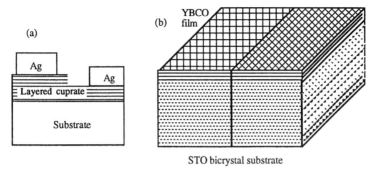

Figure 4.17. Natural HTS Josephson junctions. (*a*) Intrinsic *c*-axis junctions. (*b*) Grain boundary junction on bicrystal substrate.

by reducing the oxygen content (and T_c) to obtain clear Josephson junctions. These intrinsic junctions are always present for a number of junctions in series; it is difficult to see how to reliably isolate just one. But a relatively small number can be fabricated by etching part way into a film or crystal, as suggested in Fig. 4.17*a*. In some cases, $I_c R_n \sim 10$ mV or more for each junction have been seen.

The other kind of natural Josephson junction is a grain boundary junction that occurs when two grains are not perfectly aligned. This is the basis for the sharply reduced critical currents in nonaligned polycrystalline samples. The origin of this effect is not entirely clear but may be related to the combined effects of disordered material at the interface, the very short coherence length, and the anisotropic energy gap. A controlled way to make a grain boundary junction is to deposit a YBCO film on top of a substrate with a deliberate grain boundary. Such a "bicrystal" can be made by gluing together two single-crystal STO substrates with different orientations. For example, both can be (100) crystals but rotated in plane by some angle θ. The YBCO film grown on top will then consist of two *c*-axis films, with a grain boundary and an in-plane mismatch at essentially the same location (see Fig. 4.17*b*). The critical current falls very sharply (almost exponentially) with the mismatch angle. It is possible to produce multiple junctions along the interface of this bicrystal, but this is hardly the basis for an integrated circuit technology.

Inspired in part by the bicrystal results, several techniques have been developed to make artificial grain boundaries on a single-crystal substrate. One apporach is to etch a step in the substrate and grow a YBCO film over the step. This typically leads to two grain boundary junctions in series as the film tries to follow the contour of the step (Fig. 4.18*a*). Alternatively, a thin "seed layer" film can be deposited over part of the substrate to produce a new surface with different preferred growth pattern. When the YBCO film is deposited across this interface, a grain boundary junction is again formed (Fig. 4.18*b*).

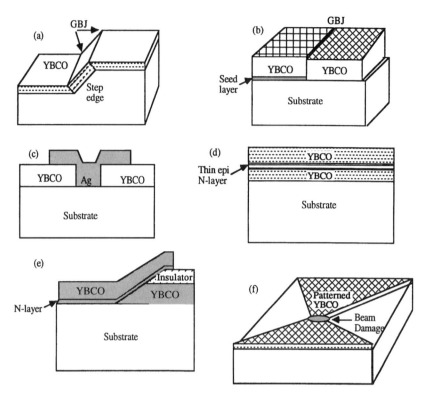

Figure 4.18. Alternative designs for fabricated Josephson junctions in HTS films. (*a*) Step-edge grain boundary junction with two series junctions indicated. (*b*) Biepitaxial grain boundary junction with seed layer film. (*c*) SNS junction across gap in film. (*d*) A *c*-axis SNS junction. (*e*) Ramp-edge SNS junction. (*f*) Beam damage SNS-like junction.

An alternative approach to the "artificial grain boundary" is to deposit an artificial barrier layer between the two superconductors. For a classic LTS superconductor, this is normally an insulator, but an HTS Josephson junction with a true insulating tunnel barrier has not yet been observed. Instead, a considerable effort has been focused on a variety of artificial barriers that consist of a normal metal or other highly resistive layer to form something like an SNS junction. The N layer might be a conventional metal such as Ag or Au or an epitaxial perovskite conductor that is not superconducting (such as $PrBa_2Cu_3O_x$) or even a disordered or doped form of YBCO. Several geometries have been investigated. In one case, the N layer bridges a narrow gap between the two pieces of superconductor (Fig. 4.18*c*). In another, there is a planar trilayer, with sequential deposition of the YBCO base electrode, an epitaxial N layer, and the YBCO counterelectrode (Fig. 4.18*d*). One geometry that has appeared promising is a ramp-edge junction, in which the YBCO base electrode is patterned by oblique ion etching into a gradual ramp (Fig. 4.18*e*) followed by deposition of the N layer and the YBCO counterelectrode.

Finally, a completely different approach has been based on producing local damage in an otherwise high-quality epitaxial YBCO film in order to produce a weak-link or Josephson junction (Fig. 4.18f). Two ways to achieve this involve narrowly focused beams of either electrons or ions. In either case, the damage should be enough to degrade the material locally but not enough to create an entire non-superconducting region. This may be similar in effect to the SNS junction above.

Despite the wide range of techniques used to fabricate HTS Josephson junctions, the junctions themselves exhibit many similar characteristics. They do not show the large low-voltage resistance that ideal LTS tunnel junctions exhibit but rather appear as if they were internally shunted by a normal resistance [a resistively shunted junction (RSJ)]. Also, there is substantial variability in I_c for nominally similar junctions, generally more than 20% variation. There is some evidence that all of these junctions are microscopically very inhomogeneous, partly because T_c can vary substantially on the scale of the coherence length, due to local variations in strain and oxygen content. This may be hard to control and might account for the variability in these junction characteristics. If this is indeed the problem, it is not clear how to solve it. However, as we will discuss in the next chapter, we can make full use of RSJs in superconducting electronics technology as long as they are reproducible. If the variation in I_c can be reduced to only a few percent, then we may have the materials basis for a technology. It is fair to say that this is the key problem standing in the way of further development of HTS microelectronics.

4.3 OTHER SUPERCONDUCTING MATERIALS AND MICROSTRUCTURES

The previous sections have focused on Nb compounds and on layered copper oxides, which provide virtually all of the practical LTS and HTS materials, respectively. In terms of microstructures, we have focused on Josephson junctions. However, it is quite possible that new materials or devices might form the basis of a new generation of superconducting technology. In this section, we will briefly survey a few of the many novel classes of superconducting materials (see Table 4.5) that appear promising in one or another respect. This illustrates the wide range of materials that display superconductivity as well as the difficulty in predicting new superconductors. We also briefly discuss one approach to the possible integration of superconductors and semiconductors together in a single microstructure.

Ternary Intermetallic Compounds (Chevrel Phase)

In the section earlier on LTS materials, we focused on intermetallic compounds containing Nb. More complex ternary intermetallic compounds have also been investigated, and one class of such compounds stands out for record-high values

Table 4.5. Selected Classes of Superconductors and Optimum Properties

Type	Compound	T_c (K)	Notable Features
Chevrel phase	$PbMo_6S_8$	14	$H_{c2} = 60$ T
Bismuthates	$Ba_{0.6}K_{0.4}BiO_3$	30	Three-dimensional isotropic oxide
Organic conductors	$(ET)_2CuN(CN)_2Cl$	13	Two-dimensional anisotropic
Fullerenes	Cs_2RbC_{60}	33	Three-dimensional

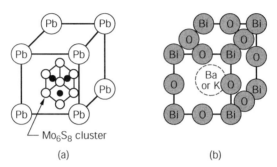

Figure 4.19. Crystal structures of novel superconductors. (*a*) Chevrel phase, $PbMo_6S_8$. (*b*) Perovskite, $Ba_{1-x}K_xBiO_3$. Unlike the cuprates, the conduction occurs via Bi–O links in three dimensions.

of critical field H_{c2}, at least before the HTS materials were discovered. These are known as "Chevrel phase" compounds and have the typical formula XMo_6S_8, where X can be any one of a number of large metallic atoms and Se can sometimes be substituted for S. The conducting properties are believed to be mostly due to the closely spaced Mo atoms. The critical temperatures of these materials are in the range of 12–15 K, but more remarkable are the upper critical magnetic fields B_{c2}, which can be as large as 60 T, for $PbMo_6S_8$. It is very difficult to make wires of this material because it is very brittle and there is no obvious ductile precursor. The crystal structure (Fig. 4.19*a*) is complex, but the superconductivity is relatively isotropic. Some success has been achieved by processing $PbMo_6S_8$ powder inside a copper can and drawing this down to wire using 'hot-isostatic pressing" in temperatures of order 1000°C. In this way, it has been possible to produce a multifilamentary composite wire with encouraging critical currents and fields at 4 K. In particular, critical currents of 10^9 A/m² have been demonstrated in fields of 20 T. Of course, with comparable efforts one can obtain superior results using the newer HTS materials such as BSCCO at low T, so that interest in these Chevrel materials has decreased somewhat. These compounds have also been fabricated by sputtering onto heated substrates, but as with HTS materials, it is difficult to envision a bulk application made using such a slow thin-film process.

Other Oxides (Bismuthates)

Although the discovery of high-temperature superconductivity in copper oxides was a surprise to many people, it was not totally without precedent. In fact, superconductivity had previously been seen in several different perovskite conductors. For example, $SrTiO_3$ becomes a conductor when it is oxygen deficient or when some Nb is substituted on the Ti site. In either case, this contributes electrons to the conduction band. Even with a rather low carrier density, this may become superconducting below about 1 K. Even more remarkable are a class of perovskites based on $BaBiO_3$ (Fig. 4.19b), which is an insulator. If some Pb is substituted on the Bi site, this contributes holes, and $BaPb_{0.7}Bi_{0.3}O_3$ has a superconducting critical temperature up to 13 K, even though the carrier density is still only $\sim 10^{27}$ m^{-3}. This brown, ceramic material, similar in certain respects to the copper oxides, was discovered a full decade before HTS. After the discovery of superconductivity in the copper oxides, the closely related hole conductor $Ba_{1-x}K_xBiO_3$ (with $x \sim 0.4$) was shown to have T_c up to 30 K! It is important to point out, however, that the bismuthate conductors are different from the cuprates is at least one key respect: They are isotropic conductors, in contrast to the layered cuprates. Both thin films and tunnel junctions have been made with BPBO and BKBO using sputtering at high temperatures and exhibit superconducting properties that may be superior to some of the HTS materials. In particular, BKBO tunnel junctions show a classic sharp energy gap, in contrast to the smeared structures seen in the copper oxides. This may be promising for potential electronic applications operating at ~ 20 K, which is too warm for Nb or NbN.

Organic Superconductors (Including Fullerenes)

We have not mentioned organic materials up to this point. Indeed, most organics (including virtually all plastics) are electrical insulators. However, there is a class of organic compounds known as "charge transfer salts" that include ringed compounds (with extended π-orbitals) and frequently incorporate metal ions. The key building block is a planar ringed structure that includes S atoms, called tetrathiofulvalene (TTF; sometimes Se substitutes for the S, to yield TSF). These form brittle crystals with highly anisotropic conductivity and superconductivity (Williams, 1992). The first such superconductors had formulas such as $(TMTSF)_2ClO_4$ and $T_c \sim 1$ K, where TMTSF is tetra-methyl-TSF (Fig. 4.20a). This is a highly anisotropic material that is a quasi-one-dimensional conductor, with current flow predominantly along a single crystalline direction. More recently, related compounds based on bis(ethylenedithio) TTF (also called BEDT-TTF, or simply ET; see Fig. 4.20b) have been studied and found to exhibit T_c up to 13 K in $(ET)_2Cu[N(CN)_2Cl]$. This is essentially a two-dimensional superconductor, analogous to the cuprates. Some related compounds exhibit antiferromagnetic spin ordering, another way that they may be analogous to the copper oxides. Several theoretical predictions for high-temperature

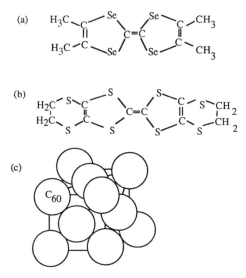

Figure 4.20. Structures of novel organic superconductors: (*a*) TMTSF; (*b*) BEDT-TTF; (*c*) face-centered cubic crystalline structure of C_{60} molecules; the alkalai atoms (K, Cs, Rb) go in interstitial spaces between the larger molecules.

superconductivity in organic conductors have been made, although it is not clear how to achieve this in real materials.

A completely different class of C-compounds is based on C_{60}, which is a quasi-spherical shell of 60 C atoms (~ 10 Å across) called buckminsterfullerene (often shortened to fullerene or "buckyball"), for its similarity in structure to the geodesic dome of architect Buckminster Fuller (or alternatively a soccer ball). These molecules can be packed together in a solid, just as if they were large atoms. The resulting material is an insulator or semiconductor and is essentially an alternative form of solid carbon to the more conventional graphite or diamond. However, fullerene can be strongly doped with electrons by adding alkali metal atoms (Na, K, Rb, Cs) that fit in the interstitial positions between the C_{60} balls, turning it into a conductor. With sufficient doping, fullerene has been found to be superconducting, with T_c up to 33 K in the compound Cs_2RbC_{60}, again above the pre-HTS record. In contrast to the other organics or the cuprates, this appears to be an isotropic three-dimensional superconductor. Fullerene remains difficult to make in significant quantities, although it has been made in thin-film form as well as in bulk.

Unidentified Superconducting Objects

Before leaving the subject of superconducting materials, let us address briefly some of the sporadic reports of superconductivity close to room temperature, for example around 240 K, or even above 300 K. These have generally been seen in mixed-phase materials that are extremely resistive at high temperatures, such as copper oxides that are far from the standard stoichiometry. Typically, there is a sharp drop in resistance in such a sample as it is cooled below some temperature T_0. Sometimes this may be accompanied by a very small change in the magnetic susceptibility. Both of these could signal a small fraction of a superconducting material in this mixture.

Unfortunately, most of these reports also involve samples that either degraded with time or could not be reproduced by the original researchers or by others. These reports have been referred to as "unidentified superconducting objects" (USOs). This does not necessarily imply that these measurements were invalid or otherwise incorrect. Given the complex crystal chemistry of the copper oxides, it is certainly conceivable that an unstable compound might form as a filament or interface among other closely related materials. Such a filament in a brittle material might well be damaged by thermal cycling and could be difficult to reproduce. However, effects completely unrelated to superconductivity can sometimes give rise to sharp changes in both resistance and magnetic properties. (The colossal magnetoresistance in perovskites containing manganese oxides is an example.) So it is better to be skeptical about these reports unless and until all four of the following criteria are met:

1. true zero resistance,
2. a true diamagnetic Meissner effect,
3. reasonable stability of the sample, and
4. quick confirmation by several laboratories.

There will certainly be more USOs in the future, and it is not impossible that one of them will ultimately be confirmed to be a true superconductor at room temperature or above.

This discussion has not exhausted all of the various classes of superconductors. It is not at all clear that their superconductivity is similar in origin. Although research is continuing, theory has not provided much of a basis for discovering new superconducting materials. However, given the very wide range of materials that exhibit superconductivity, it is not unlikely that the next breakthough will come in a completely unexpected type of material. This might even lead to a class of superconductors that operates at significantly higher temperatures. But until that happens, the best bet is to continue development of applications using Nb-based LTS and Cu-oxide-based HTS materials.

Other Superconducting Microstructures

The power and versatility of semiconductor device technology has led many researchers to explore the possibility of integrating superconductors together with semiconducting devices to form some sort of superconducting transistor or "three-terminal device" (Kleinsasser and Gallagher, 1990). We have already spoken about the flux flow transistor, which is effectively dual to the semiconductor field-effect transistor (FET), as is the SQUID. Here we will be addressing a structure more like a conventional FET, as in Fig. 4.21. The control is applied electrically (via a voltage) rather than magnetically (via a current).

When one tries to integrate superconductors with semiconductors, one faces several major hurdles. They are fundamentally quite different: A superconductor is a low-resistance, low-voltage, high-current material, whereas a classical semiconductor is a high-resistance, low-current, high-voltage material. This is in addition to the difference in typical operating temperatures. Nevertheless, heavily doped, small-gap semiconductor is more compatible with a superconductor and may even itself be induced into a weakly superconducting state. This is the basis for the device in Fig. 4.21, sometimes called a JoFET, for Josephson field-effect transistor. It is essentially a gate-controlled Josephson junction. The two superconducting electrodes represent the source and the drain of the transistor; the gate electrode can also be superconducting but is electrically isolated from the others. A voltage on the gate varies the carrier density in the channel, which in turn varies the Josephson coupling between the source and the drain, changing the critical current I_c. This permits the voltage for $I > I_c$ to be modulated.

The surface of the semiconducting channel below the gate is essentially a two-dimensional metal, but it is still not a superconductor. For there to be any Josephson current at all, the channel length must be less than several times the "normal metal coherence length" ξ_n (which is typically ~ 100 nm at low T, similar to the coherence length in a superconductor) so that a weak proximity effect from the superconductors on the two sides may lead to some coupling of Cooper pairs. Thus, this may be viewed as an SNS structure, where the gate controls I_c through its control over ξ_n. However, in many cases, there is not a good ohmic contact between the superconductor and the semiconductor. Then it may be more correct to view this as an SINIS structure, where Schottky barriers act as tunneling layers and reduce I_c further.

A number of prototype devices have been studied. These include Nb contacts on InAs, which generally avoid Schottky barriers at the interfaces. They do indeed show very substantial modulation of the critical current but generally require voltages of ~ 1 V to control a channel voltage of ~ 1 mV (of order the

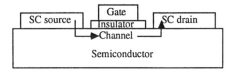

Figure 4.21. Prototype integrated superconducting–semiconducting device, the JoFET. A voltage on the gate controls the carrier density in the channel below, thereby affecting the critical current I_c of the weak induced superconductor and hence the voltage between the source and the drain for $I > I_c$.

superconducting gap voltage). Several HTS FET-like devices have also been tested but have similar limitations. For this reason, it is difficult to see how the output of one device could be used to control another one, and none of these devices is practical. So the search for a superconducting three-terminal device or transistor is continuing. However, it is worth pointing out (as we will show further in the next chapter) that the SQUID functions quite well as a superconducting transistor, with matched levels of input and output. It is not clear that an alternative device structure is necessary.

SUMMARY

- Many metallic elements (especially Nb and Pb, with T_c of 9 and 7 K) are superconducting at low T.
- Key LTS metallic alloys and compounds are Nb–Ti, NbN, and Nb_3Sn.
- Multifilamentary Nb–Ti and Nb_3Sn wires are used for high-field superconducting magnets.
- Niobium and NbN thin films and Josephson junctions are well developed for integrated circuits.
- All high-T_c copper oxide superconductors consist of conducting CuO_2 planes separated by thin insulating layers: T_c ranges up to 135 K.
- The HTSs are highly anisotropic, with anisotropic values of GL parameters λ and ξ; most applications require strong alignment of conducting planes.
- YBCO is most used for thin-film electronic applications; BSCCO is used for bulk wire.
- Prototype BSCCO multifilamentary wires in a Ag matrix have been fabricated, but vortex lattice melting may limit high-field applications to $T < \approx 30$ K.
- YBCO thin films generally require epitaxial growth on single-crystal substrates for highest J_c. Such films exhibit low rf surface resistance, with new applications to microwave filters.
- YBCO Josephson junctions are being developed using several approaches, including grain boundary and step-edge junctions.
- Organic and fullerene superconductors are examples of novel classes of superconducting materials that may offer future promise for new discoveries.

REFERENCES

L. Abelson, S. L. Thomasson, J. M. Murduck, R. Elmadjian, G. Akerling, R. Kono, and H. W. Chan, "A Superconductive Integrated Circuit Foundary," *IEEE Trans. Appl. Supercond.* **3**, 2043 (1993).

M. R. Beasley, "High-Temperature Superconductive Thin Films," *Proc. IEEE* **77**, 1155 (1989).

J. G. Bednorz and K. A. Müller, "Perovskite-Type Oxides; The New Approach to High-T_c Superconductivity," *Rev. Mod. Phys.* **60**, 585 (1988).

X. D. Chen, S. Y. Lee, J. P. Golben, S. I. Lee, R. D. McMichael, Y. Song, T. W. Noh, and J. R. Gaines, "Practical Preparation of Copper Oxide Superconductors," *Rev. Sci. Instrum.* **58**, 1565 (1987).

M. Cyrot and D. Pavuna, *Introduction to Superconductivity and High-T_c Materials* (World Scientific, Singapore, 1992).

R. Gross, L. Alff, A. Beck, O. M. Froelich, D. Koelle, and A. Marx, "Physics and Technology of HTS Josephson Junctions," *IEEE Trans. Appl. Supercond.* **7**, 2929 (1997).

A. W. Kleinsasser and W. J. Gallagher, "Three Terminal Devices," Chap. 9 in *Superconducting Devices*, Eds. S. T. Ruggeiro and D. A. Rudman (Academic, New York, 1990).

D. C. Larbalestier, "The Road to Conductors of High Temperature Superconductors," *IEEE Trans. Appl. Supercond.* **7**, 90 (1997).

J. M. Phillips, "Substrate Selection for High-Temperature Superconducting Thin Films," *J. Appl. Phys.* **79**, 1829 (1996).

B. W. Roberts, "Survey of Superconductive Materials," *J. Phys. Chem. Ref. Data* **5**, 581 (1976).

S. L. Thomasson, A. W. Moopenn, R. Elmadjian, J. M. Murduck, J. W. Spargo, L. A. Abelson, and H. W. Chan, "All Refractory NbN Integrated Circuit Process," *IEEE Trans. Appl. Supercond.* **3**, 2058 (1993).

M. Tinkham, *Introduction to Superconductivity*, 2nd ed., Chap. 9 (McGraw-Hill, New York, 1996).

R. M. White and T. H. Geballe, *Long Range Order in Solids* (Academic, New York, 1979).

J. M. Williams, Ed. *Organic Superconductors (Including Fullerenes)* (Prentice-Hall, Englewood Cliffs, NJ, 1992).

D. Zhang, G. C. Liang, C. F. Shih, R. S. Withers, M. E. Johansson, and A. Dela Cruz, "Compact Forward-Coupled Superconducting Microstrip Filters for Cellular Communications," *IEEE Trans. Appl. Supercond.* **5**, 2656 (1995).

PROBLEMS

4.1. Superconducting magnet wire.

(a) Using numbers given in Table 4.1, estimate the maximum vortex depinning current density J_p for Nb–Ti. Compare this to data in Fig. 4.3 and comment on differences. If this is made into a multifilamentary composite with superconductor covering one-third of the area, estimate the total average current density.

(b) For BSCCO, determine J_p assuming that the magnetic field points in the c direction, so that the parameters for the ab plane in Table 4.3 may be used. Comment on factors limiting the practical critical current J_c in BSCCO wires.

PROBLEMS

4.2. Microstrip transmission lines. Microstrip transmission lines are commonly used in superconducting integrated circuits both for signal transmission and as lumped inductances. Typical film thicknesses are 200 nm.

(a) Assuming all films are 200 nm thick, design a microstrip transmission line with $Z_0 = 10\ \Omega$ using NbN thin films with $\lambda = 300$ nm and an SiO_2 insulating layer with $\varepsilon_r = 4$. Determine the inductance L and capacitance C per micrometer and the width w of the line.

(b) Design a similar transmission line using YBCO films with $\lambda = 200$ nm and an epitaxial $LaAlO_3$ insulator with $\varepsilon_r = 23$.

4.3. HTS Josephson junctions. Most real HTS Josephson junctions do not exhibit classic SIS tunnel junction characteristics but rather appear to be shunted internally by a normal linear resistance.

(a) For an ideal SIS junction, $I_c R_n = \pi \Delta / 2e$, where $2\Delta \approx 3.5\ kT_c$ is the energy gap at low temperatures. Using $T_c = 92$ K for YBCO and assuming $I_c = 1$ mA, sketch the expected $I(V)$ relation for an ideal SIS junction and label the units on the axes.

(b) Now assume that this junction is internally shorted by a linear resistance with $R = 1\ \Omega$. Sketch the expected $I(V)$ relation for this case on the same plot as above. This "resistively shunted junction" is more like most of the HTS Josephson junctions that have been measured.

4.4. HTS and vortex lattice melting. Consider the case of current flow in the ab planes of BSCCO-2212 (Fig. 4.9c), with a magnetic field in the c direction. The two CuO_2 layers around the central Ca are strongly linked, as in YBCO. Use the characteristic lengths for BSCCO-2223 in Table 4.3.

(a) One way to approximate the melting of the vortex lattice in highly anisotropic HTS materials is by viewing each two-dimensional layer as an isolated superconducting thin film. Estimate roughly the two-dimensional vortex lattice melting temperature for a single layer in BSSCO. Compare to Fig. 4.12b.

(b) In the vortex liquid state, the vortex lattice is ineffective at restraining vortex motion. Estimate the flux flow resistivity for a perpendicular field of 10 T assuming an in-plane $\rho_n = 100\ \mu\Omega$-cm. How does this compare to the resistance of a good conductor like copper (with $\rho \sim 1\ \mu\Omega$-cm)? What is the significance of this for magnetic applications of BSCCO?

4.5. Critical fields of organic superconductor. The upper critical field B_{c2} has been measured for $(ET)_2Cu[N(CN)_2Br]$ close to $T_c = 11$ K, where the fields are not too large. This fit a straight line $B_{c2} = \alpha(T_c - T)$, where α equals 20 T/K for a field applied in the ac plane and 2.2 T/K for a field applied in the b direction.

(a) From these values, estimate the values of the coherence lengths $\xi(T)$ in various directions in this anisotropic material.

(b) Extrapolate to determine the low-temperature coherence lengths and compare to those of the cuprates such as YBCO and BSCCO.

5

JOSEPHSON DEVICES

Most superconducting electronic devices, apart from passive elements based on zero resistance, are based on Josephson junctions. This has been true for LTS electronics and for developing HTS systems as well. Applications range from a sensitive magnetometer, to a precision voltage standard, to a source of microwave radiation, to a digital switch based on picosecond pulses. We begin this chapter by deriving the dc and ac Josephson effects and show how a Josephson junction may be regarded alternately as a nonlinear inductor or as a voltage-controlled oscillator. This is continued with a more complete circuit model for practical junctions, including shunt resistance and capacitance. This provides the basis for the SQUID (superconducting quantum interference device), essentially just a loop with one or two Josephson junctions, with a sensitivity to weak magnetic fields that approaches the quantum limit. Finally, we will explore how arrays of junctions may be coupled together to create circuits for high-frequency oscillators, amplifiers, and pulse generators and transporters. The digital circuit applications are a sufficiently important subject that they will be treated separately in Chapter 6.

5.1 THE JOSEPHSON EFFECT

In its most general form, a Josephson junction is simply a weak connection between two pieces of superconductor, S_1 and S_2 (Fig. 5.1). Each superconductor is characterized by its complex pair wave function

$$\Psi = |\Psi|\exp(i\theta) \tag{5.1}$$

5.1 THE JOSEPHSON EFFECT

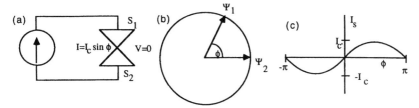

Figure 5.1. Current-biased Josephson junction linking two superconductors, illustrating dc Josephson effect. (*a*) Circuit schematic. (*b*) Phasors for complex pair wave functions Ψ_1 and Ψ_2, with phase angle ϕ between them. (*c*) Sinusoidal current-phase relation for junction.

(we are reverting here to the use of $i = \sqrt{-1}$ for complex numbers in quantum mechanics). The magnitude of the wave function can be normalized so that $|\Psi|^2 = \frac{1}{2}n_s$, the effective density of superconducting pairs, each of charge $-2e$ and mass $2m$, and θ relates in the standard way to energy $\mathcal{E}$ and momentum $\mathbf{p}$ by $\mathcal{E} = -\hbar\, d\theta/dt$ and $\mathbf{p} = \hbar\nabla\theta$. For simplicity, let us assume that the magnitude $|\Psi| = \Psi_0$ on both sides of the junction but that there is a phase difference across the junction,

$$\phi = \theta_1 - \theta_2 \tag{5.2}$$

Just as for a strong connection, a lossless dc supercurrent I_s can flow between S_1 and S_2. The Josephson effect states that

$$I_s = I_c \sin(\phi) \tag{5.3}$$

where I_c is the maximum supercurrent (the critical current of the junction) and ϕ is a constant for $V = 0$. This relation is sometimes called the "dc Josephson effect." (For $I > I_c$, we must consider parallel channels for current flow, which we will delay until the next section.)

Equation (5.3) continues to hold for $V > 0$, but now ϕ depends on the voltage across the junction by the relation

$$\frac{d\phi}{dt} = \frac{2eV}{\hbar} = \frac{2\pi V}{\Phi_0} \tag{5.4}$$

so that $\phi = 2eVt/\hbar$, which yields an alternating current

$$I_s = I_c \sin(\omega_J t) \tag{5.5}$$

which oscillates at the Josephson frequency

$$f_J = \frac{\omega_J}{2\pi} = \frac{2eV}{h} = \frac{V}{\Phi_0} \tag{5.6}$$

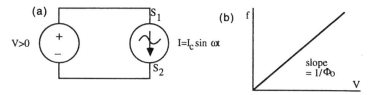

Figure 5.2. Voltage-biased Josephson junction illustrating ac Josephson effect. (a) Equivalent circuit, with junction as voltage-controlled current oscillator, with amplitude I_c and frequency ω_J. (b) Linear relation between oscillator frequency and applied voltage, with slope $2e/h = 1/\Phi_0 = 483.59$ MHz/μV.

This ac current is referred to as the ac Josephson effect. Essentially, a Josephson junction acts like a superconducting short for $V = 0$ but like an ideal voltage-controlled oscillator for $V \neq 0$ (Fig. 5.2). This may initially seem rather paradoxical, but both effects follow directly from the quantum-mechanical nature of the superconducting state.

To obtain Eq. (5.3), let us first review the formula for current density J_s in a superconductor as it relates to Ψ. For carriers of mass $2m$ and charge $-2e$, we can write

$$J_s = -n_s e v_s = \frac{-n_s e p_s}{2m} = \left(\frac{-n_s e}{2m}\right)\hbar \nabla\theta = \left(\frac{-e\hbar}{m}\right)|\Psi|^2 \nabla\theta, \quad (5.7)$$

which can be transformed to an equivalent standard form

$$J_s = \left(\frac{-e\hbar}{m}\right) Im[\Psi^* \nabla\Psi] \quad (5.8)$$

where Ψ^* is the complex conjugate of Ψ and Im is the imaginary part of a complex function. Now consider a geometry in which S_1 and S_2 are separated by a distance d in the x direction, with $x = 0$ halfway between them (Fig. 5.3). Within the junction, assume that the magnitudes of the wave functions $|\Psi|$ from both sides tail off exponentially but maintain their phases:

$$\Psi_1(x) = \Psi_0 \exp[-\alpha(x + \tfrac{1}{2}d)] \exp(i\theta_1) \quad (5.9)$$

$$\Psi_2(x) = \Psi_0 \exp[-\alpha(\tfrac{1}{2}d - x)] \exp(i\theta_2) \quad (5.10)$$

This would be the case, for example, if there were a very thin insulating layer between the two superconductors. The total wave function in the junction is just the superposition (i.e., the complex sum) of the tails from the two sides:

$$\Psi(x) = \Psi_1(x) + \Psi_2(x) \quad (5.11)$$

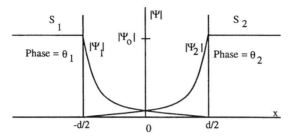

Figure 5.3. Configuration for derivation of dc Josephson relation for thin insulating layer between two superconductors S_1 and S_2. Complex wave functions from each superconducting electrode (each with its own phase θ) decay exponentially as they enter the insulating region. The total wave function is the coherent superposition of those from the two sides.

so that we also have

$$\nabla \Psi = -\alpha \Psi_1 + \alpha \Psi_2 = \alpha(\Psi_2 - \Psi_1) \quad (5.12)$$

Then from Eq. (5.8) we have

$$J_s = \left(\frac{-e\hbar}{m}\right) \text{Im}[-\alpha \Psi_1^* \Psi_1 + \alpha \Psi_2^* \Psi_2 + \alpha \Psi_1^* \Psi_2 - \alpha \Psi_2^* \Psi_1] \quad (5.13)$$

The first two terms in the brackets are real and drop out; the two cross terms at the end combine to give a purely imaginary component, leading (for any x inside the junction) to

$$J_s = \frac{2e\hbar\alpha}{m} \Psi_0^2 \exp(-\alpha d) \sin(\theta_1 - \theta_2) = J_c \sin(\phi) \quad (5.14)$$

For a junction of area A, this is the same as the dc Josephson relation in Eq. (5.3), where

$$I_c = J_c A = \frac{2eA\hbar\alpha\Psi_0^2}{m} \exp(-\alpha d) \quad (5.15)$$

Equation (5.4) follows directly from the fact that if there is a dc voltage across the junction, this raises the energy level of Cooper pairs on one side relative to the other by $\Delta\mathscr{E} = \mathscr{E}_1 - \mathscr{E}_2 = -2eV$ (Fig. 5.4). Further, Eq. (5.6) is just an expression of the basic quantum relation that $\Delta\mathscr{E} = hf$, where f is the frequency of a photon coupling two energy levels in a quantum system. An oscillating ac current is not quite a complete photon, however, since one also needs an in-phase component of ac voltage to couple net ac power out of the system. Of course, since the dc current $\langle I_s \rangle = 0$, there is no net transfer of Cooper pairs

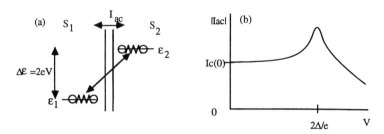

Figure 5.4. (a) Quantum picture of pair energy levels on opposite sides of a Josephson junction with voltage V across junction. The Josephson oscillation occurs at a frequency f_J such that $hf_J = \Delta\mathscr{E} = 2eV$. (b) Amplitude of Josephson oscillation as a function of voltage. Note the resonance at the gap voltage $V = 2\Delta/e$ and the rolloff above.

from one energy level to the other; they merely oscillate back and forth. Both of these are consistent with the absence of energy loss in this perfectly superconducting system (but see the section below on Shapiro steps).

The Josephson frequency–voltage relation might suggest that one can apply a 1-V dc voltage and obtain a current oscillation at ~ 500 THz, corresponding to orange light! This is *not* true, because the superconducting response rolls off for voltages greater than the gap voltage $2\Delta/e$, when direct pair-breaking can occur (Fig. 5.4b). [The BCS theory also predicts a resonance at the gap voltage.] But for frequencies much below $f_{\max} = (2e/h)(2\Delta/e) = 4\Delta/h$ (~ 1 THz for Nb), we can regard the amplitude of the ac Josephson response as essentially constant at I_c.

Note also that the critical current in Eq. (5.15) is exponentially decreasing with separation d and characteristic decay length $1/\alpha$. For a classic Josephson junction, the separation of the two superconductors is an insulator, and conduction can normally occur only by quantum-mechanical tunneling. In Chapter 2, we discussed tunneling of single electrons across such a tunnel barrier; here we are dealing with electron pairs. The wave function for a pair of electrons should decay twice as fast as that for a single electron, so that $\alpha = 2\kappa \sim 2$ Å^{-1}, where κ is the decay constant for single-electron tunneling from Eq. (2.53). Since typically $d \sim 10$ Å, J_c is reduced by orders of magnitude from the prefactor value, which is of order J_c of the superconducting film itself. It is remarkable that the formula for the normal-state electron tunneling current I_{nn} has exactly the same exponential dependence on separation d as does I_c in Eq. (5.15). In fact, the BCS theory gives the relation (for $T \ll T_c$) that

$$I_c R_n = \frac{\pi\Delta}{2e} \tag{5.16}$$

independent of the separation d. Typical values for a Nb tunnel junction might be $I_c \sim 100$ µA and $R_n \sim 20$ Ω. This accounts for the quasiparticle current above

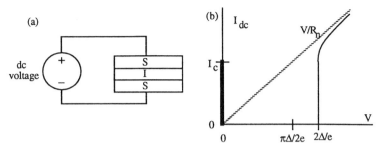

Figure 5.5. DC characteristics for voltage-biased SIS tunnel junction. (a) Circuit schematic. (b) I(V) for $T \ll T_c$, showing both Josephson current for $V = 0$ and normal-electron tunneling current for $V > 2\Delta/e$. Note that $I_c R_n = \pi\Delta/2e$.

the gap voltage $2\Delta/e$ being comparable to the superconducting critical current I_c (see Fig. 5.5). Closer to T_c, since Ψ_0^2 goes to zero linearly at T_c, I_c goes to zero in a similar fashion. The more general expression from the BCS theory is

$$I_c R_n = \frac{\pi\Delta(T)}{2e} \tanh\left(\frac{\Delta(T)}{2kT}\right) \qquad (5.17)$$

These Josephson relations are *not* limited to classic tunnel junctions. The derivation above assumed an exponential decay of $|\Psi|$ but did not specify the nature of this decay. A similar decay (with a much larger decay length $\sim \xi$) occurs in a normal-metal layer between two superconductors, forming the basis for an SNS junction that may also exhibit a similar Josephson effect. Furthermore, even a superconducting constriction (ScS) can act as weak link, particularly if its dimensions are on the order of the coherence length ξ. In fact, much of the early work on Josephson junctions was accomplished using a "point contact," a sharpened Nb wire pressed into a solid Nb block. On a microscopic level, this typically consisted of a metallic microconstriction. While this clearly does not form the basis of a technology for making reproducible junctions, it provided an easy way to make single junctions for research. Indeed, the standard symbol for a Josephson junction may have come from a similarity to a point contact. Even the $I_c R_n$ product for an SNS or ScS junction may be similar to Eq. (5.16) for an SIS in certain cases. There are distinct differences in electrical characteristics among these types of junctions (see Fig. 5.6), which we will discuss further in the next section, but these differences have more to do with the effective current paths parallel to the supercurrent than with the ideal Josephson element itself.

Josephson Inductance

What kind of electrical element is a Josephson junction? As we will show below, it can be viewed as an inductance, but one that is highly nonlinear. As a lossless

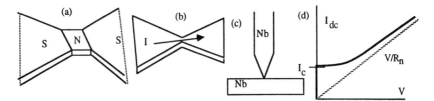

Figure 5.6. Other microstructures exhibiting Josephson effects. (*a*) SNS junction. (*b*) Thin-film constriction or microbridge. (*c*) Metallic point contact. (*d*) Typical *I*(*V*) for these junctions.

element, it can only be reactive, and it exhibits kinetic inductance in much the same way as a superconducting film. It is nonlinear in that its response can mix different frequencies; for example, a dc voltage causes an ac current.

We can see this by noting that the time integral of the voltage in Eq. (5.4) is simply the fluxoid Φ' discussed in Chapter 3 and is proportional to the phase difference ϕ across the junction:

$$\Phi' = \int V \, dt = \frac{\Phi_0 \phi}{2\pi} \tag{5.18}$$

Then we can write

$$I_s = I_c \sin\left(\frac{2\pi\Phi'}{\Phi_0}\right) \tag{5.19}$$

(Fig. 5.7*b*). For a conventional inductor or for a superconducting film below the critical current, this would be a linear dependence, where the effective inductance L is the slope of $\Phi'(I)$. Here, it is linear for currents $I_s \ll I_c$, corresponding to a Josephson inductance

$$L_{J0} = \frac{\Phi_0}{2\pi I_c} = \frac{\hbar}{2e I_c} \tag{5.20}$$

For $I_c \sim 100 \, \mu\text{A}$, this corresponds to $L_{J0} \sim 3$ pH. More generally, there is a differential inductance

$$L_J = \frac{\partial \Phi'}{\partial I_s} = \frac{L_{J0}}{\cos \phi} \tag{5.21}$$

which would correspond to the effective ac inductance with a dc current (or flux) bias and a small ac component. Note that this is *not* the same as $\Phi'/I_s = L_{J0}\phi/\sin\phi$, except in the limit that $\phi \to 0$. Also, we can evaluate the energy

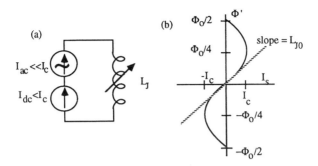

Figure 5.7. Effective inductance of Josephson junction. (*a*) Schematic of junction biased with dc $I < I_c$ and a small ac current, to give a variable nonlinear inductance. (*b*) Relation between fluxoid $\Phi' = \int V\,dt$ and current I_s in junction. The slope at a given operating point is the effective Josephson inductance L_J.

stored in the junction E_J as

$$E_J = \int I_s V\,dt = \frac{\Phi_0 I_c}{2\pi}(1 - \cos\phi) = L_{J_0} I_c^2 (1 - \cos\phi) \qquad (5.22)$$

which is of order 0.2 eV for $I_c \sim 100$ μA. Because of nonlinearity, this is not in general equivalent to $\frac{1}{2} L_J I_s^2$, except again in the linear regime that $\phi \to 0$. And of course, a dc voltage applied to a linear inductance would yield a current ramp, whereas for the Josephson junction, the current oscillates, requiring an effective inductance that oscillates as well.

A very useful analog to help understand the Josephson junction is a pendulum in a uniform gravitational field (Fig. 5.8). The angle of the pendulum from the vertical corresponds to the Josephson phase ϕ, and an applied dc voltage corresponds to rotation at a uniform rate. One revolution corresponds to a phase change of 2π, or a fluxoid change $\Delta\Phi' = \Phi_0$, a single flux quantum. The Josephson current corresponds to the gravitational torque on the pendulum,

$$T = mgl\sin(\phi) \qquad (5.23)$$

where m is the mass at the end of a rod of length l and g is the standard gravitational acceleration. This clearly changes sign depending on whether the pendulum is on the left or the right. In addition, the Josephson energy E_J corresponds to the gravitational potential energy

$$U = mgl[1 - \cos(\phi)] \qquad (5.24)$$

We will discuss the dynamics of this pendulum (including inertial and damping terms) later in the context of shunted-junction models.

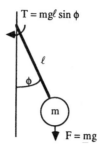

Figure 5.8. Pendulum analog of a Josephson junction, where the equilibrium position of the mass is hanging down. The angle ϕ from the vertical corresponds to the junction phase ϕ, and the gravitational torque corresponds to the supercurrent through the junction. A complete 2π rotation corresponds to a fluxoid of Φ_0.

Shapiro Steps

Instead of biasing the junction with a dc voltage, we can choose to bias it with a series combination of a dc and an ac voltage (Fig. 5.9):

$$V = V_0 + V_1 \cos(\omega_1 t) \tag{5.25}$$

Since the Josephson junction is a voltage-controlled oscillator, with frequency proportional to the voltage, by modulating the voltage, we are also modulating the frequency, which is, of course, the basis for standard frequency modulation (FM). Let us first consider the limit that $V_1 \ll V_0$ and also assume that $\omega_1 < \omega_0 = 2eV_0/\hbar$. In the standard way, this creates two side bands of the main carrier frequency, at $\omega_0 \pm \omega_1$. We can see this mathematically by integrating Eq. (5.25) to obtain ϕ, and therefore

$$I_s = I_c \sin(\phi) = I_c \sin\left[\omega_0 t + \phi_0 + \frac{2eV_1}{\hbar\omega_1}\sin(\omega_1 t)\right] \tag{5.26}$$

where $\omega_0 = 2eV_0/\hbar$. In the limit of small V_1, we can expand $\sin(A+B) \approx \sin(A) + B\cos(A)$ for small B, to obtain

$$\begin{aligned}
I_s &\approx I_c\left[\sin(\omega_0 t + \phi_0) + \frac{2eV_1}{\hbar\omega_1}\sin(\omega_1 t)\cos(\omega_0 t + \phi_0)\right] \\
&= I_c\left[\sin(\omega_0 t + \phi_0) + \frac{eV_1}{\hbar\omega_1}\sin(\omega_1 t + \omega_0 t + \phi_0)\right. \\
&\quad\left. + \frac{eV_1}{\hbar\omega_1}\sin(\omega_1 t - \omega_0 t - \phi_0)\right]
\end{aligned} \tag{5.27}$$

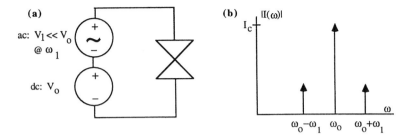

Figure 5.9. Frequency modulation in a Josephson junction. (a) Circuit schematic of a junction with both a dc and ac voltage bias. (b) Frequency spectrum of supercurrent, showing main carrier wave at ω_0 and two side bands at $\omega_0 \pm \omega_1$.

Note that the nonlinearity of the Josephson junction has effectively mixed the external voltage oscillation at ω_1 with the internal current oscillation at ω_0 to generate a current response containing sum and different frequencies at $\omega_0 \pm \omega_1$, as expected (see Fig. 5.9b).

Now, if the modulation frequency is increased until it is equal to ω_0, then the lower side band moves down to zero frequency, and we expect to see a dc contribution. From Eq. (5.27), the time-averaged value of supercurrent $\langle I_s \rangle$ takes the form

$$\langle I_s \rangle = -I_c \frac{V_1}{2V_0} \sin(\phi_0) \tag{5.28}$$

which is indicated by the vertical lines on the dc $I(V)$ at $V = \pm \hbar\omega_1/2e$ (Fig. 5.10). These are sometimes called "ac Josephson steps" or "Shapiro steps." For a typical ac frequency in the microwave range from 10 to 100 GHz, the Shapiro step will be at a voltage $V = hf/2e \sim 20\text{--}200\ \mu V$.

We can also understand this within the pendulum picture, in which we are rotating the pendulum at a reasonably constant rate but with a small periodic change in the rate. When the applied ac frequency ω_1 is synchronized to the pendulum rotation frequency $\omega_0 = 2eV_0/\hbar$, one would expect the pendulum to spend more time on one side than the other, depending on the location of the maximum rate. For example, if the rotation rate is maximum on the left and minimum on the right each period, then the pendulum will on average spend more time on the right. In terms of Josephson junction, $\langle I_s \rangle$ can have either sign, depending on the relative phase of the internal Josephson oscillator and the external ac voltage, as is indicated by Eq. (5.28).

It is also useful to view this from the point of view of power transfer. In terms of the dc I–V characteristics, an operating point in the first quadrant (positive I and V) normally corresponds to a dissipative element such as a resistor; one in the fourth quadrant (positive I and negative V) normally corresponds to an active device such as a power supply. The Josephson junction is a lossless

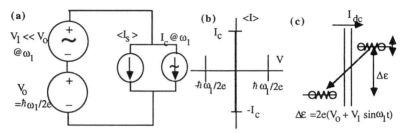

Figure 5.10. Shapiro steps in a Josephson junction. (*a*) Circuit schematic representing junction on Shapiro step as parallel combination of ac and dc current sources (the additional source at $2\omega_1$ can be neglected here). Net dc power can be transferred from the dc voltage source to the junction, while the same power is transferred back to the ac voltage source. (*b*) DC $I(V)$, showing critical current and first-order Shapiro steps. (*c*) Energy levels in Josephson junction on Shapiro step. Net pair transfer down in energy corresponds to stimulated emission of photons (net transfer up corresponds to resonant absorption).

superconducting element that conserves net power, but its nonlinearity permits it to extract power from the dc power supply and transfer it to the ac supply, or vice versa. In particular, on the Shapiro step at $V_0 = \hbar\omega_1/2e$, the dc power absorbed is

$$P_{dc} = \langle I_s \rangle V_0 = -\tfrac{1}{2} I_c V_1 \sin(\phi_0) \tag{5.29}$$

while the ac power absorbed at frequency $\omega_0 = \omega_1$ is

$$P_{ac} = \langle I_{ac}(t) V_{ac}(t) \rangle = I_c V_1 \langle \sin(\omega_0 t + \phi_0)\cos(\omega_1 t) \rangle = \tfrac{1}{2} I_c V_1 \sin(\phi_0) \tag{5.30}$$

corresponding to overall conservation of energy for all values of the phase parameter ϕ_0. This can also be seen from the viewpoint of a two-level atom, with an energy of $\hbar\omega_1 = 2eV_0$ between the levels (Fig. 5.10c). A net dc current from the lower level to the upper level ($\langle I_s \rangle < 0$) is effectively resonant absorption, with one photon absorbed for each Cooper pair crossing the junction. Similarly, $\langle I_s \rangle > 0$ corresponds to stimulated emission. (This suggests the possibility of something analogous to a laser, but the available power from a single Josephson junction is far too small to be practical. We will discuss coherent arrays of Josephson junctions in a later section.)

We can generalize these results for a larger ac voltage amplitude V_1. In this case, there is higher order mixing, with frequency components at $\omega = \omega_0 \pm n\omega_1$, where n is any integer. This follows from a determination of the harmonic content of I_s from Eq. (5.26) using a "Fourier–Bessel series":

$$I_s = I_c \sum_{n=-\infty}^{\infty} (-1)^n J_n \frac{2eV_1}{\hbar\omega_1} \sin(\omega_0 t + \phi_0 - n\omega_1 t) \tag{5.31}$$

where $J_n(x)$ is the nth-order Bessel junction and $J_{-n}(x) = (-1)^n J_n(x)$. (These same formulas come up in classic FM theory.) For small arguments the Bessel functions approach simple polynomials,

$$J_0(x) \approx 1 - \tfrac{1}{2}x^2 \qquad J_n(x) \approx \frac{(x/2)^n}{n!} \quad \text{for } n > 0 \tag{5.32}$$

but for large arguments, they oscillate and gradually damp out (see Fig. 5.11). For a given frequency ω_1, Eq. (5.31) gives rise to a dc contribution (i.e., a Shapiro step) at multiple values of voltage,

$$V_0 = \frac{\hbar\omega_0}{2e} = \frac{n\hbar\omega_1}{2e} \qquad n = \pm 1, 2, 3, \ldots \tag{5.33}$$

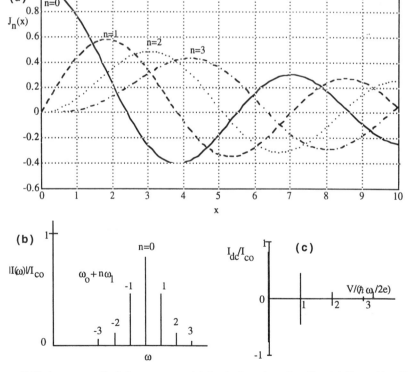

Figure 5.11. Large-amplitude frequency modulation in Josephson junction. (a) Bessel functions $J_n(x)$ for $n = 0, \ldots, 3$. (b) Frequency spectrum of current oscillations for $x = 2eV_1/\hbar\omega_1 = 1$, normalized to I_{co}, the critical current in the absence of the ac voltage. (c) Corresponding set of Shapiro steps in dc $I(V)$.

where the width ΔI_n of the nth Shapiro step is given by

$$\Delta I_n = 2I_c \left| J_n\left(\frac{2eV_1}{\hbar\omega_1}\right) \right| \tag{5.34}$$

For small arguments, this does indeed reduce to Eq. (5.28). For $n = 1$, the maximum step width is $2I_c J_1(1.8) = 1.2I_c$. Note that Eq. (5.34) also applies to the $n = 0$ case, since the critical current for $V = 0$ can be viewed as the $n = 0$ Shapiro step. Therefore, as the amplitude of the ac voltage is increased (for any frequency), the effective critical current is reduced to zero but then rises again with increasing rf voltage. The first zero of J_0 occurs for $x = 2eV_1/\hbar\omega_1 = 2.4$, which corresponds to $V_1 \sim 50\,\mu\text{V}$ for $\omega_1 = 10\,\text{GHz}$. The zeros of the various Bessel functions are in general out of phase. An example of the relative step heights for argument $x = 1$ is shown in Fig. 5.11c. Note also that from the energy level picture of Fig. 5.10c, these higher order Shapiro steps can be viewed as multiphoton absorption and emission processes, with $\Delta\mathscr{E} = n\hbar\omega_1$.

International Voltage Standard

The voltage–frequency relation of the Shapiro steps [Eq. (5.33)] is highly precise, determined by the fundamental physical constant $\Phi_0 = h/2e$. This has been developed to the stage that it now is used to define the International Standard Volt (Hamilton, Burroughs, and Benz, 1997). The old standard voltage was a particular electrochemical cell (i.e., a battery), which was difficult to stabilize to better than one part per thousand. In contrast, frequencies can be measured to almost arbitrary accuracy, essentially by counting zero-crossings. If we couple a standard frequency into a Josephson junction, we can use this to define a standard voltage.

One practical concern is that the voltages obtainable from a Shapiro step are normally rather small. Even for $f = 100\,\text{GHz}$, the voltage on the $n = 5$ step is approximately 1 mV. But using Nb-based superconducting integrated circuits, 10,000 or more identical Josephson junctions connected in series have been fabricated on a single chip (Fig. 5.12). The voltage on such an array can thus be chosen to provide any dc voltage up to 10 V or more, permitting direct calibration of secondary standards and precision digital voltmeters. Precision is typically within a few parts per *billion*. Note that this is an application that cannot be achieved using semiconductor electronics, since it requires the fundamental quantum properties of a Josephson junction.

Fluxonic Picture of Josephson Junction

Another way to regard the Josephson junction is from the point of view of fluxon motion. Consider, for example, a Josephson junction that is biased with a voltage source, as shown in Fig. 5.13. Assume for the present that the loop itself has negligible inductance and that there is no externally applied magnetic

5.1 THE JOSEPHSON EFFECT   191

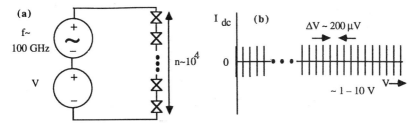

Figure 5.12. Josephson junction voltage standard with series array of ∼10,000 junctions. (a) Circuit schematic, with dc and ac voltage sources. (b) *I*(*V*) showing large number of Shapiro steps separated by $\Delta V = hf/2e \sim 200$ μV for $f \sim 100$ GHz.

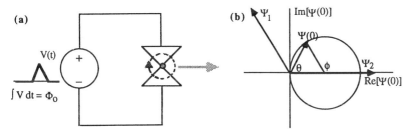

Figure 5.13. Single fluxon motion and phase slip in Josephson junction. (a) Application of single-flux-quantum voltage pulse to Josephson junction is equivalent to vortex moving transversely across junction. (b) Phasor diagram of complex pair wave function $\Psi(0)$ in the middle of a Josephson junction (Eq. 5.35), illustrating the "phase slip" that occurs once each period when $\phi = \pi$, $|\Psi(0)| = 0$ and $\Phi' = \frac{1}{2}\Phi_0$. This corresponds to the vortex center crossing the junction, as in (a).

field. If the voltage corresponds to a single-flux-quantum (SFQ) pulse of integral Φ_0, then one flux quantum is introduced into the loop and escapes across the junction. Looking at this more closely, during this process, the current first rises, then falls to zero and reverses sign, then finally returns to zero. This corresponds closely to what we would expect if we think of a vortex, with a circulating screening current, crossing the junction; the vortex is halfway across when $\Phi' = \frac{1}{2}\Phi_0$. Of course, we get the same currents if we take a vortex of the opposite sign and move it in the opposite direction; these are equivalent as far as the Josephson junction is concerned.

We can also examine the form of the complex wave function in the center of the junction at $x = 0$. From Eqs. (5.9) and (5.10), taking $\theta_2 = 0$ for simplicity, so that $\Psi(-\frac{1}{2}d) = \Psi_0 e^{i\phi}$ and $\Psi(+\frac{1}{2}d) = \Psi_0$, we have

$$\Psi(0) = |\Psi(0)|e^{i\theta(0)} = \Psi_0 \exp(-\tfrac{1}{2}\alpha d)(1 + e^{i\phi}) \tag{5.35}$$

If we plot this on a polar plot for varying ϕ, this traces out the circle shown in Fig. 5.13. For a constant voltage, the location of $\Psi(0)$ rotates around the circle at a constant rate. As shown in the figure, $\theta(0) = \frac{1}{2}\phi$ for $0 < \phi < \pi$ (this also follows from symmetry). But note that for $\phi = \pi$, $\theta(0)$ changes suddenly from $\frac{1}{2}\pi$ to $-\frac{1}{2}\pi$. This is possible only because $|\Psi| = 0$ at that point, so that θ is strictly undefined. We can also regard ϕ as changing from π to $-\pi$ at the same time, which corresponds to Φ' changing from $\frac{1}{2}\Phi_0$ to $-\frac{1}{2}\Phi_0$. And although nothing physically discontinuous happens at this point, this "phase slip" can be viewed as a leakage of a single flux quantum Φ_0 across the junction. This is also consistent with the vortex picture, since the center of a vortex also has $|\Psi| = 0$, with the supercurrent changing sign as the vortex passes by.

If we bias the Josephson junction with a dc voltage, then the current oscillates, corresponding to a periodic train of vortices crossing the junction (Fig. 5.14). The dual to this circuit is the single-electron transistor (SET) (Likharev, 1987), essentially a current-biased nonlinear capacitor, that stores charge but then releases it periodically in fixed units. This is also analogous to a leaky faucet, where the droplets correspond to flux quanta.

Alternatively, we can also think of a Josephson junction as providing a superconducting screening current that tries to maintain zero flux within the loop. As the flux introduced in the loop is increased, the junction initially acts like an inductor that opposes it. But for $\Phi' > \frac{1}{2}\Phi_0$, one fluxon has already escaped the loop, so that the effective flux is now negative ($-\frac{1}{2}\Phi_0$), and the induced circulating current now changes direction to try to increase the flux back toward zero.

We have generally been assuming in this section that a junction is biased by a voltage source or equivalently by a given fluxoid. However, when we take into account inductance in the superconducting leads, this can change the picture somewhat. In particular, the Josephson inductance of a typical junction (with $I_c \sim 100$ µA) is $L_J \sim 3$ pH. But $\mu_0 \sim 1$ µH/m $= 1$ pH/µm, so that the series

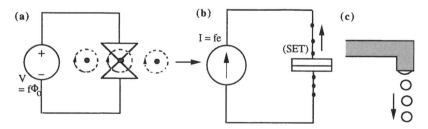

Figure 5.14. Josephson current oscillation and periodic fluxon motion. (a) Voltage-biased Josephson junction corresponding to periodic train of vortices that cross the junction transversely. (b) Dual of circuit in (a), corresponding to current-biased variable capacitor, which leaks a periodic series of discrete charges, similar to the "single-electron transistor" (SET). (c) Leaky faucet, with series of discrete droplets analogous to discrete flux quanta in (a) and electrons in (b).

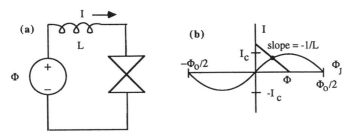

Figure 5.15. Effect of lead inductance on biasing of Josephson junction. (*a*) Circuit schematic, with ideal voltage source providing a fluxoid bias to a series combination of an inductor *L* and Josephson junction. (*b*) Load line solution for operating point of junction.

inductance L in the leads can easily be nanohenrys or greater. We can determine the effect of this by using a load line analysis similar to that for a nonlinear resistive element with a series resistance. Consider the circuit in Fig. 5.15a, where an ideal voltage source provides a flux bias Φ to the series combination of a Josephson junction and a linear inductor L. If we label the fluxoid across the junction by Φ_J, the current in the loop is given by the two expressions

$$I = \frac{(\Phi - \Phi_J)}{L} = I_c \sin\left(\frac{2\pi\Phi_J}{\Phi_0}\right) \tag{5.36}$$

where the first expression is the load line and the second is the usual sinusoidal current-phase relation. As shown in Fig. 5.15b, the intersection of these two gives the operating point. As L becomes large, this becomes less like a flux (or voltage) bias and more like a current bias. Indeed, an ideal current bias is probably a better approximation than an ideal voltage bias in most cases. To deal with a current bias, we must first address what happens when $I > I_c$, which leads into the next section on shunted-junction models.

5.2 SHUNTED-JUNCTION MODELS

The model of the Josephson junction as a nonlinear inductor is incomplete, since for a real tunnel junction there are parallel channels for current flow. In particular, there may be a normal "resistive" current I_n associated with single-electron (quasiparticle) tunneling or normal current flow in the case of an SNS or ScS junction. Furthermore, the structure of a tunnel junction is essentially that of a capacitor and parasitic capacitance is generally unavoidable, so that a "displacement current" $I_d = C\, dV/dt$ may also be important. We can represent both of these effects by shunting the ideal Josephson element with a resistor (possibly nonlinear) and a capacitor (Fig. 5.16a). If only the resistor is included, this resistively shunted junction

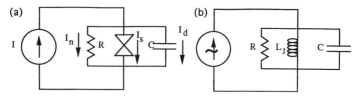

Figure 5.16. General circuit models for Josephson junction. (*a*) Capacitive resistively shunted junction (CRSJ) model. (*b*) Low-amplitude linearized approximation as parallel *LCR* resonator.

forms the RSJ model; if a capacitor is also included, this becomes a capacitive resistively shunted junction, or CRSJ (or RCSJ) model.

Junction Parameters

We can express the contributions to the total current of the CRSJ model in the following form:

$$I = I_s + I_n + I_d = I_c \sin(\phi) + \frac{V}{R} + C\frac{dV}{dt}$$

$$= I_c \sin(\phi) + \frac{\hbar}{2eR}\frac{d\phi}{dt} + \frac{\hbar C}{2e}\frac{d^2\phi}{dt^2} \tag{5.37}$$

using the standard Josephson relation that $V = (\hbar/2e)\, d\phi/dt$. We will assume for simplicity here that I_c, R, and C are constants. The presence of the $\sin(\phi)$ term makes this a second-order nonlinear differential equation, which in general does *not* have a closed-form analytic solution. But for the moment let us compare to the linearized equation for small ϕ, which can be written in the form

$$I = \frac{\Phi'}{L_{J0}} + \frac{V}{R} + C\frac{dV}{dt} \tag{5.38}$$

where $V = d\Phi'/dt$ and $L_{J0} = \Phi_0/2\pi I_c = \hbar/2eI_c$ is the Josephson inductance in this limit. This describes a current-driven parallel-*LCR* resonator (Fig. 5.16*b*), with resonant frequency

$$\omega_0 = 2\pi f_0 = \frac{1}{(L_{J0}C)^{1/2}} \tag{5.39}$$

bandwidth $B = 1/RC = \omega_0/Q$ with

$$Q = \omega_0 RC = \left(\frac{R^2 C}{L_{J0}}\right)^{1/2} \tag{5.40}$$

and maximum impedance (at resonance) $|Z| = R$. As long as $I < I_c$, the junction response will remain rather similar to that of a standard LCR resonator. Of course, the more interesting features of the junction behavior occur outside of this regime.

These same parameters also characterize the more general nonlinear response of the Josephson junction. In the literature of Josephson junctions, ω_0 is often called the "plasma frequency" of the junction, and a parameter

$$\beta_c = Q^2 = \frac{2eR^2CI_c}{\hbar} \tag{5.41}$$

is generally described instead of Q. Let us estimate typical magnitudes of the various parameters. For a tunnel junction with insulator thickness ~ 2 nm and dielectric constant $\varepsilon_r \sim 10$, the capacitance per unit area is $\varepsilon_0\varepsilon_r/d \sim 0.05$ F/m$^2 = 0.05$ pF/µm^2. For a typical junction area of 10 µm^2, we obtain $C \sim 0.5$ pF. A typical critical current density is ~ 1000 A/cm^2, which for the same size gives $I_c \sim 100$ µA. This, in turn, yields $L_{J0} = \hbar/2eI_c \sim 3$ pH. This gives the plasma frequency $f_0 = \omega_0/2\pi \sim 130$ GHz. Using the relation $I_c R_n = \pi\Delta/2e$, we can also estimate the normal-state resistance at $R_n \sim 20\,\Omega$ (for Nb), although the effective low-voltage resistance can be much larger than this (as we will discuss later). Taking $R = R_n$ for now, we have $RC \sim 10$ ps, so that $Q = \omega_0 RC \sim 8$, or $\beta_c = Q^2 \sim 60$. The area of the junction actually cancels out for most of these parameters. So for a tunnel junction, we generally expect to be in the regime that $\beta_c \gg 1$. On the other hand, other types of junctions (such as SNS or ScS) have a much larger separation between the superconducting electrodes, typically ~ 200 nm or more, corresponding to a capacitance per unit area ~ 100 times smaller than that for a tunnel junction. If I_c and R are comparable to those for the tunnel junction (as is typical), then we expect that $\beta_c \ll 1$ for these alternative junctions. Both low-Q and high-Q Josephson elements are commonly used in circuits, so that it is important to understand both their similarities and differences.

Mechanical Analogs

We can also obtain insights into this problem by extending the pendulum analogy introduced earlier. The dynamics of a pendulum of a mass m and length l (Fig. 5.17a) can be written in the form

$$T_{\text{tot}} = \mu\frac{d^2\phi}{dt^2} = T_a - \kappa\sin(\phi) - \eta\frac{d\phi}{dt} \tag{5.42}$$

The first expression is the angular form of Newton's law $F = ma$, where $\mu = ml^2$ is the angular mass (the so-called moment of inertia), $d^2\phi/dt^2$ is the angular acceleration, and T_{tot} is the total angular force (i.e., the torque) acting on the

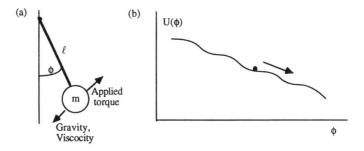

Figure 5.17. Mechanical analogs for CRSJ: (*a*) pendulum model; (*b*) tilted washboard model.

Table 5.1. Correspondences between CRSJ Model of Josephson Junction and Pendulum Analog

Josephson Junction	Pendulum
Phase difference ϕ	Angle from vertical ϕ
Voltage $V = (\hbar/2e)\, d\phi/dt$	Angular velocity $d\phi/dt$
Capacitance C	Moment of inertia $\mu = ml^2$
Conductance $1/R$	Viscocity η
Critical current I_c	Restoring constant $\kappa = mgl$
Applied current I	Applied torque T_a
Plasma frequency $\omega_0 = 1/\sqrt{L_{J0}C}$	Oscillator frequency $\omega_0 = \sqrt{\kappa/\mu}$

pendulum. The terms on the right are contributions to the total torque, namely the applied torque, the gravitational torque (with $\kappa = mgl$), and the drag associated with motion of the pendulum with angular velocity $d\phi/dt$ through a viscous medium. This corresponds *exactly* to the equation for the dynamics of the CRSJ model of the Josephson junction, if we make the identifications in Table 5.1. Essentially, the junction capacitance corresponds to the mass, and the shunt conductance $1/R$ corresponds to the viscosity. A high-Q pendulum (underdamped) is one that will oscillate or rotate a long time based only on its inertia, while a low-Q pendulum (overdamped) will move significantly only while it is being forced. Similarly, a high-Q (high-β_c) junction will oscillate or rotate many periods, while a low-Q junction will slow down as soon as the driving current is turned off, without oscillating. The boundary between these cases is known as critical damping and corresponds to $Q = \frac{1}{2}$ ($\beta_c = \frac{1}{4}$). A critically damped pendulum or junction, if given a sharp impulse, will rotate at most once and then quickly settle down without oscillation.

Another mechanical analog that is also used to describe the dynamics of a Josephson junction is the "tilted washboard model," in which a particle moves down a slope with a sinusoidally modulated height, as in Fig. 5.17*b*. This is really just another representation of the pendulum, where the angle ϕ is folded out horizontally, and the slope is proportional to the applied torque. The particle's motion is determined by the equation

$$\mu \frac{d^2\phi}{dt^2} = -\frac{\partial U(\phi)}{\partial \phi} - \eta \frac{d\phi}{dt} \tag{5.43}$$

where $\mu = ml^2$ is the moment of inertia as before, the particle is moving in a viscous medium, and the potential energy

$$U(\phi) = mgl(1 - \cos \phi) - T_a\phi \tag{5.44}$$

incorporates both the gravitational potential and the effect of the applied torque. Taking the correspondences in Table 5.1, this also describes the dynamics of the junction. This picture is particularly useful in describing a junction biased with a constant applied current, so that the slope is constant in time. For $I < I_c$, the junction will be statically located in one of the local potential minima. If the slope is sufficiently steep (i.e., if $I > I_c$), then the particle will start to roll down the slope, with a steady-state velocity corresponding to the average voltage. This velocity (or voltage) will be determined by a balance between the gravitational energy gained and the viscous energy lost. This is just another way of stating that the net dc electrical power into the junction $\langle IV \rangle$ must be balanced by net dissipation of power in the resistor R, regardless of the energy stored in the reactive elements.

RSJ Model

Let us consider first the low-β_c (low-Q) limit, where the capacitance is so small that it can be neglected; such a junction can also be called overdamped or an RSJ. This corresponds to neglecting the inertial (mass) term in the mechanical analogs, leaving the dynamics dominated by viscous loss. The dynamical equation for the junction can be written in the form

$$I = I_s + I_n = I_c \sin(\phi) + \frac{V}{R} = I_c \sin \phi + \frac{\hbar}{2eR} \frac{d\phi}{dt} \tag{5.45}$$

If the junction is biased with a dc voltage (Fig. 5.18), then we have $I_s = \sin(\omega_J t)$ and $I_n = V/R$. The dc $I(V)$ relation (including both I_s and I_n) will look just like that of a resistor, except for $V = 0$, when the usual dc supercurrent is present. If now we add an ac voltage at frequency f_1, then the supercurrent displays a set of Shapiro steps as before at $V = nhf_1/2e$, which adds to I_n.

The situation is somewhat modified if we have a constant current bias (Fig. 5.19). Before solving Eq. (5.45), let us examine what we would expect. For $I < I_c$, the resistor is shorted out, and we get the usual dc Josephson current. For

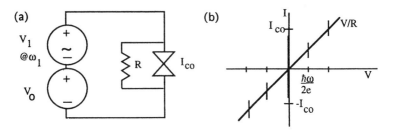

Figure 5.18. Voltage-biased RSJ. (a) Circuit schematic, including both dc and ac sources. (b) DC I–V curve, displaying normal conductance and Shapiro steps.

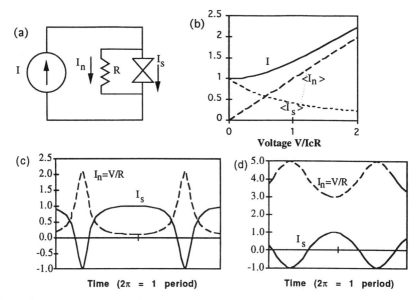

Figure 5.19. Current-biased RSJ. (a) Schematic circuit. (b) DC $I(V)$, together with dependence of dc average components of I_s and I_n. (c) Time dependence $I_s(t)$ and $I_n(t)$ for $I = 1.1 I_c$, displaying periodic sequence of SFQ pulses. (d) $I_s(t)$ and $I_n(t)$ for $I = 4 I_c$, showing almost sinusoidal current oscillations.

$I \gg I_c$, most of the current goes through the resistor (since the supercurrent can carry no more than I_c), and this effectively places a constant voltage bias of IR on the parallel Josephson element, yielding only a sinusoidal ac current. But for I just above I_c, the situation can be quite different. The current going through R produces a voltage across the junction, which produces a supercurrent oscillation, which in turn feeds back into the resistor and modulates the voltage. We can also think of this in terms of either the pendulum or the washboard. For the tilted washboard, the instantaneous velocity (i.e., the voltage) will vary, depending on which part of the cycle it is in. As the tilt is increased, the average velocity should increase smoothly. For the RSJ, over a range of currents above I_c, we would likewise expect the average voltage to provide a smooth connection between the zero-voltage line and the voltage-biased line (see Fig. 5.19b).

This can be obtained in greater detail if we solve Eq. (5.45). As a first-order differential equation, the solution can be obtained analytically. Separating variables, we have

$$\frac{d\phi}{(I - I_c \sin \phi)} = \frac{2eR}{\hbar} dt \tag{5.46}$$

We can take the integral on the left and invert to obtain $\phi(t)$, although the result is rather messy. Since we expect a result whereby ϕ changes by 2π each period, we can also use the definite integral of Eq. (5.46) to express the period T in the form

$$\int_0^{2\pi} \frac{d\phi}{I - I_c \sin \phi} = \frac{2\pi}{\sqrt{I^2 - I_c^2}} = \frac{2eR}{\hbar} T \tag{5.47}$$

This period can also be related to the dc average voltage $\langle V \rangle$, since the fundamental frequency of this (nonsinusoidal) Josephson oscillation is $f_J = \omega_J/2\pi = (2e/h)\langle V \rangle = 1/T$. This gives the particularly simple result

$$\langle V \rangle = R\sqrt{I^2 - I_c^2} \tag{5.48}$$

Note also that the current-biased dc $I(V)$ relation for the RSJ is continuous and nonhysteretic, that is, it is the same for dc current either sweeping up or sweeping down. This will change when we consider the effect of the capacitance below.

The time-dependent solutions of the RSJ model for $I_s(t)$ and $V(t)$ can be expressed in the forms (Van Duzer and Turner, 1981, p. 187)

$$I_s(t) = I_c \sin \phi(t) = I_c \frac{[I_c - I \sin(\omega_J t)]}{I - I_c \sin(\omega_J t)}$$

$$V(t) = I_n R = R[I - I_s(t)] = \frac{R[I^2 - I_c^2]}{I - I_c \sin(\omega_J t)} \tag{5.49}$$

These results are plotted in Fig. 5.19. The oscillation is indeed nonsinusoidal; if $V_0 = \langle V \rangle$, the amplitude of the mth harmonic can be written in the form

$$V_m = 2V_0 \left(\frac{I}{I_c} - \frac{V_0}{I_c R}\right)^m \tag{5.50}$$

Note that for I just above I_c, the voltage corresponds to a regular series of voltage pulses of height $\sim 2I_c R$ and width $\sim \Phi_0/2I_c R$. (For the pendulum, this corresponds to a complete 2π rotation.) The area under each such pulse is thus an SFQ Φ_0. For a typical value of $I_c R \sim 2$ mV, the pulse width is ~ 0.5 ps. For higher voltages, these pulses overlap, but there is still one flux quantum transferred across the Josephson junction for each period.

Operating points with positive values of both dc voltage and current correspond to net power dissipation in the resistor. But note that the average supercurrent $\langle I_c \rangle = I - V/R$ is also nonzero for $V > 0$, suggesting that dc power is going into the "lossless" Josephson element. The solution to this paradox is the same as that associated with power absorbed on the Shapiro step; it is due to power being transferred from dc to ac. Within each SFQ pulse, I_s and V are out of phase, indicating a source of ac power. This ac power, in turn, is being dissipated in the shunt resistor, since I_n and V are in phase. For an operating point just above I_c, the energy generated by each SFQ pulse is $\sim I_c \Phi_0$, and the rate of such pulses is $N = \langle V \rangle/\Phi_0$, corresponding to a total power transfer $N\mathscr{E} \sim I\langle V \rangle$, which is consistent with the dc power absorbed.

We have been neglecting the influence of thermal fluctuations on the dynamics of the Josephson junction. This approximation is often valid, depending on the relative size of the thermal energy $E_{\text{th}} = kT$ and the characteristic Josephson energy $E_J = L_{J0} I_c^2 = \hbar I_c/2e$. For a junction of a low-T_c superconductor with $I_c \sim 100$ μA, we have $E_J/k \sim 2500$ K, suggesting that thermal fluctuations are indeed negligible for either a LTS junction at 4 K or a HTS junction at 77 K. However, for a junction with $I_c \sim 1$ μA, the situation would no longer be quite as clear. The primary effect of thermal fluctuations for a current-biased RSJ is to initiate random phase slips even for the case where I is below I_c; this is most significant just below I_c. This shows up as a rounding of the sharp corner at I_c. In fact, this smearing effect is quite significant even for $E_{\text{th}}/E_J \sim 1/10$, leading to an effective reduction of the critical current by about 50%. For practical Josephson junctions, then, we normally require that $I_c \gtrsim 1$ μA at 4 K and $I_c \gtrsim 20$ μA at 77 K.

The current source can also include an ac component. Just as for the case of voltage bias, this leads to the presence of Shapiro steps in the dc $I(V)$ at voltages $V = n\hbar\omega_1/2e$, as shown in Fig. 5.20. Each of these is part of a continuous curve (like that near I_c), in contrast to the discrete spikes that are present in the voltage-biased case. Actually, if one switches the horizontal and vertical axes (as is often done here since I is the independent variable), these look even more like a staircase, with a series of flat steps equally spaced in the vertical direction. There is not an analytic solution to this problem in terms of standard functions, but the behavior of these steps with increasing ac amplitude I_1 has been found (both numerically and experimentally) to follow a qualitatively similar oscillating dependence to the Bessel functions described in the previous section. Also, thermal smearing can have a significant effect on the Shapiro steps if the step width is less than ≈ 1 μA (at $T = 4$ K) for the same reason described above for the critical current.

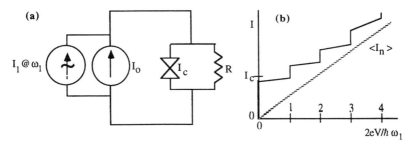

Figure 5.20. Shapiro steps in a current-biased RSJ. (*a*) Schematic circuit. (*b*) DC *I*(*V*), showing sequence of steps at $V = n\hbar\omega_1/2e$.

One interesting way to view Shapiro steps in a current-biased RSJ is in terms of frequency pulling and phase locking. If the junction is on the $n = 1$ step, then the frequency of the internal Josephson oscillator matches the applied frequency ω_1. If the operating point is not too close to the edge of the step, then if the bias current changes slightly, the junction will remain locked to the same frequency. If ω_1 itself drifts, then the junction will track this frequency, remaining locked to it if the variation is not too fast. Even for operating points off the step, the frequency is pulled toward ω_1. On the higher order steps, the junction will similarly lock onto the nth harmonic of the external signal for sufficiently large amplitude, as expected for a highly nonlinear oscillator.

Capacitance and Hysteresis

We have been addressing the characteristics of a CRSJ in the limit of small capacitance, where $Q \ll 1$. In the opposite limit $Q \gg 1$, the dynamics are quite different. There are several ways to view this effect. From the point of view of characteristic impedances in the system, recall that the characteristic time in the RSJ model is the width of the SFQ pulse, $\Delta t \sim L_{J0}/R$, and the characteristic frequency $\omega_R = 1/\Delta t = 2eI_cR/\hbar$. At this frequency, the impedance of the shunt capacitor is

$$|Z_c(\omega_R)| = \frac{1}{\omega_R C} = \frac{R}{Q^2} = \frac{R}{\beta_c} \qquad (5.51)$$

where Q is the same quantity as defined in Eq. (5.41). So if $Q \ll 1$, then the capacitive shunt is virtually an open and can be ignored. In the other limit $Q \gg 1$, however, the capacitor is virtually a short at ac frequencies. For a voltage-biased junction, this has no effect on the dc current. The only effect on the Shapiro steps is that the ac voltage source will draw much more ac current, possibly making it difficult to couple sufficient power into the junction.

For a current-biased junction, the effect of capacitance is somewhat more subtle. For $I > I_c$, the dc current will produce a voltage across R, which in turn will generate a sinusoidal current in the Josephson element. In the low-Q case, this ac current couples back to R, in turn generating an ac voltage that acts back on the junction, causing a nonsinusoidal $I_s(t)$. In the present high-Q case, however, this ac current is shunted across C without producing a significant ac voltage. The voltage across the junction, then, is pure dc, reducing this to the voltage-biased case. So when $I > I_c$, the voltage jumps sharply to the normal resistance at $V = I_c R$ (Fig. 5.21). The more subtle part occurs when I is now reduced to below I_c. The voltage remains on the voltage-biased branch of the curve, rather than jumping back down to $V = 0$. The only way to return to $V = 0$ is to reduce the current to $I = 0$. This discontinuity and hysteresis are characteristic features of high-Q junctions.

One can gain additional insight into this hysteresis via the mechanical analogs. The high-Q limit describes the situation in which the pendulum is dominated by inertial effects rather than viscous loss. For the tilted washboard model, $I > I_c$ corresponds to a tilt that is enough to initiate the ball rolling down the corrugated slope. Hysteresis corresponds to the situation where the tilt angle is reduced, but the ball continues to roll down the slope. For the ball to be "retrapped" in one of the periodic dips, it must first slow down and come to a stop. In this limit, the dissipation is so low that the steady-state velocity is very large, barely modulated by the corrugations in the washboard.

One does not have to reduce I quite all of the way to zero in order to return to the zero-voltage state (Fig. 5.21b). If we reduce V by a factor of Q^2 from the onset of $I_c R$, then the frequency will also be reduced by the same factor, and the impedances of R and C are now comparable. For $I < I_c/Q^2$, the capacitor should no longer be dominant, and the junction should return to $V = 0$. A more careful analysis (Tinkham, 1996, p. 209) has yielded a somewhat larger criterion for the return current I_r:

$$I_r \approx \frac{4}{\pi Q} I_c \quad \text{for} \quad Q \gg 1 \tag{5.52}$$

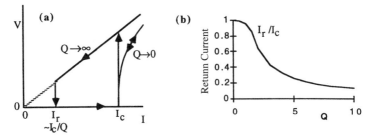

Figure 5.21. Hysteresis in high-Q current-biased CRSJ. (a) $V(I)$ for high-Q and low-Q junctions (the V and I axes have been switched from earlier figures). Note that the return current $I_r \sim I_c/Q$ for the high-Q case. (b) Return current I_r versus Q.

For an ideal CRSJ at $T = 0$ for any nonzero value of Q, there is always some hysteresis (i.e., $I_r < I_c$). However, once effects of thermal fluctuations are taken into account, critical current hysteresis typically disappears for $Q \lesssim 1$.

Hysteretic I–V curves are often a disadvantage, as in the case of SQUIDs (to be discussed in the next section). Furthermore, generating a single SFQ pulse (as for digital applications to be discussed in the next chapter) can be easily achieved only with a low-Q junction. It is common to produce a low-Q junction from a high-Q superconducting tunnel junction (e.g., based on Nb) by fabricating a thin-film normal resistive shunt externally across the inputs of the junction, thus producing a true RSJ (Fig. 5.22). It is also notable that most HTS Josephson junctions (including all of those discussed in the previous chapter) behave very much like an ideal RSJ, with a low Q and no hysteresis. Furthermore, they typically exhibit $I_c R < 1$ mV, much less than the expected gap voltage of ~ 20–60 mV. This, too, is consistent with a resistor shunting the intrinsic Josephson element. On a microscopic level, this may reflect an inhomogeneous junction region that consists of mixed superconducting and normal domains. While this may be less than ideal for some purposes, a variety of applications that make use of the RSJ behavior of HTS Josephson junctions are developing.

For general nonzero values of Q, the CRSJ model does not have an analytic solution. However, let us briefly mention a similar model that illustrates the same points and does have a simple solution. Consider an ideal Josephson element that is shunted by a resistance R_0 at dc and a different resistance R_1 at high frequencies of order the Josephson oscillations. If $R_1 < R_0$, this would be similar to the sort of capacitive shorting effect that occurs in real Josephson junctions. The solution to this takes the form of $I(V, R = R_1)$ for the usual current-biased RSJ, with a correction for the dc conductance of R_0:

$$I = \sqrt{\frac{V^2}{R_1^2} + I_c^2} + V\left(\frac{1}{R_0} - \frac{1}{R_1}\right) \quad (5.53)$$

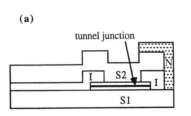
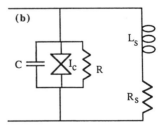

Figure 5.22. Josephson junction with external resistive shunt. (a) Cross-sectional view of thin-film layers with normal layer (N) connected across superconducting inputs (S_1 and S_2) of tunnel junction. The insulating I layer is used to define the "window" for the junction. (b) Schematic of circuit with intrinsic CRSJ in parallel with a smaller resistance R_s and a (normally very small) series inductance.

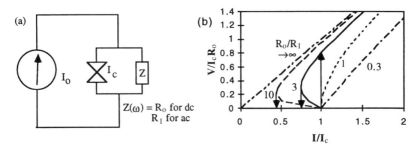

Figure 5.23. (a) Modified RSJ model, where the resistive shunt is R_0 at dc and R_1 at the Josephson frequency and harmonics. (b) $V(I)$ relation from Eq. (5.53), exhibiting hysteresis for $R_1 < R_0$, (similar to hysteresis in CRSJ model). Values of R_1/R_0 from left to right, are ∞ (voltage-biased limit), 10, 3, 1 (usual RSJ), and 0.3.

As Fig. 5.23 shows, for $R_1 < R_0$, the curves move closer to the voltage-biased line and also exhibit double-valued behavior as a function of voltage. For dc current biasing, this is a direct indication of discontinuity and hysteresis (the back-bending solutions may not be stable). Note also that for $R_1 > R_0$, the curve moves even farther from the voltage-biased solution, to the right of the standard current-biased RSJ. This might correspond to the real physical situation of an external resistive shunt with a significant series inductance, giving an enhanced high-frequency impedance.

Finally, the CRSJ model does not require that the resistance R be a constant, independent of voltage. For an ideal superconducting tunnel junction at $T \ll T_c$, the effective conductance $1/R(V)$ is very small for $V < V_g = 2\Delta/e$, at which point the conductance rises sharply toward the normal-state value $1/R_n$. In our mechanical analogs, this might correspond to motion through a medium where the viscous resistance may be negligible until one reaches a critical velocity. For the junction, this means that on exceeding I_c, the voltage would quickly rise until it reached this critical value, at which point steady state could be achieved. A typical superconducting tunnel junction is also a high-Q structure, so that if it is biased with a constant current, it will exhibit substantial hysteresis, as indicated in Fig. 5.24.

Josephson Junction as Radiation Source

The ac Josephson effect provides a precision voltage-controlled current oscillator. This suggests that a Josephson junction may be used as a tunable source of high-frequency radiation in the gigahertz range up to the gap frequency $4\Delta/h \sim 1000$ GHz. We can represent this most simply as an RSJ coupled to an electrical load with an effective load resistance R_L (Fig. 5.25a). The physical output circuit near the junction might actually consist of a transmission line, waveguide, or antenna, but these, too, could be represented by the appropriate characteristic impedance. This circuit also acts like an RSJ, but with a shunt

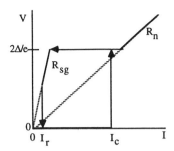

Figure 5.24. $V(I)$ for high-Q superconducting tunnel junction biased with a constant current showing nonlinear voltage-dependent resistance $R(V)$ (with large subgap resistance R_{sg} and smaller resistance R_n above the gap voltage $2\Delta/e$, equivalent to the normal-state resistance) and hysteresis in the critical current.

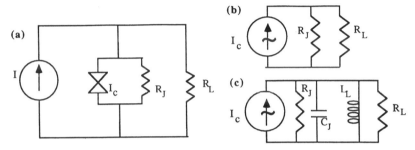

Figure 5.25. Coupling radiation out of a Josephson junction. (*a*) External load resistor R_L connected across current-biased RSJ, representing detector, transmission line, or antenna. (*b*) Equivalent ac circuit, representing junction as ac current source with amplitude I_c, delivering power to parallel combination of junction and load resistances. The optimum power transfer to the load occurs for $R_L = R_J$. (*c*) Equivalent ac circuit for CRSJ, showing matched load that cancels out shunt reactance of capacitor at resonant frequency $\omega = 1/\sqrt{L_L C_J}$.

resistance R_s that is the parallel combination of the junction resistance R_J and R_L, that is,

$$R_s = \frac{R_J R_L}{R_J + R_L} \tag{5.54}$$

In some cases, the load might be blocked at dc, which will affect the dc I–V relation (as suggested by Fig. 5.23), but the same value of R_s will apply at high frequencies.

Let us estimate how much optimum power one can expect to couple out of a typical Josephson junction at a given frequency $f = 2eV/h$. Presumably, we would like the Josephson current oscillation to be reasonably sinusoidal, so that most of the power is in the fundamental frequency. For the RSJ model, that requires that $V \gtrsim I_c R_s$. In this case, the junction can be viewed as an ac current source of amplitude I_c (Fig. 5.25b). Then, the ac power coupled into the load is approximately

$$P_{\text{out}} = \tfrac{1}{2} I_L V_L = \tfrac{1}{2} I_L^2 R_L \tag{5.55}$$

where I_L and V_L here refer to the ac quantities at frequency f. As for conventional power supplies, the optimum power transfer occurs when the source impedance and load impedance are matched, so that $R_L = R_J$. In that situation, half of the ac current (of amplitude I_c) is diverted to the load, so that

$$P_{\text{out}} = \tfrac{1}{2}(I_c/2)^2 R_J = \tfrac{1}{8} I_c^2 R_J \tag{5.56}$$

If we estimate $I_c R_J \sim 1$ mV, $R_J \sim 10\ \Omega$, and $I_c \sim 100\ \mu\text{A}$, then $P_{\text{out}} \sim 0.1\ \mu\text{W}$. This is detectable and has indeed been measured, but it is not really enough for a practical power source. We can increase I_c to ~ 1 mA, and then P_{out} rises to $\sim 1\ \mu\text{W}$, but this requires an output impedance of $\sim 1\ \Omega$, somewhat smaller than optimum for many transmission structures. So it would seem that the available microwave power is severely limited. However, as we discuss later in the chapter, it is possible to obtain coherent radiation from an array of many Josephson junctions, and this may well turn out to be practical.

Note also that the presence of a large shunt capacitance can be a serious problem for coupling out radiation, particularly at higher frequencies. For example, given the estimate earlier of $C \sim 1$ pF for a tunnel junction, this yields a shunt impedance at $f = 1$ THz of $|Z| = 1/2\pi f C = \tfrac{1}{6}\ \Omega$. A load impedance would have to be comparable to this in order to couple out significant power. It is possible in principle to "tune out" a given capacitance using a load that has a reactance of the opposite sign (Fig. 5.25c), but normally this can be accomplished only at a single fixed frequency (i.e., the resonant frequency $1/\sqrt{L_L C_J}$). Then, we no longer have a continuously tunable radiation source, unless the load reactance can also be tuned. This significantly complicates both the device and the analysis, so that it is generally preferable to deal with junctions with $Q \sim 1$ or less.

Simulation of Josephson Junction Circuits

We have already described mechanical analogs based on the pendulum and the tilted washboard that provide for reasonably accurate simulation of the dynamics of Josephson junctions. These are very useful for qualitative visualization, but less so for quantitative measurement, particularly if more than one junction is required. Somewhat more useful are electrical circuits that act as analogs of the CRSJ equations but have characteristic frequencies of order 10 kHz rather than 100 GHz, allowing the results to be displayed on an oscilloscope. A number of these have been developed, some of them based on phase-locked loops and voltage-controlled oscillators.

In recent years, however, Josephson circuit simulation programs have been developed for many standard computer platforms. These include applications of specially designed systems for superconducting circuits (such as PSCAN) and others based on standard (non-superconducting) circuit simulation tools such as SPICE. These digital simulations generally carry out solutions in the time domain, since the nonlinearity of the Josephson junction does not lend itself to frequency-domain solutions. Circuits with up to hundreds of Josephson junctions and other elements can be simulated using current computer workstations. The application of SPICE to Josephson junctions is discussed in greater detail in Appendix B.

5.3 SQUIDs AND MAGNETIC DETECTION

In the field of superconducting devices, the word "squid" refers not to the undersea mollusk, but rather to the acronym SQUID, which was coined in the mid-1960s for superconducting quantum interference device. Now the term is used to refer to any device consisting of a superconducting loop containing one or two Josephson junctions (usually resistively shunted junctions, or RSJs) with a way to introduce magnetic flux Φ into the loop. This will give rise to a circulating current $I_{\text{cir}}(\Phi)$ that is periodic in the flux quantum $\Phi_0 = h/2e$. If we can measure the characteristics of the junction(s), we can in principle provide a sensitive measurement of Φ to a small fraction of Φ_0. In a two-junction SQUID (Fig. 5.26), the junctions can be biased in parallel with a dc current and operated above the critical current in the voltage state. Hence, the two-junction SQUID is better known as the "dc SQUID." In contrast, the one-junction SQUID (Fig. 5.27) cannot be dc-biased, since it is shorted out by the loop. However, its state can be determined by coupling to a driven rf resonator, giving the more common name "rf SQUID" or sometimes "ac SQUID." Both the dc SQUID and the rf SQUID are widely used for sensitive magnetic measurements. We will first address the dc SQUID and deal with the rf SQUID afterward.

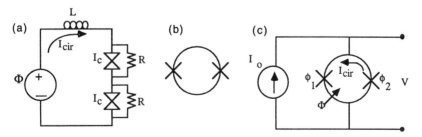

Figure 5.26. The dc SQUID. (*a*) Circuit schematic of flux-biased superconducting loop with two matched RSJs. (*b*) Standard symbol. (*c*) Typical dc biasing scheme with junctions in parallel.

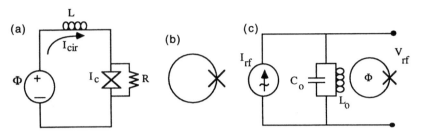

Figure 5.27. The rf SQUID. (*a*) Circuit schematic of flux-biased superconducting loop with one RSJ. (*b*) Standard symbol. (*c*) Typical rf-biasing scheme with current-driven *LC* resonator inductively coupled to rf SQUID.

Two Junctions and Superconducting Quantum Interference

Consider two junctions with identical critical currents I_c in the flux-biased circuit of Fig. 5.26a. This flux is represented here by an ideal voltage source with a time-integrated voltage pulse $\Phi = \int V dt$ in the past. In an actual circuit, it might be introduced by applying an external magnetic field perpendicular to the loop or by inductive coupling to another line. As we showed earlier, the flux Φ in a superconducting loop corresponds to a phase difference around the loop of $\Delta\theta = 2\pi(\Phi/\Phi_0)$. Assume for the present that the inductance L of the loop itself is negligible, compared to the (nonlinear) Josephson inductance of the junctions $L_J \sim \Phi_0/2\pi I_c$. Then the flux Φ must split evenly between the two identical inductances, yielding a circulating current

$$I_{\text{cir}} = I_c \sin \phi = I_c \sin \frac{\pi \Phi}{\Phi_0} \tag{5.57}$$

where ϕ is the (identical) phase drop across each junction. Note that for $\Phi = \frac{1}{2}\Phi_0$, $I_{\text{cir}} = I_c$.

Now consider the same dc SQUID loop embedded in the current-biased circuit of Fig. 5.26c. This initially looks like a simple parallel combination of two RSJs with total critical current $2I_c$, but the presence of I_{cir} complicates this picture. Let us first address the situation in the zero-voltage state, where the situation is static and we can ignore the shunt resistors. The applied current I_0 must by symmetry split evenly between the two junctions, so that

$$I_1 = \tfrac{1}{2}I_0 + I_{\text{cir}} \qquad I_2 = \tfrac{1}{2}I - I_{\text{cir}} \tag{5.58}$$

but we *cannot* simply substitute in Eq. (5.57) to obtain I_1 and I_2; the principle of superposition does not apply to this nonlinear system. Instead, we have the constraint that the difference in phases between the two junctions is proportional to the applied flux:

$$\phi_1 - \phi_2 = \Delta\phi = \frac{2\pi\Phi}{\Phi_0} \tag{5.59}$$

We can see this graphically using the construction of Fig. 5.28a. Here, the currents for the two junctions correspond to the ends of the line segment of width $\Delta\phi$, but the segment is free to move horizontally. The total current is the sum of those for the two junctions, and the circulating current is half the difference. Two examples are shown corresponding to $I_0 = 0$ and to the maximum total current I'_c. Working out the detailed dependence using trigonometric identities, we have

$$I_0 = I_1 + I_2 = I_c(\sin\phi_1 + \sin\phi_2) = 2I_c\sin[\tfrac{1}{2}(\phi_1 + \phi_2)]\cos[\tfrac{1}{2}(\phi_1 - \phi_2)]$$
$$= 2I_c\cos\left(\frac{\pi\Phi}{\Phi_0}\right)\sin[\tfrac{1}{2}(\phi_1 + \phi_2)] = I'_c\sin\phi' \tag{5.60}$$

where the effective critical current of the parallel junctions is

$$I'_c = 2I_c\left|\cos\left(\frac{\pi\Phi}{\Phi_0}\right)\right| \tag{5.61}$$

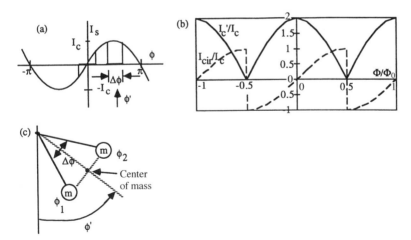

Figure 5.28. DC SQUID in zero-voltage state: currents and pendulum picture. (*a*) Supercurrent versus phase ϕ for two junctions in SQUID; the line segment $\Delta\phi$ is fixed by applied flux ϕ. Examples shown correspond to total current $I_0 = 0$ and I'_c, the effective critical current of the SQUID. (*b*) Effective critical current I'_c (solid line) and circulating current I_{cir} (for $I_0 = 0$—dashed line) versus flux Φ in loop for negligible loop inductance L. (*c*) Compound pendulum analog of dc SQUID, which acts like a single pendulum with mass $2m$ at the center of mass, and phase ϕ'.

and the effective phase is $\phi' = \frac{1}{2}(\phi_1 + \phi_2)$. So the two junctions do indeed act like a single junction but one that may have a reduced critical current; indeed, I'_c can be reduced all the way to zero.

The dependence of I'_c is periodic in Φ_0 (see Fig. 5.28b) and is known as superconducting quantum interference, for reasons that will be discussed below. Note in particular that the critical current goes to zero for $\Phi = (n + \frac{1}{2})\Phi_0$. Also, the circulating current is

$$I_{\rm cir} = \tfrac{1}{2}(I_1 - I_2) = \tfrac{1}{2}I_c(\sin \phi_1 - \sin \phi_2) = I_c \sin\left(\frac{\pi\Phi}{\Phi_0}\right)\cos \phi' \qquad (5.62)$$

which goes smoothly to zero as I_0 goes to I'_c (i.e., $\cos \phi' = 0$ when $\sin \phi' = 1$) for all values of Φ. This ensures that both junctions start to develop a voltage at the same current, which of course is necessary for junctions in parallel. The dependence of $I_{\rm cir}$ for $I_0 = 0$ is also periodic in Φ_0 (Fig. 5.28b). Note the sudden change in sign at $\frac{1}{2}\Phi_0$; the basis for this will be made clearer below.

Let us note that a magnetic field can also depress the critical current of a single Josephson junction, for essentially the same reason as described for a SQUID. However, if the junction is small enough so that the flux inside the junction is much less than Φ_0, this effect can be neglected. This will be discussed in more detail in Section 5.4 on distributed Josphson junctions, but for the time being, we will assume that this is not a concern.

We can gain insight into these equations by considering again the pendulum analogy introduced earlier for the single Josephson junction, but let us now consider two pendula with the same mass, hanging from the same pivot point (Fig. 5.28c). The condition (5.59) corresponds to a fixed angle $\Delta\phi = \phi_1 - \phi_2$ between the two. This is equivalent to a single compound pendulum, with its full mass $m' = 2m$ located at the center of mass halfway between the two individual masses, which is a distance $l' = l\cos(\frac{1}{2}\Delta\phi)$ from the pivot point. Alternatively, we can view the net gravitational torque (or effective critical current) as the magnitude of the vector sum of those from the two pendula. In the absence of an applied torque, this compound pendulum hangs with the center of mass pointing down, with the individual pendula at an angle $\frac{1}{2}\Delta\phi$ on either side. If a torque is applied, this rotates as a unit, with its effective phase angle halfway between the two [$\phi' = \frac{1}{2}(\phi_1 + \phi_2)$], and its effective gravitational torque $m'gl' = 2mgl\cos(\frac{1}{2}\Delta\phi)$, exactly analogous to the case with the junctions. When $\phi' = \frac{1}{2}\pi$, the torque (and hence the supercurrent) is a maximum, corresponding to the total critical current I'_c. Note that for $\Delta\phi = \pi$ (which corresponds to $\frac{1}{2}\Phi_0$), the center of mass is at the pivot point, so that the gravitational torque (corresponding to I'_c) is 0.

Furthermore, the circulating current $I_{\rm cir}$ corresponds to half the difference of the horizontal projections of the two pendula. As Eq. (5.62) indicates, for fixed $\Delta\phi$, this goes as $\cos(\phi')$. If $\Delta\phi > \pi$, then the center of mass is actually above the pivot point, so that the stable configuration is with the pendulum flipped over, and the effective angle between the pendula is really $2\pi - \Delta\phi$. This then reverses the sign of I_{cir}, corresponding to the sudden change at $\Phi = \frac{1}{2}\Phi_0$ in Fig. 5.28b.

Thus far we have considered only the zero-voltage state for $I < I'_c$. However, the same relations continue to be valid above the critical current. In terms of Fig. 5.28a, the horizontal line segment moves to the right for larger values of ϕ, leading to oscillation of supercurrents. Similarly, the compound pendulum rotates around, still maintaining the fixed angle $\Delta\phi$. We can see how the other parameters of the model transform from either the equation for the total current or the schematic itself; the shunt capacitances add, as do the shunt conductances:

$$I = I_1 + I_2 = \left(I_c \sin \phi_1 + \frac{V}{R} + C\frac{dV}{dt}\right) + \left(I_c \sin \phi_2 + \frac{V}{R} + C\frac{dV}{dt}\right)$$

$$= I'_c \sin \phi' + \frac{V}{R'} + C'\frac{dV}{dt} \tag{5.63}$$

where I'_c and ϕ' are exactly as obtained above, $R' = \frac{1}{2}R$, $C' = 2C$, and $V = (\hbar/2e)d\phi'/dt$. Thus, for fixed ϕ, the dc SQUID acts just like a single Josephson junction with these parameters in the voltage state as well. This includes ac Josephson oscillations at high voltage and Shapiro steps if an ac current or voltage is added to the dc bias. We will focus on the nonhysteretic limit, where we can essentially neglect the capacitance. In this case, the dc $V(I)$ relation (Fig. 5.29a) is exactly that given earlier within the RSJ model:

$$V(I) = \begin{cases} 0 & I < I'_c \\ R'\sqrt{I^2 - I'^2_c} & I > I'_c \end{cases} \tag{5.64}$$

In addition, through the dependence of I'_c on the flux Φ, this also gives an implicit dependence $V(\Phi)$ for fixed I (Fig. 5.29b):

$$V(\Phi) = I_c R \sqrt{\left[\left(\frac{I}{2I_c}\right)^2 - \cos^2\left(\frac{\pi\Phi}{\Phi_0}\right)\right]} \qquad I > 2I_c \tag{5.65}$$

This is periodic in Φ with period Φ_0, and the modulation of the voltage is greatest at $I = 2I_c$, where the voltage goes between 0 and $I_c R$ as the flux goes from 0 to $\frac{1}{2}\Phi_0$. More generally, the range of voltage covered (for fixed current) lies between the current-biased and voltage-biased RSJ curves. This range decreases quickly for larger currents, so that for greatest sensitivity, the SQUID is normally operated just above $2I_c$. Since the junctions must be nonhysteretic in this range, this requires low-Q junctions. If high-Q tunnel junctions are used, they are typically shunted by a normal resistor to bring Q down to $\lesssim 1$.

For $V \gg I_c R$, the dc voltage signal becomes small, but there is still an ac current at the Josephson frequency $2eV/h$ of amplitude I'_c (now almost

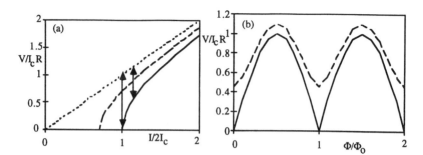

Figure 5.29. DC SQUID in voltage state. (*a*) DC $V(I)$ for several values of $\Phi(\frac{1}{2}\Phi_0, \frac{1}{4}\Phi_0, 0)$ from left to right. (*b*) Dependence $V(\Phi)$ for fixed $I = 2I_c$ (bottom) and $I = 2.2\,I_c$ (top).

sinusoidal) generated by the Josephson junctions, which for the current-biased case gives rise to an ac voltage of amplitude

$$V_{ac} = I'_c R' = I_c R |\cos(\tfrac{1}{2}\Delta\phi)| \qquad (5.66)$$

where $\Delta\phi = 2\pi\Phi/\Phi_0$ is the phase delay between the two Josephson oscillators. Also, there is an oscillating circulating current in the loop at the same frequency of amplitude

$$I_{cir} = I_c \sin(\tfrac{1}{2}\Delta\phi) \qquad (5.67)$$

This high-voltage limit is identical to a familiar standard problem, that of the "interference" of two coherent wave sources with a phase delay between them. Depending on whether the waves add in phase or out of phase, the relative amplitude may vary $\pm 100\%$ from that of the individual sources. For example, this occurs in the two-slit interference pattern in optics and in the radiation pattern of two coherent antennas in the rf regime. This is the basis for the term "superconducting quantum interference"; the quantum aspect here is that the phase delay is due to the effect of magnetic flux on the phase of the superconducting quantum wave function. For the more general interference problem, it is common to represent the sources using phasors (Fig. 5.30). In the present case, the two Josephson oscillators must be oscillating at the same frequency since they have the same voltage across them, and their phase delay is constrained by the applied flux. Let us label the two current sources by phasors I_c and $I_c \exp(j\,\Delta\phi) = I_c \angle \Delta\phi$, both driving current into the shunt resistance $R' = \tfrac{1}{2}R$. So the ac voltage amplitude must take the form

$$|V_{ac}| = R'|I_c + I_c \exp(j\Delta\phi)| = R' I_c |\exp(-\tfrac{1}{2}j\Delta\phi) + \exp(\tfrac{1}{2}j\,\Delta\phi)|$$

$$= I_c R |\cos(\tfrac{1}{2}\Delta\phi)| \qquad (5.68)$$

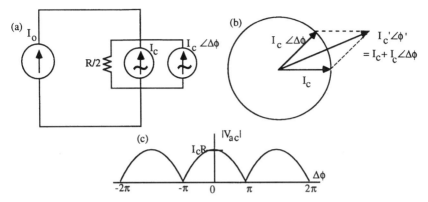

Figure 5.30. Interference in dc SQUID for $I \gg 2I_c$. (*a*) Schematic with junctions represented as ac current sources with phase delay $\Delta\phi$. (*b*) Phasors for two current sources and resultant net current I'_c. (*c*) Net ac voltage amplitude versus phase delay $\Delta\phi$.

the same as that obtained by a different approach in Eq. (5.66) above. This "Josephson phasor" approach can also be applied when the current oscillation is not sinusoidal or even in the zero-voltage state. This permits us to easily see the effect of having Josephson junctions with unmatched values of I_c. From the interference picture, this will reduce the maximum modulation of the critical current (or the voltage), since the full range can only go from $|I_{c1} - I_{c2}|$ to $I_{c1} + I_{c2}$. We can also generalize to a circuit with more than two junctions.

Loop Inductance and Screening Current

Thus far, we have assumed that we can ignore the loop inductance L, compared to the Josephson inductances $L_J \sim \hbar/2eI_c$. However, this is *not* generally a good assumption. First, the typical Josephson inductances are rather small; for $I_c \sim 100$ μA, $L_J \sim 3$ pH. Given that $\mu_0 \sim 1$ pH/μm, $L \ll L_J$ would require an unreasonably small loop $\lesssim 1$ μm. But Φ_0 for such a small loop would correspond to $B \sim 1$ mT (10 G), which would not make for a sensitive detector of magnetic fields. It is possible to increase L_J somewhat by decreasing I_c, but if $I_c \lesssim 1$ μA at 4 K or if $I_c \lesssim 20$ μA at 80 K (i.e., if the Josephson energy $\hbar I_c/2e \lesssim kT$), thermal fluctuations begin to destroy the required coherence. On the other hand, we do not want the loop inductance L to be too large. If $L \gg L_J$, then the ac circulating currents will be blocked, and each junction will behave separately as a current-biased RSJ. In other words, the critical current of the SQUID would be simply the sum of those of the two junctions, independent of the applied flux. So it should be no surprise that the optimum situation occurs for $L \sim L_J$, that is, $LI_c \sim \Phi_0$.

We can also consider this from the point of view of the superconducting screening current. If this loop were simply a strongly superconducting wire with no Josephson junctions, then it would be expected to oppose the applied flux

Φ_a and try to maintain the enclosed magnetic flux Φ_{int} at an integral number of flux quanta (depending on the initial condition), regardless of Φ_a. This corresponds to the large-L limit, since strong coupling implies a large I_c and a small $L_J \sim 1/I_c$. The screening flux is $\Phi_s = LI_{cir}$, so that the net flux in the loop is

$$\Phi_{int} = \Phi_a - LI_{cir} \tag{5.69}$$

Since $I_{cir} \leq I_c$, when $LI_c \gtrsim \Phi_0$, screening of a flux quantum will be significant. This is essentially the same condition as $L \gtrsim L_J \sim \Phi_0/2\pi I_c$. We have been dealing in the other limit, when the maximum screening flux $\sim LI_c \ll \Phi_0$ and can be ignored. But with practical SQUIDs, this is no longer valid, and the earlier treatment must be modified somewhat.

Let us examine this situation in more detail. Consider the circuit in Fig. 5.31a, which shows the inductance L split evenly between the two parallel branches, as is the applied flux Φ_a. Since the Josephson junction is essentially a nonlinear inductor, we can regard the applied flux $\frac{1}{2}\Phi_a$ in each branch as being divided across a series combination of $\frac{1}{2}L$ and L_J. This is directly analogous to the standard problem of biasing a nonlinear resistor (such as a transistor) using a voltage source with a series resistance, which is typically addressed using load line analysis. In the present case, the same current I must flow in both the junction and the series inductor, so we have

$$I = I_c \sin \phi = I_c \sin\left(\frac{2\pi\Phi_J}{\Phi_0}\right) \qquad I = [(\tfrac{1}{2}\Phi_a) - \Phi_J]\left(\frac{2}{L}\right) \tag{5.70}$$

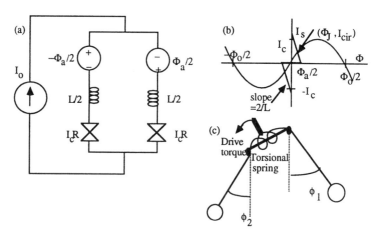

Figure 5.31. DC SQUID with loop inductance included. (a) Schematic of dc SQUID with symmetric loop inductance $\frac{1}{2}L$ in each branch. (b) Load line analysis of current and flux relation for junction with zero total bias current. (c) Compound pendulum analog for dc SQUID, showing torsional spring between pendula to simulate effect of loop inductance.

where Φ_J is the "flux drop" across a junction. In general, this must be solved numerically, but the solution can be given graphically by the intersection of the two lines from Eq. (5.70), as shown in Fig. 5.31b. We have shown here the load lines for both junctions of the SQUID for zero applied current. The distance between the intercepts on the flux axis is Φ_a; the corresponding separation between the operating points is somewhat less. Note the similarity between this figure and Fig. 5.28a; the only difference (apart from the trivial factor of $2\pi/\Phi_0$ in going from ϕ to Φ) is that the load lines now lean at an angle. When we vary the applied current, the segment on the flux axis moves as a unit, and similarly in the voltage state. However, the phase delay between the two Josephson junctions $\Delta\phi$ is no longer a constant, and the solution must be obtained numerically.

We must also modify the pendulum analog to account for the series inductance. The inductor can be represented by a torsional spring between the two separate pendula on the rotation axis (Fig. 5.31c). The flux bias is equivalent to a net twist in this spring. A loose spring corresponds to a large inductance. Within this picture, we would expect that the phase angle $\Delta\phi$ between the pendula will be somewhat reduced when they are hanging down but would actually be increased when they are pointing up. The drive torque is applied to the middle of this spring, rather than on one or the other of the pendula directly. This would cause the phases of the junctions to lag behind the driving torque, again consistent with the construction in Fig. 5.31b.

As L increases, things become even more complicated. When the slope of the load line $2/L$ decreases below $2\pi I_c/\Phi_0$ (i.e., when $\frac{1}{2}L > L_J = \Phi_0/2\pi I_c$), there can be two intersection points of the load line with the sine wave, leading to hysteretic behavior, which would cause problems for the operation of this device. This is also apparent in a plot of Φ_J versus Φ_a (Fig. 5.32a), determined using

$$\Phi_a = 2\Phi_J + LI_c \sin\left(\frac{2\pi\Phi_J}{\Phi_0}\right) \qquad (5.71)$$

In general, we want to design a dc SQUID with as large a loop inductance L as possible that does not cause flux hysteresis. This must be solved numerically, and simulations have shown that $L \sim 2L_J$ still retains about a 50% reduction in I_c for $\Phi = \frac{1}{2}\Phi_0$. This, in turn, leads to a voltage modulation that is only slightly reduced from the value $I_c R$ that was calculated earlier ignoring the effect of L (Fig. 5.32b). So dc SQUIDs are typically designed with $LI_c \sim \frac{1}{2}\Phi_0$.

The flux resolution of a dc SQUID is always much less than one flux quantum and is generally limited by thermal noise. For an optimally designed SQUID, the noise level at 4 K is typically less than $1 \times 10^{-5}\ \Phi_0/\sqrt{\text{Hz}}$. Given that a flux of $\frac{1}{2}\Phi_0$ yields a voltage $\sim I_c R$, this corresponds (for $I_c R \sim 100\ \mu\text{V}$) to a voltage noise $\sim 2\ \text{nV}/\sqrt{\text{Hz}}$. A noise spectrum of this form, with a noise power $\sim V^2$ being proportional to the effective bandwidth of the measurement, is characteristic of "white noise." In addition, SQUID systems also exhibit an additional

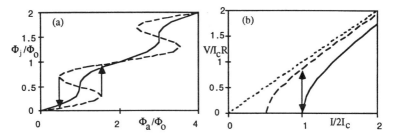

Figure 5.32. Effect of screening current on dc SQUID characteristics. (a) Net flux Φ_J across junction in dc SQUID versus applied flux Φ_a for zero current bias and with $LI_c = \Phi_0/\pi$ (solid line) and Φ_0 (dashed line). Note that the latter value causes hysteretic behavior. (b) Typical $V(I)$ for dc SQUID with $LI_c = \Phi_0/\pi$ for $\Phi_a = 0$ (solid) and $\frac{1}{2}\Phi_0$ (dashed). In contrast to Fig. 5.29, the maximum suppression of I_c is only about 50%. (The straight line $V = \frac{1}{2}IR$ is included for comparison.)

"$1/f$-noise" that is typically dominant at low frequencies in the kilohertz range and below. Some implications of this for optimum measurements are discussed later.

RF SQUID

The rf SQUID has only one junction in a superconducting loop and no bias current, so one might initially expect it to be even simpler than a dc SQUID. It is based on similar principles, but the readout is somewhat indirect. This turns out to make its analysis significantly more complicated, and the device is generally also somewhat less sensitive. Nevertheless, the early commercial SQUIDs made use of a point contact junction that was difficult to make in matched pairs (as required for the dc SQUID), so that a device based on a single junction had a clear advantage. The descendants of these early rf SQUIDs continue to be widely used. As shown earlier in Fig. 5.27, the readout of an rf SQUID is based on measuring the effective resonant impedance of a driven LC resonator inductively coupled to the SQUID. For an appropriate rf current level, this rf impedance turns out (not surprisingly) to be periodic in applied flux with period Φ_0.

The resonator is typically a standard rf LC "tank circuit" with a resonant frequency of 20–30 MHz and $Q \lesssim 100$; a high-Q superconducting resonator is not necessary or even preferable. This is inductively coupled to the SQUID loop, and the resonator is driven at its resonant frequency. The output signal is the amplitude of the rf voltage for a fixed amplitude of rf current. We will not be solving this in detail, but we can obtain a qualitative understanding of the primary phenomena using a similar picture to that for the dc SQUID (Tinkham, 1996, p. 229). For a single junction in a loop with inductance L, the relation between the applied flux Φ_a and the flux across the junction Φ_J (Fig. 5.33a) is

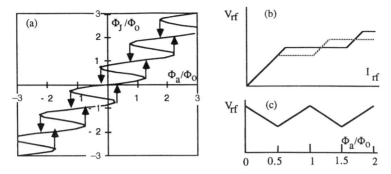

Figure 5.33. Operating characteristics of rf SQUID. (*a*) Hysteretic dependence of internal flux Φ_J on applied flux Φ_a for $LI_c = \Phi_0$. (*b*) Approximate rf current–voltage characteristic for resonant circuit inductively coupled to SQUID loop for $\Phi_a = n\Phi_0$ (solid line) and $(n + \frac{1}{2})\Phi_0$ (dashed line). The plateaus are associated with onset of hysteresis losses in (*a*). (*c*) Typical relation $V_{rf}(\Phi_a)$, showing characteristic triangular periodic behavior, with I_{rf} biased on the first plateau.

given in the quasi-static limit by

$$\Phi_a = \Phi_J + LI_c \sin\left(\frac{2\pi\Phi_J}{\Phi_0}\right) \qquad (5.72)$$

almost the same as Eq. (5.71). In this situation Φ_J is also the same as the internal (screened) flux. The applied flux Φ_a is the sum of the static flux and the ac flux from the resonator. This becomes hysteretic for $LI_c > \Phi_0/2\pi$, as can also be seen using a load-line construction similar to that in Fig. 5.31*b*. In contrast to the dc SQUID, where magnetic hysteresis is to be avoided, the rf SQUID is typically operated with $LI_c \approx \Phi_0$, well into the hysteretic regime. However, the amplitude of the rf flux induced in the loop is normally sufficiently large that an entire hysteresis loop is swept out during a cycle. In this way, the average V_{rf} is itself not hysteretic in the static applied flux.

The typical dependence of V_{rf} on the drive current amplitude I_{rf} exhibits a series of flat steps connected by ramps, as indicated roughly in Fig. 5.33*b*. These are not Shapiro steps or anything similar. Instead, they reflect the fact that having the junction traverse a hysteresis loop absorbs energy (delivered to the shunt resistor), and this energy loss reduces the Q of the resonant circuit. Since the impedance at resonance is proportional to Q, a contribution to energy loss will show up as a characteristic feature in V_{rf}. The first plateau is associated with the hysteresis loop near Φ_0; the second plateau occurs when I_{rf} is large enough that the second jump near $2\Phi_0$ can also occur. The dc flux bias affects this by changing the threshold at which the jump can start. By choosing an appropriate value of I_{rf}, a triangular dependence of V_{rf} versus Φ_a with period Φ_0 is normally obtained (Fig. 5.33*c*). The flux sensitivity of an rf SQUID (limited by thermal noise in the SQUID and in the rf electronics) is typically of order 1×10^{-4}

$\Phi_0/\sqrt{Hz}$, about a factor of 10 less sensitive than the dc SQUID. On the other hand, it is far more sensitive than any other competing technology for the measurement of weak magnetic fields.

Issues in Practical SQUIDs

We have established that SQUIDs can resolve a very small magnetic flux but that is not quite the same as a small magnetic field. The difference, of course, is the area of the SQUID loop, which is normally rather small. We would like to increase the effective area of the loop without increasing the inductance L above the value $\sim \Phi_0/I_c$. In fact, a SQUID is usually placed inside a magnetic shield, with external magnetic flux coupled in via an input circuit. The typical input circuit is a closed superconducting loop known as a "flux transformer" that consists of a pickup coil in the ambient magnetic field and an input coil inductively coupled to the SQUID loop (Fig. 5.34a).

The flux transformer is based on the conservation of flux in a superconducting loop maintained by a screening current I_s. If the loop initially contains flux $\Phi_{tot} = 0$, then if a magnetic flux density Φ_p is applied to the pickup loop of inductance L_p, the total flux will be

$$\Phi_{tot} = 0 = \Phi_p - (L_p + L_i)I_s \tag{5.73}$$

where the inductance L_i of the input coil implicitly includes the effect of the mutual inductance of the SQUID loop. Furthermore, the effective applied flux coupled into the SQUID is

$$\Phi_a = MI_s = \frac{M\Phi_p}{L_p + L_i} \tag{5.74}$$

where M is the mutual inductance between the input coil and the SQUID loop L, given by the transformer relation

$$M = k\sqrt{LL_i} \tag{5.75}$$

where k is the usual transformer coupling constant, which ideally approaches 1. The optimum condition for flux transfer is for $L_p = L_i$, so that

$$\Phi_a \approx \left(\frac{M}{2L_i}\right)\Phi_p = \sqrt{\frac{L}{L_i}}\frac{\Phi_p}{2} \tag{5.76}$$

Since we normally have $L_p \gg L$, this configuration would appear to decrease the flux sensitivity of the device. This is true, but because the area of the pickup loop A_p is also much larger than that of the SQUID loop (by a factor of order $(L_p/L)^2$), this can give rise to a substantial increase in the sensitivity of the

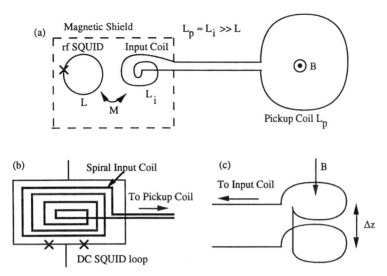

Figure 5.34. Coupling coils for SQUID magnetometers. (*a*) Flux transformer coupled to rf SQUID (inside of magnetic shield) showing large pickup coil and multiturn input coil. (*b*) Multiturn thin-film input coil on top of dc SQUID loop. (*c*) Gradiometer pickup coil with two matched counter-wound coils that couple a flux proportional to the magnetic field gradient dB/dz.

magnetic field in the pickup loop $B = \Phi_p/A_p$. In order to match the input coil to the large coupling coil, while at the same time coupling it strongly with the small SQUID coil, the input coil typically has multiple turns ($L_i \sim n^2$). None of these coils has a magnetic core (it would be too lossy), but reasonably strong coupling ($k \to 0.9$) is possible even in thin-film geometry with a "washer type" coil overlaying the SQUID loop (Fig. 5.34b). In this way, we can use a pickup coil on the centimeter scale, with L_p approaching 1 µH, to match to a SQUID loop on the micrometer scale with L on the pH level. Sensitivity to magnetic fields as low as 1 fT has been achieved with a dc SQUID and ~ 100 fT with an rf SQUID.

One problem with such a SQUID magnetometer is that it is too sensitive to ambient fields and external magnetic noise. There are two ways to deal with this. One is to place the entire system inside magnetic shielding, either superconducting (if possible) or room-temperature ferromagnetic shielding (such as mumetal). An alternative approach is to extend the input flux transformer to include two adjacent pickup coils of the same dimensions, coupling flux with opposite sign (Fig. 5.34c). If these coils are exactly matched, then a uniform magnetic field will introduce no net flux into the loop. In practice, this has the effect of drastically reducing the magnetic signal from any distant source. More precisely, this configuration will effectively measure the gradient of the magnetic field and so is generally called a "gradiometer." A "second-order" gradiometer is also possible using three (or four) properly balanced coils; this will null out both a constant field and its first-order gradient. Using these gradiometer systems, we can effectively measure extremely weak magnetic fields (on the pT scale and below) from close sources, without being completely saturated by much larger background fields. The gradiometer principle has also been used for a SQUID susceptometer, in which a weakly magnetic object is placed inside one of the two coupling coils of a gradiometer. In the presence of a uniform external magnetic field, only the induced magnetic moment of the object (proportional to its magnetic susceptibility) will give rise to a signal, since the field itself is already nulled out.

For best performance, a dc SQUID is normally not operated strictly at dc but rather at a frequency of ~ 100 kHz or higher, since both the SQUID and the room-temperature electronics tend to exhibit some excess $1/f$ noise and drift at low frequencies. Even though 100 kHz is certainly in the rf range, this is still called a dc SQUID, since the name rf SQUID is reserved for the single-junction SQUID. In addition to the static magnetic flux Φ_{dc} supplied by the input coil, a sinusoidal flux $\Phi_{ac}(t) = \Phi_1 \cos(2\pi f_0 t)$ is also supplied to the SQUID via a second coil. The voltage across the SQUID can be expressed in a Taylor series expansion in Φ:

$$V(t) = V[\Phi(t)] = V(\Phi_{dc}) + \left(\frac{\partial V}{\partial \Phi}\right)\Phi_{ac}(t) + \left(\frac{\partial^2 V}{\partial \Phi^2}\right)\left(\frac{\Phi_{ac}^2}{2}\right) \cdots$$

$$= V_0 + V_1 \cos(2\pi f_0 t) + V_2 \cos(4\pi f_0 t) \cdots \quad (5.77)$$

We showed earlier that $V(\Phi)$ is periodic with period Φ_0, but it is clear that the same can also be said for the derivatives. Note also that $\partial V/\partial \Phi$ has a maximum value of order $(I_c R)/(\frac{1}{2}\Phi_0)$, so that for $\Phi_1 \approx \frac{1}{4}\Phi_0$ (giving the maximum response), the second term varies between $\approx \pm \frac{1}{2} I_c R$, very similar in range to that of the first term. So by measuring the first harmonic response at 100 kHz, we can avoid the low-frequency noise and obtain higher resolution. This measurement is generally carried out using a lock-in amplifier (Fig. 5.35), operating at room temperature with conventional electronics, which mixes the voltage $V(t)$ from the SQUID with the reference voltage from the oscillator, to obtain a phase-sensitive dc contribution proportional to the amplitude at f_0. A transformer-coupled input is often used to better match the low output impedance of the SQUID with the high input impedance of the lock-in amplifier or pre-amplifier.

There is one additional important aspect of SQUID measurement techniques. Both dc and rf SQUID magnetometers are generally operated using feedback in the "flux-locked loop" mode. Consider the flux-modulated SQUID circuit of Fig. 5.35, with $\Phi = 0$ initially in the SQUID, so that the dc output voltage $V_{out} = 0$. If we then apply a flux Φ_a to the input coil, that will generate a nonzero V_{out}, which can be used to drive a feedback current that can be coupled back to the SQUID. This, in turn, will induce a dc flux in the SQUID that opposes the applied flux. With sufficient gain in the feedback loop (including both the

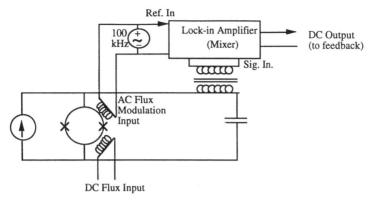

Figure 5.35. DC SQUID magnetometer operated in ac flux-modulation mode. The 100-kHz ac oscillator and lock-in amplifier are generally conventional room-temperature electronic systems. A transformer couples the SQUID output to the lock-in amplifier to provide additional gain and impedance matching. The dc output of the lock-in amplifier may also be used to provide a feedback current that couples flux back into the SQUID loop.

SQUID and the lock-in amplifier), the total flux in the SQUID will remain close to $\Phi_{tot} = 0$. There will, however, be a small error flux proportional to the applied flux Φ_a. The measurement signal is now the feedback current, which is therefore linear with applied flux even much larger than Φ_0. In this way, one obtains both a large dynamic range and high precision for this linear flux-to-voltage amplifier.

In designing circuits using SQUIDs, it is often useful to consider the principle of duality, as discussed earlier in Section 3.4, where one exchanges the roles of current and voltage, electric charge and magnetic flux. One example presented in that section was that of the superconducting flux-flow transistor (SFFT), in which a current in a control line modulates the resistance of a superconducting film by introducing flux into the film. A SQUID can be viewed in a very similar manner: Current in the input circuit introduces flux into the SQUID loop, thereby modulating the resistance of the SQUID. So like the SFFT, the SQUID is functionally dual to the semiconductor field-effect transistor (FET), in which a voltage on the gate introduces charge into the channel, thereby modulating its conductance. These analogies can be carried further by comparing a SQUID magnetometer to an FET electrometer. The former is a low-impedance device (typically a few ohms) that essentially measures magnetic flux; the latter is a high-impedance device ($\gtrsim G\Omega$) that essentially measures charge. For the magnetometer, a flux transformer circuit is used to carry flux from an input coil to the inductively coupled SQUID. For the electrometer, a "charge transporter" circuit is used to carry charge from an input capacitor to the capacitively coupled gate of the FET. In these and many other respects, the relevant circuits associated with these two systems are also dual to one another, and insights from one technology can be used to guide circuit design in the other.

High-quality SQUIDs have been fabricated using both LTS and HTS materials. Most LTS SQUIDs are now made using Nb tunnel junctions with an external resistive shunt to eliminate hysteresis in the junction, combined with thin-film Nb wiring and coupling coils, and are designed to operate at 4 K in liquid helium. The HTS SQUIDs are generally made from YBCO, with step-edge or grain-boundary junctions that are intrinsically overdamped without the need for an external resistive shunt. In addition, multiturn flux transformers have been fabricated using two levels of YBCO films separated by an epitaxial insulating layer (such as strontium titanate). The performance of the best LTS SQUIDs remains somewhat better than that of HTS SQUIDs (see Table 5.2) for several reasons. First, the higher operating temperatures for HTS materials increase the effects of thermal noise. Furthermore, HTS films and junctions tend to exhibit relatively high levels of $1/f$ noise, associated in part with inhomogeneous flux trapping within the films and junctions; this problem is gradually improving as HTS materials technology matures. Nonetheless, for applications where cooling to 4 K is impractical, HTS SQUIDs (cooled to 60–80 K) are already much better than competing technologies for measurement of weak magnetic fields.

In comparing the performance of different SQUIDs, one important figure of merit is the noise energy per unit bandwidth, defined by $\delta\mathscr{E} = \langle\delta\Phi\rangle^2/2L$, where L is the inductance in the SQUID loop. Analysis has shown (Clarke, 1989) that the thermal limit of this noise for a dc SQUID is given approximately by $\delta\mathscr{E} \approx 9kTL/R$ (where R is the shunt resistance of the junctions), and indeed, many dc SQUID systems exhibit noise near this level, at least in the high-frequency regime where the $1/f$ noise is no longer dominant. Typical values in optimized systems at 4 K are of order 10^{-30} J/Hz. Note that the units of $\delta\mathscr{E}$ (J/Hz) are the same as those of Planck's constant ($h \approx 10^{-34}$ J-s), and indeed, further analysis has shown that there is a quantum limit at low temperatures, $\delta\mathscr{E} \sim h$. In fact, careful experiments with optimum dc SQUIDs cooled to near $T = 0$ have achieved noise levels of just a few times this quantum limit.

Table 5.2. SQUID Instrumentation and Typical Sensitivity or Resolution

Magnetometers	
HTS DC SQUID	100 fT
LTS RF SQUID	100 fT
LTS DC SQUID	<10 fT
Scanning SQUID microscope	100 μm
Helmet array (~100 SQUIDs)	1 mm
Susceptometer ($\chi = \mu_0 M/B$)	10^{-11}
Voltmeter	1 pV
Position sensor, strain $\delta L/L$	10^{-18}

Applications of SQUIDs

As we have discussed, a SQUID is an extremely sensitive detector of magnetic flux, and with the proper pickup coil, it also provides a magnetometer of unsurpassed sensitivity (Clarke, 1989, Gallop, 1991). SQUID magnetometer systems are being made commercially by several companies worldwide, and others are under development. A magnetometer is often configured to measure the magnetic moment (in A-m^2) in a region on the centimeter scale; sensitivities down to $\sim 10^{-13}$ A-m^2 can be obtained using a SQUID. We can also express this in terms of equivalent field sensitivities of order 100 fT (10^{-13} T), with sensitivities better than 10 fT for the best LTS dc SQUID systems. For comparison, the field of the earth is about 50 µT, which must generally be reduced or canceled out for maximum sensitivity. Apart from systems designed to be used in a magnetically shielded room, either a gradiometer configuration of the input coil must be used or, alternatively, the background field can be subtracted off digitally in a multi-SQUID system. In fact, a properly designed SQUID magnetometer can measure the field or magnetic moment of a small object even in the presence of a uniform field of order 10 T produced by a superconducting solenoid! This sort of performance is orders of magnitude beyond what can be achieved using a non-superconducting magnetometer.

Magnetic field mapping (in two or three dimensions) can be obtained either with an array of SQUID sensors or alternatively with one or more sensors that scan across the region of interest. For example, a multi-SQUID "helmet" with ~ 100 SQUID sensors can achieve resolution of magnetic sources on the millimeter scale within the human brain. Also, a scanning SQUID microscope has exhibited spatial resolution on the 100-µm scale if the object to be imaged is located in the cryogenic environment with the superconducting pickup loop. A closely related instrument is a SQUID susceptometer, which measures the small induced magnetization in the presence of a moderately large magnetic field.

Furthermore, a SQUID can also provide a sensitive measurement of any other physical quantity that can be transformed to a magnetic flux. For example, a current can be passed through a mutual inductor to couple flux into the SQUID loop; sensitivities to 1 pA or less are possible. Similarly, we can generate such a current from a voltage source, so that a SQUID voltmeter with sensitivity to ~ 1 pV (or less, depending on the source resistance) can be achieved. Since the SQUID output is also a voltage, this makes it clear that the SQUID can also be viewed as a low-noise, high-gain amplifier. This is not limited to dc, but can also be extended well into the rf range, even using the dc SQUID. Another example is an electromechanical position transducer, since a very small motion of a magnet or coupling coil can couple flux into the SQUID loop. This has been applied to measuring extremely small strains in a metal bar (of order 10^{-18}) as part of a research project to observe gravitational waves.

SQUIDs are also being applied to the biomedical and geophysical fields as well as to nondestructive testing (NDT) (Wikswo, 1995). In the biomedical area, the human body is generally nonmagnetic, but very small magnetic fields are generated by low-frequency electrical currents. These currents are due to ionic transport across cell membrances and are associated with both muscular contractions and nerve impulses. These are the same currents that are responsible for the more commonly known electrical voltages due to the heart (electrocardiogram) and brain (electroencephalogram). The corresponding magnetic signals are often referred to as the magnetocardiogram (MCG) and magnetoencephalogram (MEG), with field magnitudes on the picotesla and femtotesla scale (see Table 5.3). The magnetic signals can be detected using SQUID systems, with appropriate gradiometers and cancellation schemes to remove the much larger background. In addition to the spontaneous magnetic output of the brain, a magnetic response may also be evoked by given external stimulus (visual, auditory, tactile, etc.), and this can also be measured. For medical purposes, the key question is whether the magnetic signal provides additional information that is unavailable using more conventional room-temperature voltage measurements. Although these technologies are still being developed, MEG seems to be much better than EEG at localizing a current source in the brain. This is because the presence of an electrically insulating skull distorts the electrical image of such a source but does not affect the magnetic image. Thus, "magnetic source imaging" can be used to determine the location of various functional regions of the brain prior to (or instead of) exploratory brain surgery. Large arrays of SQUIDs in a helmet configuration (to fit around the head) are being developed to address this potential market. There are other developing technologies that may have some similar capabilities (such as functional MRI and positron emission tomography), but SQUID magnetometry is still competitive.

Table 5.3. Some Applications of SQUID Magnetometers and Magnitudes of Typical Sources

Category	System or Measurement	Typical Field
Biomedical	Heart (magnetocardiogram)	1–100 pT
	Brain (magnetoencephalogram)	0.1–1 pF
	Evoked brain response	10 fT
Geophysical	Magnetotellurics	100 fT
	Rock magnetism	$\mu \sim$ pA-m^2
Nondestructive testing (NDT)	Corrosion in metals	1 nT
	Stress and microcracks in steel	1 nT

Note: For comparison, the earth's magnetic field is 50 μT.

SQUIDs have also been applied to geophysical magnetometry (Clarke, 1983). In one application, samples of rock containing magnetic inclusions or impurities are brought into the laboratory and measured in a SQUID magnetometer. These can provide information on the prior magnetic history of the rock, including the polarity of the earth's magnetic field when the rock was formed. Another class of applications is conducted out in the field and provides "magnetic sounding" information that can be used in the search for groundwater and for mineral and petroleum deposits. The key distance here is the frequency-dependent electromagnetic skin depth $\delta = 1/\sqrt{\pi\mu_0\sigma f}$, which is typically of order 1 km for $f = 1$ Hz for a ground conductivity $\sigma \sim 1$ S/m. There are naturally occurring ultra-low frequency "magnetotelluric waves" originating in ionospheric disturbances, which propagate toward the ground. A SQUID magnetometer system (or several such systems operating at different locations) can be used to measure the amplitude of these waves, which enables one to unfold the depth-dependent effective conductivity $\sigma(z)$ of the ground. The same thing can be done at higher frequencies using artificial radio-wave sources. Although the need for cryogenics in distant locations might seem to be a barrier to application, low-loss dewars and superior sensitivity may give SQUID magnetometers an edge in this developing application.

Finally, there are a variety of applications in NDT where SQUID magnetometers may develop a niche. One such application is in the detection of corrosion, which as a chemical process requires ionic transport between different materials (as in a battery). This gives rise to small electrical currents, which in turn produce small magnetic fields. Laboratory experiments have shown that SQUID gradiometers are effective at detecting such corrosion-induced fields (of order 1 nT or less), even ~ 1 cm underneath other layers. Research has also indicated that defects (such as microcracks and regions under stress) may have a characteristic magnetic signature in a material such as steel. In some cases, in either magnetic or nonmagnetic steel, the stray magnetic field at the surface of such a defect would be very small, and a SQUID magnetometer mapping out this field may be useful in detecting such a defect before it leads to a more serious structural failure.

5.4 DISTRIBUTED JUNCTIONS AND ARRAYS

We have been regarding a Josephson junction as a lumped element, but in many cases the finite length of the junction is significant, even if it is typically on the micrometer scale. This is particularly true when we consider the effect of a magnetic field. The supercurrents in a junction are much smaller than those within a strong superconductor, but in some cases these Josephson currents can also screen a small magnetic field and permit a larger field to penetrate in the form of quantized vortices. This leads to Josephson flux-flow devices analogous to those for a homogeneous thin film. Finally, some of these same principles can also be applied to devices based on coupled arrays of Josephson junctions, with application to microwave oscillators and flux-transfer devices. Some of these will also form the basis for logic circuits to be described in Chapter 6.

Josephson Penetration Depth

As for the case of a superconductor itself, the junction can be treated as a distributed element within a transmission line picture. Consider a classic tunnel junction with geometry as shown in Fig. 5.36a, with two superconducting electrodes of thickness t, length l, and width w separated by an insulating barrier of thickness d. Assume uniformity along the width. This is the same geometry as the superconducting transmission line that we discussed earlier in Section 2.3, except that there may now be tunneling currents through the barrier; this is sometimes known as a Josephson transmission line. We assume that a magnetic field points along the y direction, a voltage is applied in the x direction, and a wave may propagate in the z direction. Then we can represent the junction using the modified transmission line picture of Fig. 5.36b, where the shunt elements are simply the distributed components of the CRSJ model (the Josephson current together with the shunt conductance and capacitance), and the series inductance per unit length L (in H/m) is the same as that for the transmission line:

$$L = \frac{\mu_0(2\lambda + d)}{w} \qquad (5.78)$$

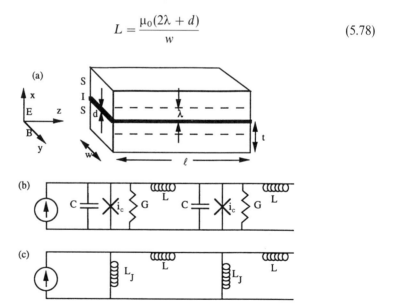

Figure 5.36. Model for long Josephson junction. (*a*) Junction geometry similar to superconducting transmission line but with small electrode separation $d \sim 1$ nm to permit Josephson currents between electrodes. (*b*) Distributed circuit equivalent of Josephson transmission line. (*c*) Low-frequency linearized equivalent of (*b*) for currents well below the critical current, which gives rise to Josephson penetration depth $\lambda_J = \sqrt{L_J/L}$.

where we are assuming here that the film thickness $t \gg \lambda$, the magnetic penetration depth. The distributed Josephson element is given by a current per unit length $i = i_c \sin \phi$ where $i_c = J_c w = I_{c0}/l$ is the critical current per unit length, assumed here to be uniform. The series resistance in the superconducting electrodes (within a two-fluid picture) could also be included, but the loss in the junction is normally dominant.

Before addressing the general solution, let us focus first on the very low frequency properties for small bias currents. Then we can approximate the distributed junction by its equivalent Josephson inductance per unit length (in H-m) (Fig. 5.36c)

$$L_J = \frac{\hbar}{2e J_c w} \qquad (5.79)$$

This is the same sort of series/parallel inductance network that gave rise to the magnetic penetration depth in Section 2.1, and the same relation can be obtained here. For a series impedance per unit length $Z = j\omega L$ and a shunt admittance per unit length $Y = 1/j\omega L_J$, the propagation constant is

$$\gamma = \sqrt{ZY} = \sqrt{\frac{L}{L_J}} = \frac{1}{\lambda_J} \qquad (5.80)$$

where the effective "Josephson penetration depth" λ_J is given

$$\lambda_J = \sqrt{\frac{L_J}{L}} = \sqrt{\frac{\Phi_0}{2\pi J_c \mu_0 (2\lambda + d)}} \qquad (5.81)$$

independent of frequency down to dc. To estimate a typical value, let us assume that $\lambda = 100$ nm (so that $d \sim 1$ nm is negligible) and $J_c = 10^7$ A/m² = 10 µA/µm². Then $\lambda_J \approx 10$ µm, and indeed we have been discussing junctions only a few micrometers across.

As with the usual superconducting penetration depth, λ_J determines the length of the junction associated with superconducting screening currents and magnetic flux penetration. The reason that $\lambda_J \gg \lambda$ is that the Josephson currents are so much weaker than the screening currents in the superconductor itself. In the usual "short-junction" case with $l, w \ll \lambda_J$, our previous analysis as a lumped element should generally continue to remain valid.

Field Suppression of Critical Current in Short Junctions

There is one situation in which the distributed nature of the Josephson junction is important even for a short junction, and that is in the effect of an applied magnetic field on the critical current of the junction. In this case, the external magnetic field will penetrate the entire junction uniformly (Fig. 5.37a) since

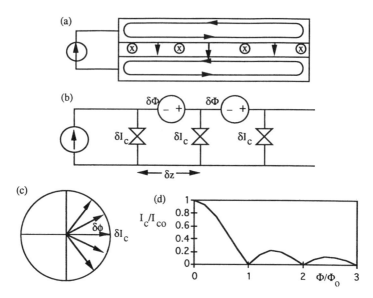

Figure 5.37. Model for short Josephson junction in magnetic field. (*a*) Configuration of fields and currents. (*b*) Distributed circuit model with uniform flux $\delta\Phi$ in each differential loop. (*c*) Josephson phasor picture with *N* differential elements with phase $\delta\phi$ between each element. (*d*) Dependence of total junction critical current I_c on total flux Φ in junction.

screening currents are ineffective. Following our earlier analysis of the effect of a magnetic field on a dc SQUID loop, we can model the system with the distributed diagram of Fig. 5.37*b*. Here, for a differential length δz, an equal flux $\delta\Phi$ is applied to each loop:

$$\delta\Phi = B(2\lambda + d)\,\delta z \approx 2\lambda B\,\delta z \tag{5.82}$$

This, in turn, implies a phase difference between each adjacent pair of junctions,

$$\delta\phi = \frac{2\pi}{\Phi_0}\delta\Phi = \frac{4\pi\lambda B}{\Phi_0}\delta z \tag{5.83}$$

with a total phase difference across the junction,

$$\Delta\phi = \frac{2\pi}{\Phi_0}2\lambda l B \tag{5.84}$$

corresponding to total flux $\Phi = 2\lambda l B = N\delta\Phi$ in the junction, where N is the total number of differential junctions. Since the junctions along the length will not all be in phase, the maximum supercurrent will be reduced from the value $I_{c0} = J_c w l$ that it would have in the absence of the applied field. Clearly, this effect will become significant when $\Delta\phi$ approaches 1.

We can view this more quantitatively (Fig. 5.37c) by generalizing our "Josephson phasor" picture used earlier for the dc SQUID (see Fig. 5.30). Each differential length δz will have a critical current $\delta I_c = J_c w\, \delta z$. The total current can be expressed as

$$I(\Phi) = \delta I_c \sum_{i=1}^{N} \sin \phi_i = \delta I_c \operatorname{Im}[e^{j\phi_1}(1 + e^{j\delta\phi} + \cdots + e^{jN\delta\phi})] \quad (5.85)$$

The total critical current I_c is then the maximum value of this sum:

$$I_c(\Phi) = \delta I_c |1 + \exp(j\delta\phi) + \exp(2j\delta\phi) + \cdots + \exp(Nj\delta\phi)| \quad (5.86)$$

This has the form of a standard geometric series, which has closed-form sum:

$$S = 1 + x + x^2 \cdots + x^N$$
$$xS = x + x^2 \cdots + x^N + x^{N+1}$$

so that

$$S = \frac{1 - x^{N+1}}{1 - x} \quad (5.87)$$

In the present case, $x = \exp(j\delta\Phi)$, so that

$$I_c(\Phi) = \frac{\delta I_c |1 - \exp(Nj\delta\phi)|}{|\exp(j\delta\phi)|} = \frac{\delta I_{c0} |\sin(\Delta\phi/2)|}{|\sin(\delta\phi/2)|} \quad (5.88)$$

In the limit that δz goes to zero, so does $\delta\Phi$, and we can express I_c in the form

$$I_c(\Phi) = \frac{I_{c0}|\sin(\Delta\phi/2)|}{|\Delta\phi/2|} = \frac{I_{c0}|\sin(\pi\Phi/\Phi_0)|}{|\pi\Phi/\Phi_0|} \quad (5.89)$$

This functional dependence (shown in Fig. 5.37d) is the same as the antenna pattern for a linear phased array or the optical intensity pattern from diffraction through a single slit; in all of these cases, we have a linear array of sources with a phase delay between them. This result can also obtained by transforming the sum in Eq. (5.85) to an integral:

$$I_c(\Phi) = \left| \int_0^l w J_c(z) \exp[j\phi(z)]\, dz \right| \quad (5.90)$$

where $\phi(z) = kz = z(4\pi\lambda B/\Phi_0)$. In this form, it is clear that this is the spatial Fourier transform of the function $i_c(z) = J_c w$ inside the junction and zero

230 JOSEPHSON DEVICES

outside. In a sense, we can think of the magnetic field as modulating the phase in space (with wavelength $2\pi/k = \Phi_0/2\lambda B$) in much the same way that a voltage across the junction modulates the phase in time (with Josephson period Φ_0/V).

This function has its first zero for $\Delta\phi = 2\pi$, which corresponds to one flux quantum Φ_0 in the junction. It continues to oscillate, but the amplitude rolls off as $1/B$ for larger fields. For typical values $\lambda = 100$ nm and junction scale $l = 2$ μm, the first zero corresponds to a field $B = \Phi_0/2\lambda l \sim 5$ mT, as compared to the earth's ambient field ~ 50 μT.

We assumed above that J_c is constant, but J_c may sometimes vary significantly across a real junction, particularly with the newer high-T_c superconductors. In that case, $I_c(\Phi)$ is proportional to $|F(i_c(z))|$, the one-dimensional Fourier transform of the critical current per unit length. Because of the absolute value, this cannot be uniquely inverted (the phase information has been lost), but significant information on the spatial distribution of $i_c(z)$ may be obtained from $I_c(\Phi)$. Also, this analysis is not limited to a rectangular junction; more generally, the junction width $w(z)$ would be a function of the position z along the junction, and $i_c(z) = J_c(z)w(z)$.

Consider, for example, a superconducting transmission line that has Josephson junctions near the two ends (Fig. 5.38a). This is actually a dc SQUID, even though the coupling loop is rather small. Each junction is of length l, and the two junctions are a distance L apart. We can model this as a single junction with a critical current distribution (Fig. 5.38b)

$$i_c(z) = \begin{cases} 0 & |z| < \tfrac{1}{2}(L-l) \\ J_c w & \tfrac{1}{2}(L-l) < |z| < \tfrac{1}{2}(L+l) \end{cases} \tag{5.91}$$

The Fourier transform of this distribution is then

$$F[i_c(z)] = J_c w \left\{ \int_{(L-l)/2}^{(L+l)/2} dz \, \exp(jkz) + \int_{(-L-l)/2}^{(-L+l)/2} dz \, \exp(jkz) \right\}$$

$$= 2J_c w \int_{(L-l)/2}^{(L+l)/2} dz \, \cos(kz) = \frac{2J_c w}{k} [\sin(\tfrac{1}{2}k)(L+l) - \sin(\tfrac{1}{2}k)(L-l)]$$

$$= \frac{4J_c w}{k} \cos(\tfrac{1}{2}kL)\sin(\tfrac{1}{2}kl) \tag{5.92}$$

So we have

$$I_c(\Phi) = \frac{2I_{c0}|\cos(\pi\Phi_L/\Phi_0)| \, |\sin(\pi\Phi_l/\Phi_0)|}{|\pi\Phi_l/\Phi_0|} \tag{5.93}$$

where $\Phi_L = 2\lambda LB$ is the flux in the main loop, $\Phi_l = 2\lambda lB$ is the flux in each junction, and $I_{c0} = J_c wl$ is still the critical current of each junction in zero

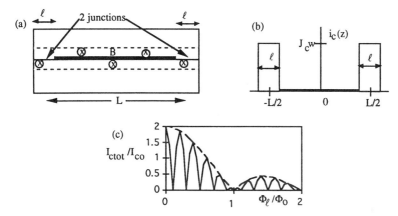

Figure 5.38. Model for magnetic flux dependence of dc SQUID with extended junctions. (a) Geometry of superconducting transmission line with Josephson junctions (of length l) at two ends separated by distance L. Critical current distribution $i_c(z)$ in equivalent single junction. (c) Fourier transform $|F(i_c(z))|$ of (b) corresponding to flux dependence of critical current of SQUID for $L = 5l$. The envelope (dashed line) shows the critical current of the junctions.

magnetic field. This relation is shown in Fig. 5.38c for the case where $L = 5l$. This is clearly the product of the two-junction SQUID interference modulation and the single-junction diffraction modulation. [This result would follow more directly from the convolution theorem of Fourier transforms, since $i_c(z)$ is actually the convolution of two narrow junctions with a broadening function for each junction.] The more general SQUID loop would be opened up much more than this, but the principle is the same. As long as the the field is not large enough to significantly suppress I_c of an individual junction (i.e., $\Phi_l \ll \Phi_0$), the earlier treatment of a SQUID should remain valid.

Incidentally, a magnetic field that suppresses the critical current of a Josephson junction may have little effect on the energy gap of the superconducting electrodes, or on the quasiparticle tunneling current, since these do not depend on the phase of the pair wave function. For example, the field ~ 100 mT from a close permanent magnet can provide strong suppression of I_c of a junction, while not significantly affecting the sharp rise in the current at $V = 2\Delta/e$. A similar effect is evident in a granular superconductor, that is, a superconductor that consists of a packed array of superconducting grains weakly coupled via Josephson junctions. A relatively modest magnetic field can act to kill the macroscopic superconductivity, even while the separate grains themselves may remain strictly superconducting. This is particularly notable for ceramic HTS samples; despite the very large upper critical field of the grains, the transport critical current of the assembly can be remarkably small.

We have been describing the situation for a short junction below I_c, but there are similar effects in the voltage state above the critical current. Here, in addition to the supercurrent J_s, there is also a normal current J_n associated with the shunt

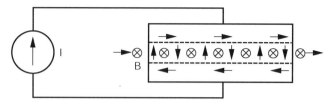

Figure 5.39. Josephson flux flow in a short junction with flux $\Phi \approx 3\Phi_0$ in the junction and magnetic field pointing into the page. For applied current $I > I_c$, the Josephson vortices move to the right due to the Lorentz force ($J \times B$) of the dc current pointing down.

conductance. For a fixed dc voltage, J_s at each position oscillates sinusoidally at the Josephson frequency. For a junction containing one or more flux quanta, the pattern of spatially alternating supercurrent in the junction will move along the junction in a direction that is consistent with the Lorentz force on vortices (Fig. 5.39). For this reason, it is sometimes helpful to think of these supercurrent patterns as a close-packed linear array of "Josephson vortices," even though the magnetic field is uniform in the junction. This is analogous to flux flow in a type II superconductor in a large magnetic field (near H_{c2}). This is also consistent with the fluxonic picture of a Josephson junction introduced earlier in Section 5.1, where one fluxon enters and leaves the junction traversely each Josephson period.

Long Josephson Junction and Josephson Solitons

As noted above, a long Josephson junction has a transverse dimension $l \gg \lambda_J$. As with an ordinary superconductor thicker than λ, such a junction can screen out a magnetic field by surface screening currents. In addition, the transport current will tend to flow near the outer perimeter of such a junction, on the scale of λ_J. This means, for example, that the effective critical current I_c of such a junction will be less than $J_c lw$, by a factor of order λ_J/l. (Since the quasiparticle current I_n and displacement current I_d are unaffected, the shunt conductance G and capacitance C would be unchanged.) In fact, this relation puts an effective upper limit on the critical current of a Josephson junction, at least if we want to take full advantage of uniform supercurrent flow. If we estimate that the optimum junction will be circular with radius λ_J, then from Eq. (5.79) we have

$$I_{c,\max} \approx \pi \lambda_J^2 J_c I_c = \frac{\Phi_0}{8\pi\mu_0\lambda} \approx 2 \text{ mA} \tag{5.94}$$

for $\lambda \approx 100$ nm. We cannot improve on this either by increasing the size of the junction or by increasing J_c.

However, a long Josephson junction has another mode of operation that may offer some possibilities for application. Like a type II superconductor, a long Josephson junction can reduce its energy by permitting flux penetration into its interior via discrete quantized vortices, above a characteristic lower critical field $H_{c1} = B_{c1}/\mu_0$. A more complete analysis (Orlando and Delin, 1991, p. 436) shows that in this case $B_{c1} = 4\mu_0 J_c \lambda_J/\pi \approx \Phi_0/\pi^2 \lambda_J \lambda$, similar to the formula for an anisotropic type II superconductor. For typical values $J_c \sim 1000$ A/cm^2 and $\lambda \sim 100$ nm, this yields $B_{c1} \sim 200$ µT = 2 G, a rather small field. Furthermore, we will show below that such a Josephson vortex concentrates the magnetic flux within a distance $\sim \lambda_J$ along the junction and λ into the electrodes and will again respond to a transverse Lorentz force and propagate in a long junction, producing flux-flow resistance.

To determine how such a traveling Josephson vortex solution can arise in a long junction, let us go back to the general Josephson transmission line picture of Fig. 5.36b. First consider the regime below the critical current, where the Josephson element can be approximated by a simple linear inductance per unit length (in H-m) $L_J \approx \Phi_0/2\pi w J_c$ and let us neglect the lossy shunt conductance. At low frequencies, as we indicated earlier, an electromagnetic wave cannot propagate down such a line. Only at frequencies above the shunt $L_J C$ resonance will the shunt admittance look capacitive and a wave can again propagate. This characteristic frequency (the same as ω_0 for a short Josephson junction) is called the Josephson plasma frequency

$$\omega_J = \frac{1}{\sqrt{L_J C}} \approx \sqrt{\frac{2\pi d J_c}{\varepsilon \Phi_0}} \tag{5.95}$$

which for typical values gives a frequency $f_J = \omega_J/2\pi \sim 100$ GHz. This is analogous to the situation in a metal, which becomes transparent for frequencies above the ultraviolet [see, e.g., Fig. 2.1 and the note above Eq. (2.5)]. The difference in magnitude reflects the much lower effective density of carriers in the junction. At high frequencies $f \gg f_J$, waves can propagate as on a regular superconducting transmission line, with a velocity

$$u = \frac{1}{\sqrt{LC}} = \lambda_J \omega_J \approx \frac{c}{\sqrt{\varepsilon_r}} \sqrt{\frac{d}{2\lambda}} \tag{5.96}$$

where $c = 1/\sqrt{\mu_0 \varepsilon_0}$ is the speed of light and $\varepsilon_r = \varepsilon/\varepsilon_0$ is the relative dielectric constant. Because of the very small value of d (typically ~ 2 nm), this velocity is a factor of 10 below the usual electromagnetic velocity on a transmission line with a similar dielectric; typical values are $u \sim 10^7$ m/s.

But we must consider the complete nonlinear equation to account for the propagating Josephson vortex. The circuit equations for the Josephson transmission line can be written in terms of the voltage $V(z)$ across the junction and the current $I(z)$ along the electrodes:

$$\frac{\partial I}{\partial z} = C\frac{\partial V}{\partial t} + GV + i_c \sin\phi + i \tag{5.97}$$

$$\frac{\partial V}{\partial z} = L\frac{\partial I}{\partial t} \tag{5.98}$$

where Eq. (5.97) has been generalized to permit a distributed feed current i per unit length into the junction. The voltage, phase, and flux are related in the usual way by

$$V = \frac{d\Phi}{dt} = \frac{\Phi_0}{2\pi}\frac{d\phi}{dt} \tag{5.99}$$

Then, Eq. (5.98) can be integrated and rewritten as

$$\frac{\partial \Phi}{\partial z} = \frac{\Phi_0}{2\pi}\frac{\partial \phi}{\partial z} = LI$$

which can then be substituted into Eq. (5.97) to yield the following second-order nonlinear partial differential equation in $\phi(z, t)$:

$$\lambda_J^2 \frac{\partial^2 \phi}{\partial z^2} = \omega_J^{-2}\frac{\partial^2 \phi}{\partial t^2} + \sin\phi + \frac{i}{i_c} + \frac{Q}{\omega_J}\frac{\partial \phi}{\partial t} \tag{5.100}$$

where $Q = \omega_J C/G$ is the same factor defined for the lumped Josephson junction in Section 5.2.

The only nonlinear term in Eq. (5.100) is $\sin\phi$, and this is precisely what gives rise to the discrete Josephson vortex, also known as a "soliton." The simplest nonlinear solution occurs in the static case, where there is no drive current or dissipation, so that

$$\lambda_J^2 \frac{\partial^2 \phi}{\partial z^2} = \sin\phi \tag{5.101}$$

A solution to this takes the form (Orlando and Delin, 1991, p. 433)

$$\phi(z) = 4\arctan\left[\exp\left(\frac{z}{\lambda_J}\right)\right] - \pi \tag{5.102}$$

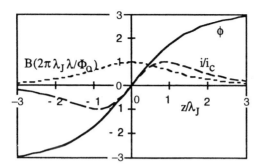

Figure 5.40. Static soliton in long Josephson junction showing spatial dependence of phase difference ϕ, supercurrent i/i_c, and magnetic field in junction $B(2\pi\lambda_J\lambda/\Phi_0)$, from Eqs. (5.102), (5.103), and (5.104).

which can be confirmed by direct substitution. This is plotted in Fig. 5.40, together with the supercurrent $i(z)$ and the magnetic field $B(z)$, given by

$$i(z) = i_c \sin\phi = 2i_c \operatorname{sech}\left(\frac{z}{\lambda_J}\right)\tanh\left(\frac{z}{\lambda_J}\right) \tag{5.103}$$

$$B(z) = \frac{\partial\phi}{\partial z}\frac{\Phi_0}{4\lambda} = \operatorname{sech}\left(\frac{z}{\lambda_j}\right)\frac{\Phi_0}{2\pi\lambda_J\lambda} \tag{5.104}$$

As the figure suggests, this represents an isolated vortex at $z = 0$, with currents and fields localized in a region of order λ_J on either side. Note that ϕ changes by 2π in crossing the soliton and $B(0)(2\lambda)(2\lambda_J) \approx \Phi_0$, as one would expect for a vortex. In contrast to an Abrikosov vortex in a type II superconductor, there is no "normal core" on the scale of the coherence length. Note that the amplitude of this solution is *not* arbitrary, unlike the situation for a linear equation.

Now we can go one step further and include the time dependence:

$$\lambda_J^2 \frac{\partial^2\phi}{\partial z^2} = \omega_J^{-2}\frac{\partial^2\phi}{\partial t^2} + \sin\phi \tag{5.105}$$

This equation is sometimes known as the "sine-Gordon" equation and can represent other physical systems as well as the long Josephson junction. A solution to this can be obtained by modifying the static solution Eq. (5.102) to a traveling-wave solution (Pedersen, 1991):

$$\phi(z,t) = 4\arctan\left[\exp\left(\frac{\gamma(z-vt)}{\lambda_J}\right)\right] \tag{5.106}$$

where

$$\gamma = \frac{1}{\sqrt{1 - v^2/u^2}} \qquad (5.107)$$

One can think of this fluxon as a particle moving with an arbitrary velocity v, as is consistent with the argument $z - vt$. The γ-factor has the same form as the Lorentz contraction of a relativistic particle, and indeed, $u = \lambda_J \omega_J$ is the speed of light in the tunnel barrier. This has the effect of limiting the maximum speed of this soliton to u in very high-Q junctions. We can also take the derivative of Eq. (5.106) to obtain the equivalent voltage pulse:

$$V(z,t) = \frac{\Phi_0}{2\pi} \frac{\partial \phi}{\partial t} = \frac{\Phi_0}{\pi} \frac{\gamma v}{\lambda_J} \operatorname{sech}\left[\frac{\gamma(z - vt)}{\lambda_J}\right] \qquad (5.108)$$

This represents an SFQ pulse moving with velocity v, with characteristic time scale of order $\lambda_J/u \sim 1/\omega_J \sim 1$ ps for a typical junction (see Fig. 5.41a). For large γ, the pulse becomes both narrower and higher, maintaining the same intergrated flux Φ_0.

This result was somewhat oversimplified, since we ignored both the drive current and the resistive loss. The drive current will tend to accelerate the fluxon, and the shunt resistance will slow it down. Taken together, these will lead to a steady-state fluxon velocity that is a function of the drive current $V(I)$ but is

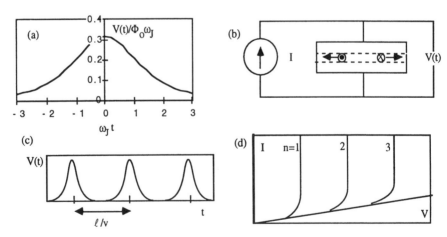

Figure 5.41. Soliton motion in long Josephson junction. (*a*) Single-flux-quantum voltage pulse $V(t)$ from soliton moving with velocity $v = u/\sqrt{2}$, where u is the velocity of light in the junction. (*b*) A propagating fluxon can reflect from the end of the junction as an antifluxon (and vice versa), corresponding to a voltage pulse of the same polarity. (*c*) Periodic pulse train corresponding to trapped fluxon. (*d*) Schematic dc *I–V* curves for low-loss junction showing zero-field steps (multiple hysteretic branches) corresponding to one or more fluxons trapped in junction.

otherwise similar to the SFQ pulse obtained above. In discussing the above solution, we have implicitly assumed that the junction is infinitely long, or at least that the ends can be ignored. It is possible to construct a toroidal junction, so that the fluxon can continue to circle the junction indefinitely. More commonly, there is a rectangular junction with a finite length $l \gg \lambda_J$. When a fluxon hits the open end of a junction, it can reflect as a antifluxon (i.e., with the opposite magnetic flux) moving in the opposite direction. This bouncing back and forth will continue indefinitely as long as the drive current continues. Each traversal (in either direction) corresponds to an SQF voltage pulse of the same polarity (Fig. 5.41b) at a repetition frequency $f = v/l$, leading to a dc average voltage $V = \Phi_0 v(I)/l$. A schematic $V(I)$ relation for a low-loss junction is shown in Fig. 5.41c and shows a characteristic saturation effect for large currents. This is due to the vortex velocity saturating at the light-speed velocity u, as described above. For $u \sim 10^7$ m/s and $l \sim 100$ μm, the saturation voltage $V \sim 200$ μV. A similar set of characteristics are present for multiple fluxons simultaneously present in the junction (see Fig. 5.41d); the saturation voltage is proportionately larger. These branches of $V(I)$ are sometimes called "zero-field steps" since they may be present in the absence of an applied magnetic filed.

We had earlier introduced a pendulum analog to describe a single Josephson junction and a dc SQUID, and a long junction can likewise be represented by a linear array of pendula coupled by torsional springs (Fig. 5.42). The Josephson soliton is then a localized region over which there is a 360° rotation in the pendulum angle, and such a feature can clearly propagate down the line and reflect back and forth off the ends of the line. This also suggests that a parallel array of discrete short Josephson junctions might show similar effects. We will address this further in the discussion below on Josephson junction arrays.

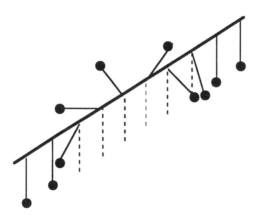

Figure 5.42. Representation of Josephson soliton in long junction as a linear array of pendula coupled by torsional springs. The angle of the pendulum from the vertical corresponds to the local phase difference across the junction. The soliton corresponds to the 360° phase shift, which may move along the array.

Applications of Long Josephson Junctions

Several potential applications have been proposed for long Josephson junctions, and prototypes are under development. These fall into two major classes: oscillators and amplifiers. The initial motivation for oscillator development (Pedersen, 1991) follows essentially from the periodic pulse train in Fig. 5.41c. This can have a significant component in the fundamental frequency, which for typical parameters ($u \sim 10^7$ m/s and $l \sim 100$ μm) is of order 100 GHz. Analysis has shown that the total amount of power that may be coupled out of such a long junction is low, typically on the order of nanowatts. This is not much more than can be obtained from a short Josephson junction, which was discussed earlier. However, the long junction has an advantage in producing a much narrower radiation linewidth, which is a key problem for the short junction.

A detailed analysis (Likharev, 1986) has shown that a broadened frequency spectrum for either type of junction follows from fluctuations in low-frequency bias current, which in turn produces fluctuations in voltage and (by the ac Josephson relation) frequency. This analysis further indicates that the linewidth Δf is proportional to R_d^2, where $R_d = dV/dI$ is the dynamic resistance at the operating point. A resistively shunted junction (RSJ) biased above the critical current may have $R_d \sim 10\ \Omega$; in contrast, a long junction in the velocity–saturation regime (see the constant-voltage steps in Fig. 5.41d) may have $R_d \ll 1\ \Omega$. For this reason, the linewidth Δf at 100 GHz for might be >10 MHz for the short junction and <10 kHz for the long junction. This latter value is highly promising for application to a narrow-band superconducting on-chip microwave receiver, such as that based on an SIS mixer. The frequency may also be tuned somewhat by application of a magnetic field. An alternative approach to obtaining a narrow linewidth is available using a coupled array of discrete short junctions, as discussed in the section below.

Another focus for the application of long Josephson junctions has been the attempt to develop a simple, compact, three-terminal superconducting device that exhibits significant current gain, essentially an amplifier (Nordman, 1995). Such a device could be used as a versatile building block in superconducting circuits, in much the same way as a transistor (either bipolar or field effect) is used in semiconductor circuits. We pointed out earlier that a SQUID properly configured can indeed be an amplifier, but the search for alternatives has continued. One approach using long junctions is similar to the superconducting flux-flow transistor (SFFT) introduced earlier in the context of Abrikosov vortices in a superconducting film. Here, current in a control line injects Josephson vortices (i.e., solitons) into a long junction (Fig. 5.43). There are several regimes of operation of such a device, depending in part on the Q of the junction, with names such as "current-injection transistor" (CIT) and "Josephson vortex flow transistor" (JVFT). Under optimum device geometries and conditions, the current gain can approach $l/2\lambda_J$ (which can be $\gg 1$), and

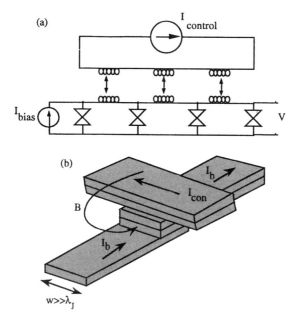

Figure 5.43. Schematic (a) and simplified layout (b) of Josephson flux-flow transistor. The magnetic field produced by control line couples flux (in the form of Josephson vortices) into the long Josephson junction.

the gain–bandwidth product can approach the Josephson plasma frequency $f_J \sim 100$ GHz. Both LTS and HTS devices have been demonstrated, but the effort to develop improved devices continues.

Discrete Josephson Array Oscillators

Once we have a technology that can make large numbers of Josephson junctions reproducibly and reliably (thus far, mostly with LTS), it becomes feasible to consider circuits that link many of these into a single device structure. We already mentioned one such application earlier in Section 5.1 (Fig. 5.12): the Josephson voltage standard, in which thousands of junctions locked on a Shapiro step are connected up in series to achieve a single (much larger) dc voltage. In the next chapter, we will show how Josephson junctions can be linked together to make digital logic circuits. But in the remaining section here, let us illustrate several less complex circuits, mostly analog, that can be achieved using regular arrays of Josephson junctions.

One potential application is for a microwave power source (Lukens, 1990). The Josephson junction is an ideal voltage-controlled oscillator, but as we described above, a single junction (even a long junction) has far too small an output power for many applications. If we can achieve coherent radiation of a large number (N) of Josephson functions, then this will help in two ways. First, if we can connect them in series, then their voltage amplitudes V_i will add up in phase, leading to a power $N^2 V_i^2/2R_L$ delivered to a load resistance R_L. This should be valid until the total series resistance NR_i approaches R_L (these are generally RSJs, with resistance R_i). Since the effective resistance of individual junctions can be quite small compared to a typical load resistance or antenna structure, this can lead to quite a substantial increase in total power, from <1 nW to >1 µW. For example, an array of $N = 50$ junctions, each with $I_c = 1$ mA and $R_i = 2\,\Omega$ (corresponding to superconducting Nb), might be coupled to a matched transmission line or antenna with $R_L = 100\,\Omega$. The power coupled to the load would be expected to be [cf. Eq. (5.56) for a single junction]

$$PL = \frac{(\tfrac{1}{2}NI_c R_i)^2}{2R_L} = \tfrac{1}{8}NI_c^2 R_L \sim 10\ \mu\text{W} \tag{5.109}$$

An additional advantage of such an array is that if the noise sources causing linewidth broadening are incoherent, then the radiation linewidth Δf should also narrow with increasing N. A more detailed analysis has shown that $\Delta f \sim 1/N$, which can bring the linewidth down to ~ 100 kHz or less for a 100-GHz source.

In order to achieve the required coherence among all the junctions of an array, careful attention must be paid to the biasing scheme. The most obvious scheme (although not the most robust) is to bias all the junctions in a series linear array, as indicated in Fig. 5.44a. If all of the junctions are identical, then they will all exhibit the same voltage drop and hence oscillate at the same Josephson frequency. Their relative phases would still tend to be random; an external phase reference is necessary to produce coherence. This could be achieved in principle using a single "feedback impedance" Z_f across the entire array, large enough so as not to divert much of the bias current away from the junctions (or to shunt the rf oscillation from the load). The circulating ac feedback current can provide the necessary phase reference to phase-lock the junctions.

However, in the more likely case that the junctions exhibit some variation in I_c and R, the common bias current will lead to a range of voltages (and hence frequencies) for the various junctions. The feedback current will tend to pull the frequencies together, but it may not be sufficient. An alternative biasing scheme has all of the junctions in parallel, as shown in Fig. 5.44b. This has the advantage that all of the junctions will then have identical voltages, so that they will oscillate at the same frequency, even with a dispersion of junction parameters. On the other hand, an important disadvantage is that the effective rf impedance of this array is reduced by a factor of N, making it difficult to couple significant power into a typical load resistance $R_L \sim 100\,\Omega$. Furthermore, an additional complication is that each pair of junctions is actually a dc SQUID loop. In the voltage state, the relative phases of the oscillations of adjacent junctions depend on the flux in the loop.

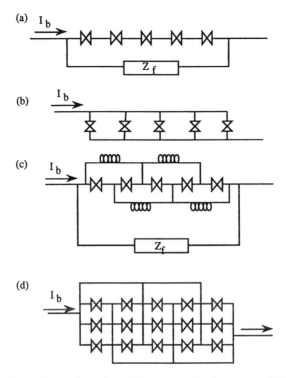

Figure 5.44. Biasing schemes for coherent Josephson junction arrays. (*a*) Series biasing of linear array with feedback impedance across array to maintain phase coherence. (*b*) Parallel biasing of linear array to assure oscillation at the same frequency; note that each adjacent pair is a SQUID loop. (*c*) Parallel–series biasing of linear array: the dc bias is in parallel, but the rf voltages are effectively in series. (*d*) Parallel–series biasing of two-dimensional array.

An alternative series–parallel scheme that combines the advantages of series and parallel biasing is often preferable. As illustrated in Fig. 5.44*c*, the junctions are arranged along a single line corresponding to a series arrangement. However, there is also a high-inductance superconducting line around each adjacent pair of junctions corresponding to alternating dc voltage across adjacent junctions. For an odd total number of junctions, this is equivalent to parallel dc biasing, but the rf voltage is coupled out in series (it is blocked by the inductors). The junctions will now all oscillate at the same frequency, although not necessarily in phase. If an appropriate feedback impedance is added, they will tend to phase-lock as well (although magnetic field sensitivity in the SQUID loops is still a concern). Using this type of design, up to hundreds of junctions have been integrated into a single microwave source, producing up to ~ 1 μW at frequencies ~ 100 GHz and above.

One important issue that we have not discussed is how this radiation is coupled out to a potential load. This can be accomplished using either a transmission line on-chip, or alternatively an antenna if the radiation is to be transmitted through free space. We have implicitly assumed that the Josephson array is much smaller than a wavelength, but this may not always be the case as the frequency increases. For $f = 100\,\text{GHz}$, for example, the wavelength $\lambda = 3\,\text{mm}$ in free space, and less in typical dielectrics. If the array is larger than about $\frac{1}{4}\lambda$, the phase shifts along the array must be taken into account in a proper design if full coherence is to be maintained.

Two-dimensional Josephson arrays have also been fabricated, although the conditions for coherent radiation are then even more complex. A series–parallel biasing scheme may also be used in this case (see Fig. 5.44d). The necessary phase reference here may be achieved by coupling to an appropriate resonance of the radiation field. Such two-dimensional arrays with hundreds of junctions have also yielded up to $\sim 1\,\mu\text{W}$ in the 100–300-GHz range, in some cases radiated into free space with an antenna structure.

Discrete Josephson Transmission Line

As a final device in this section, let us consider the discrete analog of the long Josephson junction. This is a parallel array of Josephson junctions, each with critical current I_c, linked by SQUID loops (each with inductance L), and is sometimes called the discrete Josephson transmission line. These are generally non-hysteretic (resistively shunted) junctions, with $Q \lesssim 1$. The two key parameters here are N, the number of junctions, and β_L, the ratio of the loop inductance L to the Josephson inductance L_J:

$$\beta_L = \frac{L}{L_J} = \frac{L}{\Phi_0/2\pi I_c} = \frac{2\pi L I_c}{\Phi_c} \qquad (5.110)$$

If $\beta_L \ll 1$ ($L \ll L_J$), then any Josephson vortex will be spread out over many junctions, and this behaves virtually the same as the continuous junction. Screening will occur over a Josephson penetration depth, which corresponds to the distance over which the series and parallel inductances balance. For the discrete case, this is typically expressed in terms of the number of junctions n_J (the Josephson penetration number) rather than a distance, so that $n_J L = L_J/n_J$, or

$$n_J = \sqrt{L_J/L} = 1/\sqrt{\beta_L} \gg 1 \qquad (5.111)$$

This looks like Eq. (5.81) for λ_J, but here the inductances are discrete rather than distributed. If $N < n_J$, then the array will behave as a short Josephson junction; if $N \gtrsim 2n_J$, it will behave as a long junction.

In the other limit $\beta_L \gg 1$ ($L \gg L_J$), the effective Josephson penetration number n_J would be less than 1, which only means that the continuum picture is no longer valid. The magnetic behavior is dominated by superconducting screening and fluxoid conservation in each of the loops, as in the corresponding limit in

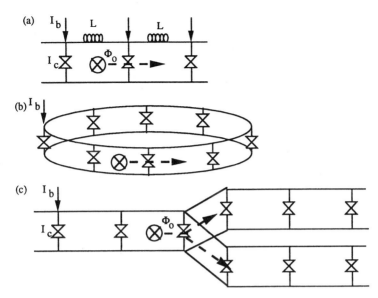

Figure 5.45. Discrete Josephson transmission line or flux shuttle. (*a*) Parallel array of Josephson junctions (each with Josephson inductance $L_J = \Phi_0/2\pi I_c$) connected by inductors L. In the regime $LI_c \sim 1$, a single loop may contain a single flux quantum Φ_0 that can propagate down the line. (*b*) Josephson ring oscillator with SFQ circulating around ring. (*c*) Split Josephson transmission line with one SFQ pulse moving to the right, generating two pulses as the line splits.

the dc SQUID. But there is also an intermediate regime with $\beta_L \sim 1$–10 (or $LI_c \sim 1$, typical of SQUID operation), which is particularly interesting. In this case, a single flux quantum Φ_0 is effectively localized in one of the loops (Fig. 5.45*a*). In the presence of an appropriate current bias, it can be "shuttled along" from one loop to the next, in a way that is quite similar to the propagation of a Josephson vortex in the long continuous junction, so that this configuration is sometimes called the "flux shuttle." For appropriate parameters, the flux contained in a given cell is either zero or Φ_0; multiple flux quanta may coexist in adjacent cells, but not in the same cell. This lends itself naturally to binary encoding (0 or 1) in a digital circuit, as we will discuss in greater detail in the next chapter.

Such a flux shuttle could also be connected in a ring geometry, in which case one or more flux quanta could travel around the ring indefinitely (Fig. 5.45*b*). This would produce an SFQ voltage pulse each time one of the quanta passed a given part of the circuit. This could function as an oscillator or, in the digital context, as a clock. The characteristic time for a pulse to be transferred across one cell is of order $\Phi_0/I_cR \sim 1$ ps, so that it is straightforward to produce such a sequence of pulses on the 100-GHz scale or above. Note that this SFQ pulse will not disperse or attenuate as it propagates down the discrete Josephson transmission line, even in the presence of resistive loss. Like the case of the soliton in the long Josephson junction, its shape is regenerated at each junction in the line. Futhermore, such a pulse can also be split and sent down two (or more) lines (Fig. 5.45c). These attributes are particularly useful in connection with digital logic circuits, as will be discussed further in Chapter 6.

SUMMARY

- A Josephson junction (JJ) exhibits quantum interference between two superconducting wave functions across a weak link between them, leading to a lossless current $I_c \sin \phi$, where the phase difference $\phi = 2\pi\Phi'/\Phi_0$, and $\Phi' = \int V \, dt$ is the fluxoid across the junction.
- A JJ is a source of microwave radiation at frequency $f_J = V/\Phi_0$; this corresponds to one fluxon crossing the junction each period. An array of many junctions can produce microwatts of power in the 100-GHz range and above.
- The dc $V(I)$ of a JJ in an rf voltage at frequency f exhibits a series of flat Shapiro steps at $V = nf\Phi_0$; an array of such junctions serves as the official voltage standard.
- A JJ can be viewed as a nonlinear inductance $\sim L_J = \Phi_0/2\pi I_c$, typically on the pH scale.
- A JJ is typically shunted by a resistance and a capacitance, leading to the RCSJ model, with $Q = \omega_0 RC$, where $\omega_0 = 1/\sqrt{L_{J0}C}$.
- An overdamped JJ (small Q) generates a series of SFQ pulses for $I > I_c$; an underdamped JJ (large Q) exhibits a hysteretic $V(I)$.
- A SQUID consists of one or two JJs in a loop with inductance $L \sim L_J$, sometimes called "rf SQUID" and "dc SQUID," with $V(I)$ strongly modulated by flux $\sim \frac{1}{2}\Phi_0$.
- A SQUID is also sensitive to a small fraction of Φ_0, which permits quantum-limited detection of small magnetic fields and ultra-low-noise amplification.
- The Josephson penetration depth is given by $\lambda_J = \sqrt{L_J/L}$, where L_J and L are distributed inductances in the junction and superconducting electrodes.
- In a short JJ ($l < \lambda_J$), magnetic flux is unscreened, and $\Phi \sim \Phi_0$ suppresses I_c.
- In a long JJ ($l > \lambda_J$), magnetic flux penetrates in discrete Josephson vortices (or solitons), which can flow down the junction, generating a single-flux-quantum (SFQ) voltage pulse.
- The discrete analog of a long JJ is a Josephson transmission line, which can also propagate an SFQ pulse, and provides a basic element of SFQ digital logic circuits.

REFERENCES

J. Clarke, "Geophysical Applications of SQUIDs," *IEEE Trans. Magn.* **19**, 288 (1983).

J. Clarke, "Principles and Applications of SQUIDs," *Proc. IEEE* **77**, 1208 (1989).

J. C. Gallop, *SQUIDs, the Josephson Effects, and Superconducting Electronics* (Adam Hilger, Bristol, 1991).

C. A. Hamilton, C. J. Burroughs, and S. P. Benz, "Josephson Voltage Standard: A Review," *IEEE Trans. Appl. Supercond.* **7**, 3756 (1997).

K. K. Likharev, *Dynamics of Josephson Junctions and Circuits* (Gordon and Breach, New York, 1986).

K. K. Likharev, "Single-Electron Transistors: Electrostatic Analogs of dc SQUIDs," *IEEE Trans. Magn.* **23**, 1142 (1987).

J. E. Lukens, "Josephson Arrays as High Frequency Sources," Chap. 4 in *Superconducting Devices*, Ed. S. T. Ruggiero (Academic, New York, 1990).

J. E. Nordman, "Superconductive Amplifying Devices Using Fluxon Dynamics," *Supercond. Sci. Technol.* **8**, 681 (1995).

T. P. Orlando and K. A. Delin, *Foundations of Applied Superconductivity*, Chaps. 8 and 9 (Addison-Wesley, Reading, MA, 1991).

N. F. Pedersen, "Fluxon Electronic Devices," *IEEE Trans. Magn.* **27**, 3328 (1991).

M. Tinkham, *Introduction to Superconductivity*, 2nd ed., Chap. 6 (McGraw-Hill, New York, 1996).

T. Van Duzer and C. W. Turner, *Principles of Superconductive Devices and Circuits*, Chaps. 4 and 5 (Elsevier, New York, 1981).

J. P. Wikswo, "SQUID Magnetometers for Biomagnetism and Nondestructive Testing," *IEEE Trans. Appl. Supercond.* **5**, 74 (1995).

PROBLEMS

5.1. Josephson radiation. An ideal Josephson element (with no resistive shunt) with critical current I_c is coupled into an external resistive load R_L and is biased with a current $I \gg I_c$. Estimate the following quantities in terms of these parameters.

(a) The dc voltage across the junction.

(b) The fundamental frequency of the ac signal coupled into R_L and the amplitude of the ac current at this frequency.

(c) The time-averaged rf power coupled into the load at this frequency.

(d) Determine the dc average power entering the Josephson element and compare it to (c). Comment on the significance of this.

5.2. Thermal modulation of Josephson junction. Consider a Josephson junction with a temperature-dependent critical current, which close to T_c can be written $I_c(T) = I_0(1 - T/T_c)$. The junction is coupled thermally, but not electrically, to a small heater with a resistance R and a thermal conductance K to the junction;

that is, the junction will be heated up by $\Delta T = I_{\text{con}}^2 R/K$. Assume a control current is $I_{\text{con}} = I_1 \cos \omega t$.

(a) Determine the effect of the heater current on the dc critical current of the junction.

(b) Will this produce any steps on the I–V characteristics, in analogy with Shapiro steps produced by an rf bias current? At what voltage(s)?

(c) If such steps exist, determine their amplitudes in terms of the above parameters. How does this compare to step widths for Shapiro steps?

5.3. Granular superconductor in magnetic field. Consider a random assemblage of weakly coupled superconducting grains, where the typical grain size a is smaller than the magnetic penetration depth λ. Assume that the grains are coupled by Josephson junctions with critical current density J_{c0}. In this case, the magnetic properties are dominated by the junctions.

(a) Estimate the effective magnetic penetration depth λ_g for the assemblage.

(b) Estimate the lower and upper critical fields for the assemblage, B_{c1} and B_{c2}, using λ_g and a for the GL parameters λ and ξ.

(c) For a superconductor with $\lambda = 300$ nm, $a = 100$ nm, and $J_{c0} = 10^8$ A/m^2, estimate these parameters.

5.4. Suppression of critical current of junction. There are some applications of SIS tunnel junctions (such as the SIS quasiparticle mixer) where it is necessary to suppress the Josephson current almost to zero but without applying a large magnetic field. This can be achieved using a rectangular Josephson junction with the field parallel to one of the axes, but only if the flux in the junction is exactly $n\Phi_0$. Otherwise, I_c increases again. So it is also desirable that the envelope of $I_c(B)$ drops as fast as possible.

(a) Plot the interference pattern $I_c(B)$ for a square junction 1 μm × 1 μm, with $\lambda = 100$ nm, with a magnetic field applied parallel to an axis of the square.

(b) Determine the interference pattern for the same square junction but with the field pointing in the diagonal direction in the junction plane. [*Hint*: This is the Fourier transform of the one-dimensional projection $i_c(x)$ in the direction perpendicular to the field.] Show this dependence on the same plot as (a). Would this be an improvement?

(c) Is there a shape where $I_c(B)$ will decrease to zero without oscillating? For the same total junction area, compare $I_c(B)$ to those for (a) and (b). (*Hint*: Consider the Gaussian.)

5.5. RSJ model.

(a) Show that the solutions for the current-biased RSJ model (with $C = 0$) given in Eq. (5.49) satisfy both Eq. (5.45) and the relation $V = (\hbar/2e)\, d\phi/dt$.

(b) Show by direct integration over a Josephson period that $V(t)$ is consistent with one flux quantum crossing the junction each period.

(c) From Eq. (5.50), plot the power spectrum of the voltage oscillation for $V_0 = \langle V \rangle = \frac{1}{5}I_cR$ and I_cR and also plot the amplitude in the first harmonic as a function of V_0.

(d) In which regime would one want to operate an RSJ as an oscillator and in which regime as a pulse generator?

5.6. DC SQUID design. Design a dc SQUID magnetometer based on Josephson junctions with $I_c = 100$ µA.

(a) Estimate the optimum loop inductance, taking $LI_c = \frac{1}{2}\Phi_0$. If this inductance consists of a loop without a ground plane, estimate the diameter d of the loop using the expression $L \approx \mu_0 d$.

(b) If the pickup coil is 1 cm in diameter, design a matched multiturn input coil that couples optimally to the SQUID loop. How many turns does this have?

(c) Assuming a flux sensitivity for the SQUID of $10^{-5}\,\Phi_0/\sqrt{\text{Hz}}$ (and ideal coupling to the input coil), what is the magnetic field sensitivity of the SQUID magnetometer for fields at the pickup coil?

5.7. Josephson soliton propagation.

(a) Show that the traveling-wave solution in Eq. (5.106) satisfies the sine-Gordon equation (5.105).

(b) What is the speed of light $u = \lambda_J \omega_J$ in the tunnel barrier for an insulator with thickness $d = 1$ nm and $\varepsilon_r = 10$ and superconducting electrodes with $\lambda = 100$ nm?

(c) What is the saturation voltage on the $n = 1$ step for a junction that is 50 µm long? What is the corresponding fundamental frequency?

5.8. Linear array Josephson oscillator. Consider a linear array of N RSJs ($N \gg 1$), each with $I_c = 200$ µA and $R = 1\,\Omega$, which is being used as a coherent oscillator. The entire array is coupled to a 50-Ω transmission line.

(a) Assume that each junction is biased with $I = 300$ µA and that all of the junctions are oscillating in phase, with their voltages adding in series. Determine the fundamental oscillation frequency, and estimate the voltage amplitude across the array at this frequency.

(b) Determine the power coupled into the transmission line as a function of N and determine the optimum value of N and the maximum power that can be coupled out.

5.9. Superconducting point contact. Much of the early work on Josephson junctions was actually carried out using a "point contact," a sharpened metal point in contact with another metal piece, with an extremely small contact area. When the size of this contact is smaller than the coherence length ξ, this can form a Josephson junction that obeys the ideal relation $I_cR_n = \pi\Delta/2e$.

(a) Estimate roughly the normal-state resistance R_n of such a point contact assuming a metallic constriction of radius $r = 30$ nm and a resistivity $\rho = 10$ µΩ-cm.

(b) Estimate I_c for this junction at low T using parameters appropriate for Nb. Also, estimate the characteristic Josephson inductance L_{J0}.
(c) Estimate the capacitance of this junction.
(d) Estimate the Q of the point contact. Would you expect the $V(I)$ to be underdamped (hysteretic) or overdamped (like the RSJ)?

6

SUPERCONDUCTING DIGITAL CIRCUITS

The superconducting circuits that we have described thus far have all been applied to analog systems. But, of course, the most revolutionary impact of semiconductor microelectronics in the past few decades has been in the area of digital computers and related systems. In the present chapter, we will describe the ways in which superconducting circuits based on Josephson junctions may be able to make a contribution here as well, notwithstanding the continuing remarkable progress in silicon integrated circuits. Optimized superconducting circuits are both faster and lower power than their silicon counterparts, and medium-scale superconducting integrated circuits (with many thousands of junctions) and prototype systems have demonstrated performance at clock speeds in excess of 20 GHz, with reliable projections of at least a factor of ten faster as the technology improves. Two distinct approaches to Josephson logic circuits have been developed, sometimes known as voltage-state logic and flux-state (or single-flux-quantum) logic; we will address both of these. This work has been accomplished primarily with LTS Nb-based circuits, but HTS circuits have also recently been demonstrated. Although the ultimate goal may be general-purpose supercomputing, the near-term projects have generally been in special-purpose signal processors, including analog-to-digital (A/D) converters, digital filters, counters, and switching arrays.

6.1 FAST SWITCHES AND MEMORIES

There are two basic elements required for any digital computing technology: a fast switch and a memory element for bit storage. For semiconductor integrated circuits, the transistor is the switch, and the capacitor is the storage element. So by the principle of duality, it should not be surprising that the corresponding superconducting elements are the Josephson junction or SQUID as the switch and the superconducting inductor for bit storage. These provided the basis for the development of a "Josephson computer" in the 1970s and 1980s, which greatly advanced the technology of Josephson junctions even if it did not ultimately lead to a practical computer. We will discuss some of these issues in later sections.

The Cryotron

But even before the discovery of the Josephson junction, a superconducting digital device called the "cryotron" was proposed as the switch, and a great deal of early development work was carried out in the late 1950s and 1960s. The cryotron consists of two superconducting wires or thin films (Fig. 6.1), where the control line has a higher critical magnetic field H_c than the gate line; material pairs such as Nb–Ta and Pb–Sn were typically used. The control current produces a magnetic field large enough to exceed H_c of the gate ($\sim 100\,G$), driving it into the normal state. With appropriate design parameters, this normal gate can switch enough current to control another gate. These cryotrons could be easily combined to produce various logic gates. The memory cell in the cryotron technology was essentially a closed superconducting loop, that could store a current (and equivalent flux) of either sign in the persistent-current mode, thus representing one bit. The flux is of course quantized, but a large number of flux quanta were typically used. If part of this closed loop is also a cryotron gate, then driving this into the normal state with the control current permits flux to be transferred into or out of the loop, that is, writing or reading the stored bit.

The key problem with this cryotron technology was that switching was limited by L/R times, which even for optimized thin-film structures were of order 10 ns or more. Furthermore, localized self-heating in the normal state often retarded the ability of the gate to return to the superconducting state. By the

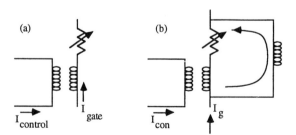

Figure 6.1. Basic switching and memory elements for the (obsolete) cryotron. (*a*) Schematic of cryotron switch, in which a magnetic field from a control line drives a superconducting gate into the resistive normal state. (*b*) Schematic of cryotron memory cell with a persistent current in a superconducting loop and a cryotron switch for reading or writing.

early 1960s, it was clear that this technology would be unable to compete with the new semiconducting integrated circuits, and cryotron development was generally abandoned.

Josephson Switches and Memories

However, some of the same concepts were carried over to the early development of the "Josephson cryotron" soon after the Josephson effect was discovered. The critical current of a Josephson junction can be suppressed by a magnetic field that is typically much smaller than the critical field of the superconductor. An even smaller field can suppress I_c of a SQUID, a parallel array of two junctions that in the digital arena has sometimes been called an "interferometer" (Fig. 6.2a). Thus, a much smaller control current (and a smaller inductance) can be used to drive a Josephson junction into the resistive "normal state" than was needed for the cryotron. Furthermore, as we will discuss in more detail below, the switching time for a Josephson junction can be orders of magnitude smaller, approaching 1 ps. An alternative to inductive coupling to a junction or SQUID is current injection (Fig. 6.2b), which is sometimes known as "overdrive." This is ultimately rather similar in effect to inductive coupling; when the critical current is exceeded, the device switches to the resistive state with $V > 0$.

A similar SQUID loop can also be used as a storage element, with flux being transferred in or out across one of the Josephson junctions. The more sensitive control permits the use of the natural binary bit of a superconducting loop, namely a single flux quantum (SFQ) Φ_0. A proper choice of element and bias parameters enables the SQUID to be switched between zero and one flux quantum in the loop, as will be discussed in further detail later in the chapter. On the other hand, this very sensitivity to small magnetic fields creates a potential problem for these SFQ memory elements and to some extent for the switching circuits as well. As we discussed previously, a superconductor tends to exclude a weak magnetic field when cooled below its critical temperature.

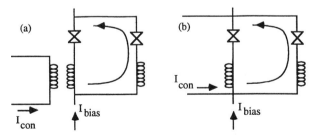

Figure 6.2. Simplified switching and memory elements based on Josephson junctions and SQUIDs. (a) Inductively coupled switch or memory cell; the control current I_{con} decreases the critical current below the bias current I_b. (b) Current injection overdrive switch or memory cell; the control current causes the total current to exceed the critical current.

Ambient magnetic fields and those produced by current lines may lead to the presence of accidental "trapped flux" in holes or inhomogeneities in the circuit, which in turn may couple magnetic fields into SQUID loops in other parts of the circuit. This may prevent proper operation of the circuit, until this trapped flux can be eliminated by heating the system above T_c and cooling again in the absence of the field. For this reason, Josephson digital circuits are typically operated inside one or more layers of magnetic shielding, composed of some combination of high-permeability magnetic and superconducting materials. In addition, thin-film Josephson integrated circuits are generally designed with "magnetic moats," slots in the superconducting ground plane where any residual magnetic flux is shunted away from the most sensitive regions of the circuit.

Characteristic Times in Josephson Junction

Although the actual devices typically have two (or even three) Josephson junctions in an "interferometer loop," one can understand the basic switching process by considering the dynamics of a single Josephson junction within the lumped CRSJ model. As we obtained earlier, the intrinsic Josephson junction can be considered a nonlinear inductor, with inductance (in the limit of zero current)

$$L_J = \frac{\Phi_0}{2\pi I_c} \tag{6.1}$$

Although this is certainly a highly nonlinear system, to a first approximation the characteristic times are the same as for the linear LCR oscillator:

$$\tau_1 = RC \quad \tau_2 = \frac{L_J}{R} = \frac{\Phi_0}{2\pi I_c R} \quad \tau_0 = \frac{1}{\omega_0} = \sqrt{L_J C} = \sqrt{\tau_1 \tau_2} \tag{6.2}$$

where ω_0 is sometimes referred to as the Josephson plasma frequency. The key dimensionless parameter is the Q of the junction, or equivalently $\beta_c = Q^2$:

$$Q = \frac{\tau_1}{\tau_0} = \sqrt{\frac{\tau_1}{\tau_2}} = R\sqrt{\frac{C}{L_J}}, \quad \beta_c = \frac{\tau_1}{\tau_2} = \frac{R^2 C}{L_J} \tag{6.3}$$

Note also that Q depends on T through the temperature dependence of I_c (and hence L_J); Q will decrease close to T_c.

As we will show below, the switching speed of a Josephson junction is limited by the slower of the times τ_1 and τ_2; τ_0 is the geometric mean of these two and is always between them. Hence, for $Q > 1$, the switching time is given approximately by $\tau_1 = RC$, whereas for $Q < 1$, it is given by $\tau_2 = L_J/R$. For $Q = 1$, all three times are the same.

For an ideal Josephson junction, the intrinsic quasiparticle resistance is nonlinear but can be approximated in many cases by the normal-state resistance R_n, where for an ideal Josephson junction well below T_c, without an external resistive shunt,

$$I_c R_n = \frac{\pi \Delta}{2e} \tag{6.4}$$

where 2Δ is the superconducting energy gap. Hence, the expression for τ_2 can be rewritten in the form

$$\tau_2 = \frac{\hbar}{\pi \Delta} \sim 0.15 \text{ ps} \tag{6.5}$$

where the energy gap for Nb, $2\Delta \approx 3$ meV, has been used here. For a typical Josephson junction in digital applications, $I_c \sim 100$ μA and the transverse scale is $a \sim 3$ μm, corresponding to $L_J \sim 3$ pH, $J_c \sim 10^7$ A/m^2, and $R_n \sim 20\ \Omega$ for an ideal junction.

Although in principle I_c can be varied by orders of magnitude (since J_c for a tunnel junction goes exponentially with the width of the insulating barrier), there are important reasons why I_c is ~ 100 μA. First, thermal fluctuations give rise to noise currents δI such that $(\hbar/2e)\,\delta I \sim kT$, so that for operation at 4 K, $\delta I \sim 0.2$ μA. If a junction is biased at 80% of I_c (80 μA), then it is extremely unlikely that thermal fluctuations will cause it to exceed I_c and switch prematurely into the resistive state. But there are also good reasons why I_c should not be too much larger than 100 μA. The loop inductance in the SQUID loops is also typically of the same order as L_J, or equivalently $LI_c/\Phi_0 \sim 0.1$–1. A small inductance can be produced using a superconducting microstripline, for which the inductance per square is $L_s = \mu_0 d_{\text{eff}} = \mu_0(d + 2\lambda)$. For a superconductor such as Nb with $\lambda \approx 100$ nm and an insulator with $d \sim 100$ nm, $L_s \sim 0.4$ pH. Thus, an inductance of ~ 3 pH can be easily designed using a line ~ 8 squares long. But if I_c were a factor of 10 larger, then the loop inductance would be ~ 0.3 pH for a line that is less than one square long. This would make the layout much more difficult.

To determine the value of Q, we must estimate the capacitance, which is normally dominant on this scale, at least for classic SIS tunnel junctions. For an insulator of thickness $d \sim 2$ nm and $\varepsilon_r \sim 10$ (such as Al$_2$O$_3$ used in most Nb tunnel junctions), one obtains a capacitance

$$C = \frac{\varepsilon_0 \varepsilon_r a^2}{d} \sim 0.5 \text{ pF} \tag{6.6}$$

for $a \sim 3$ μm. This yields the values

$$\tau_1 = RC \sim 10 \text{ ps} \tag{6.7}$$

and

$$\tau_0 = \sqrt{LC} \sim 1.2 \text{ ps} \tag{6.8}$$

which combine with the earlier estimate $\tau_2 \sim 0.2$ ps to yield

$$Q \sim 8 \qquad \beta_c \sim 60 \tag{6.9}$$

So a Nb junction on this scale is naturally underdamped, and the limiting response time $\tau_1 \sim 10$ ps.

But consider what happens if we decrease the junction scale while maintaining $I_c \sim 100$ μA (and therefore $R_n \sim 20$ Ω). This can be achieved by decreasing the barrier thickness very slightly. In this case, since τ_2 is constant, $Q \propto \tau_1 \propto C \propto a^2/d$. For $a \lesssim 0.4$ μm, we would obtain $Q < 1$, changing to the overdamped regime. The required value of $J_c \sim 6 \times 10^8$ A/m^2 is relatively large but is still much smaller than the critical current density in the superconductor itself ($\sim 10^{11}$ A/m^2). The present integrated circuit (IC) fabrication technology for Josephson junctions is still generally on the 3-μm scale (as of 1998), lagging somewhat behind the state of the art for semiconductor ICs, but these junctions are certain to get smaller as the technology matures in the next decade. This has the key advantage that the limiting characteristic time is reduced to $\tau_2 \sim 0.2$ ps as Q approaches 1, as illustrated in Fig. 6.3a.

Even without reducing the junction scale, it is also possible to turn an underdamped junction into an overdamped junction by adding an external resistive shunt, as we described earlier in the section on SQUIDs. For example, if we shunt the 20-Ω junction described above with a resistance of 3 Ω, the total shunt resistance of 2.6 Ω leads to $Q \approx 1$, with all of the above characteristics times $\tau \approx 1.3$ ps. The only assumption here is that the loop inductance associated with this shunt must be very small compared to L_J, which as we have seen

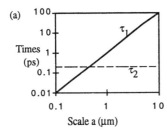
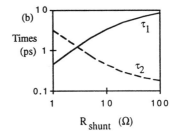

Figure 6.3. Characteristic times $\tau_1 = RC$ and $\tau_2 = L_J/R$ for typical Nb Josephson junctions assuming a constant critical current $I_c = 100$ μA. (a) For junction without external shunt resistor, dependence on junction scale a. (b) For shunted junction, dependence on shunt resistance R_s. Note that the slowest time (which sets the basic switching speed) is minimized for $Q = \sqrt{\tau_1/\tau_2} \approx 1$.

requires careful layout. In general, choosing an external shunt resistance $R_s \sim R_n/Q$ will move the junction out of the underdamped regime. As shown in Fig. 6.3b, shunting a high-Q junction with a small resistance will speed up the limiting time by reducing $\tau_1 = RC$ until $Q = 1$ is achieved. Decreasing R_s further will start to *increase* the limiting time, which is now $\tau_2 = L_J/R$. Thus, the optimum speed will be achieved for $Q \approx 1$.

For HTS Josephson junctions, ideal classic tunnel junctions are not (yet) available, and the technology is still being refined. Let us speculate on the properties of an ideal hypothetical HTS junction. For example, let us assume $I_c R \sim 10$ mV, which is conceivable given the energy gap $2\Delta \approx 30$ meV or greater. Also, let us choose $I_c \sim 1$ mA, to permit tolerance of thermal fluctuations at $T \sim 40$ K and above. This reduces the required loop inductance to $L_J \sim 0.3$ pH, which is difficult to achieve given a sheet inductance $\mu_0(2\lambda + d) \sim 0.6$ pH ($\lambda \sim 200$ nm) but may still be feasible. Then, we have $R \sim 10\,\Omega$ and $\tau_2 \sim 30$ fs. If we assume that $Q = 1$, then $\tau_1 = RC \sim 30$ fs, so $C \sim 3$ fF. This might correspond to a junction with $a \sim 1$ µm, $d \sim 3$ nm, and $\varepsilon_r \sim 12$. With such a junction, switching times of ~ 30 fs would be achievable at temperatures above 40 K.

Unfortunately, at least as of 1998, such HTS junctions are far from reality. Virtually all of the YBCO junctions under investigation act as if they have internal resistive shunts already included, so that they are in the overdamped limit. In fact, the $I_c R_n$ product for reproducible junctions is typically < 1 mV. For example, if we take $I_c R_n \sim 200$ µV and consider a value $I_c \sim 100$ µA, then $R_n \sim 2\,\Omega$. If we assume a junction scale $a \sim 3$ µm and a specific capacitance of ~ 50 fF/µm^2, one obtains $Q \sim 0.8$. The capacitance is likely to be somewhat smaller than this (due to a larger electrode separation), so that Q would be even smaller, well into the overdamped regime. The limiting characteristic time would then be $\tau_2 = L_J/R \sim 1.5$ ps. This is not yet faster than the best LTS junctions but has the possibility of operating at higher temperatures. Prototype digital circuits, based on overdamped Josephson junctions with similar parameters, have recently been fabricated and demonstrated.

Switching Dynamics of Underdamped and Overdamped Junctions

The switching behavior of a Josephson digital circuit is quite different depending on whether the junctions are underdamped or overdamped, and there are distinct logic families optimized for each regime. Let us first consider an intrinsic (unshunted) tunnel junction of the type described above, with critical current $I_c = 100$ µA, normal resistance $R_n = 20\,\Omega$, and capacitance $C = 0.5$ pF. This is well into the underdamped regime, with $Q \approx 8$. The dc I–V curves for such a junction are strongly hysteretic; once the junction goes into the voltage state, it remains there until the current bias is reduced almost to zero. This is illustrated in Fig. 6.4a for two different models for the quasiparticle resistance. The simple linear model assumes that $R = R_n$ for all voltages, whereas the more correct nonlinear model takes $R = R_n$ for $V > 2\Delta$, but $R = R_{sg}$ for subgap voltages,

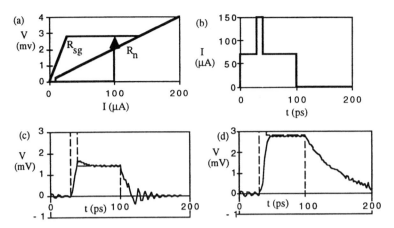

Figure 6.4. Transient response to fast current pulse of underdamped Josephson junction with $I_c = 100\ \mu\text{A}$, $R_n = 20\ \Omega$, and $C = 0.5\ \text{pF}$. (a) Hysteretic dc $V(I)$, for linear resistance with $R = R_n$ and for nonlinear resistor model with $R_{sg} = 100\ \Omega$ for $V < 2\Delta/e = 2.8\ \text{mV}$. (b) Current bias showing initial subcritical bias current followed by 10-ps supercritical current pulse, driving junction into voltage state, followed by current turn-off to reset junction to zero-voltage state. (c) Transient (solid line) and quasi-static (dashed line) voltage responses for junction with linear resistor model. (d) Transient (solid) and quasi-static (dashed) voltage responses for nonlinear resistor model. Note that the return to $V = 0$ in the "reset" period is much slower than for the linear resistor.

where $R_{sg} \gg R_n$. For the plot in Fig. 6.4a, $R_{sg} = 100\ \Omega$ is used, with an energy gap $2\Delta = 2.8\ \text{mV}$.

For either resistor model, the junction is first biased below the critical current (at $0.7I_c$) and then subjected to an additional current pulse of amplitude $0.8I_c$ and width 10 ps, followed at the end by turning off the bias current entirely (Fig. 6.4b). If this were a slower pulse, then the voltage response would be the quasi-static curve $V_{dc}[I(t)]$ indicated by the dashed lines in Figs. 6.4c and 6.4d (for the two resistor models). Note that because of the hysteresis, the pulse causes the junction to "latch" (i.e., lock) into the voltage state. The more correct transient solution (using SPICE; see Appendix B) is also shown and exhibits both RC time delays and oscillation typical of the underdamped limit. Note that the characteristic delay for switching to the voltage stage is of order $R_n C = 10\ \text{ps}$ for either resistor model. Both also exhibit "Josephson plasma oscillations" of decreasing amplitude well after the current is turned off. For the linear LCR resonator, which is relevant for the small-amplitude plasma oscillations, the period of oscillations is $2\pi\tau_0 \sim 8\ \text{ps}$, and the damping time is $2\tau_1 = 2RC$. So for the linear resistor model, the oscillations die out in a characteristic time of order 20 ps, while for the nonlinear resistor mode, the damping occurs in a much longer time, $\sim 2R_{sg}C \sim 100\ \text{ps}$. The slow damping during the "reset" of underdamped junctions will turn out to seriously limit the speed of digital circuits incorporating these junctions.

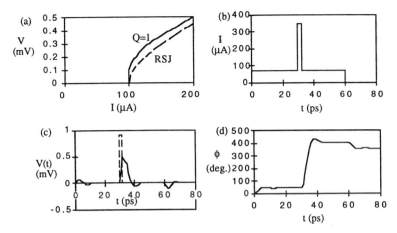

Figure 6.5. Transient response to fast current pulse of critically damped ($Q = 1$) Josephson junction with $I_c = 100$ µA, $R = 2.6\,\Omega$, and $C = 0.5$ pF. (a) Nonhysteretic dc $V(I)$, together with analytic RSJ solution $V = R\sqrt{I^2 - I_c^2}$. (b) Current bias, showing initial subcritical bias current, followed by 2-ps supercritical current pulse, driving junction temporarily into voltage state, followed by current turn-off. (c) Transient (solid) and quasi-static (dashed) voltage response corresponding to SFQ voltage pulse with characteristic pulse width of ~4 ps. (d) Evolution of junction phase $\phi(t)$ showing 360° phase rotation corresponding to SFQ pulse.

We also show the corresponding transient response for a junction near critical damping with $Q = 1$. As described earlier, this can be achieved by shunting the intrinsic junction above with a shunt resistance $R_s = 3\,\Omega$, giving all of the characteristic times of ~1.2 ps. Then, the form of the subgap resistance makes little difference. The dc I–V characteristic is now essentially nonhysteretic and resembles the simple RSJ dependence $V = R\sqrt{I^2 - I_c^2}$ that is appropriate for the overdamped limit (see Fig. 6.5a). A narrow current pulse (Fig. 6.5b) will cause the voltage to go momentarily into the voltage state, but it will *not* latch. Instead, the voltage returns quickly to zero, forming a narrow pulse (see the pulse at $t = 30$ ps in Fig. 6.5c), limited by the characteristic time $\tau_2 = L_J/R$. The minimal clearly defined pulse is one that is associated with the transfer of an SFQ across the junction, so that $\int V\,dt = \Phi_0$. In the overdamped limit, the pulse height would correspond to $2I_cR$ and the effective width $\Delta t = \Phi_0/2I_cR = \pi\tau_2 \sim 4$ ps. Because of the shunt resistance, this pulse is reduced in amplitude from that for the unshunted junction in Fig. 6.4. This also corresponds to the junction phase ϕ rotating 360°, as illustrated in Fig. 6.5d. In contrast, the phase change corresponding to Fig. 6.4 would be many flux quanta.

For $Q = 1$, there is a small overshoot following the main pulse, but it dies out quickly. The nature of the response is essentially unchanged if the junction resistance R_s is further reduced to cause the junction to be strongly overdamped. There is still a similar SFQ pulse, but it is somewhat slower (and lower in amplitude), since the width will scale as $\tau_2 = L_J/R$. This is in agreement with the analysis summarized in Fig. 6.3; the fastest pulse will be limited by the slower of τ_1 and τ_2 and will be optimized for $Q = 1$. Note that there is also a much smaller positive and negative voltage pulse associated with turning on and off the subcritical bias current.

One can better understand the dynamics of underdamped and overdamped Josephson junctions in terms of the pendulum analog. In the high-Q (underdamped) limit, if the pendulum is given a large enough impulse to cause it to rotate around, it will continue to rotate indefinitely. Furthermore, even after any drive torque is removed, the pendulum will continue to rotate for a time (depending on how quickly the energy is removed by viscous dissipation), followed by oscillation around the bottom, until eventually this dies out. Since the rotation rate is the analog of the voltage, this corresponds to a hysteretic I–V characteristic. In contrast, in the low-Q (overdamped) case, there is really only a single relevant time; τ_2 is the characteristic time for the pendulum to relax back to its equilibrium position. Since this corresponds to zero voltage, this case exhibits no hysteresis. Furthermore, if given a narrow impulse, the pendulum will rotate at most once, without further oscillation, corresponding to an SFQ. This continues to apply for $Q \approx 1$.

These two regimes form the basis for the two major classes of Josephson junction logic circuits. In voltage-state logic, or "latching logic," based on hysteretic high-Q junctions, the 'OFF' condition is the $V = 0$ state, while the 'ON' condition corresponds to the resistive state with $V \approx 2\Delta/e$. Given the hysteretic characteristic, the junction will remain in the ON state until it is reset. In contrast, in flux-state logic, which uses nonhysteretic junctions with $Q \sim 1$, the binary information is encoded in the passage of an SFQ pulse, corresponding to a 360° rotation of the phase. This is a form of nonlatching logic, since the junction automatically resets to the zero-voltage state. The SFQ circuits can be much faster than voltage-state circuits, but timing issues associated with the motion of the SFQ pulse may make design of integrated circuits rather more difficult. In succeeding sections, both classes of logic circuits will be discussed.

Speed–Power Diagram

One of the key reasons that superconducting digital circuits are of potential interest is that they can be extremely fast. As we have discussed, junction switching times approaching 1 ps are possible in some cases. But equally important is the very low power dissipation associated with these devices, so that they can be packed very close together. Otherwise, the time delays associated with ideal transmission at the speed of light (~ 100 μm/ps) can dominate the delays from the devices themselves. A Josephson junction dissipates no power at all in its zero-voltage state, and even in the voltage state, both the voltage and the current are very small. The narrowest voltage pulse associated with switching of a junction is an SFQ pulse. The peak power of such a pulse is of order $2I_c^2 R \sim 50$ nW for the parameters given above for the shunted junction for a time $\pi\tau_2 \sim \Phi_0/2I_c R \sim 4$ ps, corresponding to a total energy

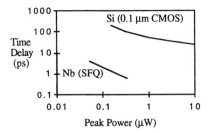

Figure 6.6. Speed–power trade-offs for state-of-the-art superconducting and silicon devices. The superconducting figures represent an SFQ pulse on a Nb Josephson junction with scale varying from 3 μm down to 0.4 μm operating at 4 K. The silicon figures are for each stage of a 100-stage CMOS ring oscillator operating at room temperature with a gate length of 0.1 μm and source–drain voltage varying from 0.4 V up to 2 V as the gate delay decreased (Taur et al., 1995).

$\sim I_c \Phi_0 \sim 2 \times 10^{-19}$ J (0.2 aJ). As we described earlier, reductions in size are likely to lead to a further decrease in the switching time to less than 1 ps, although the total energy dissipated would likely remain about the same. This trade-off curve is shown in Fig. 6.6 in a plot of gate delay versus power dissipation. This is merely a single junction, but such junctions can be combined together to form discrete Josephson transmission lines as well as other structures with digital applications that we will discuss in the next sections. A switch constructed from a hysteretic Josephson junction would be somewhat slower ($\sim 2R_n C$), and the peak power $\sim I_c^2 R_n$ would also be somewhat larger, perhaps 2 μW and 20 ps for junctions on the 3-μm scale.

Although there have been phenomenal advances in both speed and packing densities of semiconducting integrated circuits, superconducting circuits are still ahead in certain respects and are likely to remain so for the foreseeable future. To make a fair comparison, we should probably take the fastest, most advanced results from semiconductor device research. A recent report on a Si CMOS ring oscillator (Taur, et al. 1995), with 0.1 μm channel length (and operating at room temperature), gave figures on the power-delay trade-off in this technology (see Fig. 6.6). The devices could be very fast (as low as 20 ps/stage) or very low power (less than 200 nW/stage), but not under the same conditions. So optimum superconducting devices are likely to remain at least a factor of 10 faster and lower in power than their silicon counterparts. This establishes the motivation for pursuing superconducting circuits; we must still prove that large-scale Josephson integrated circuits can achieve the same functionality that we have grown to expect from silicon.

6.2 VOLTAGE-STATE LOGIC

The switching dynamics of a Josephson junction are quite distinct depending upon whether it is damped or underdamped, as was discussed in some depth in the previous section. Although a damped junction with $Q \sim 1$ can switch much faster, the early development work on Josephson logic was focused almost exclusively on underdamped, hysteretic junctions. This was partly because LTS junctions on the several-micron scale are naturally underdamped, but in addition, these underdamped junctions appeared to provide a more robust basis for practical realization of Josephson logic circuits. This was certainly the viewpoint in the 1970s, when a major research and development program was initiated by the IBM Corporation to develop a prototype "Josephson computer" (Anacker, 1980).

This Josephson computer was based on high-Q hysteretic Josephson junctions, which "latch" into the resistive branch of the $V(I)$ characteristic when a magnetic field is applied or a supercritical current injected; the bias current must be reduced to zero to reset the junction into the superconducting state. These two states of the junction, superconducting = "0" and resistive = "1", define the binary encoding and correspond to voltage levels $V = 0$ and $V \sim V_\Delta = 2\Delta/e$ (~ 3 mV for Nb), where 2Δ is the superconducting energy gap. This is similar to the way that voltage levels define the binary bit in ordinary transistor logic schemes. Even after IBM terminated this project in 1983 (due partly to project difficulties and partly to tremendous advances in the semiconducting competition), this same general approach of "voltage-state logic" or "latching logic" continued to be pursued by other groups, particularly in Japan (Hasuo, 1993; Hasuo and Imamura, 1989; Hayakawa, 1990). More recently, by the late 1980s, schemes to harness the ultrafast speed of damped (nonhysteretic) junctions were introduced, and these alternative approaches to Josephson logic appear to be superceding the earlier voltage-state logic in many potential applications. Nevertheless, several major technical accomplishments were demonstrated using voltage-stage logic, and it still provides the basis for comparison that any would-be alternative technology must exceed. In the present section, we will describe some of the key features of logic circuits based on hysteretic junctions; the newer nonlatching logic approaches will be addressed in Section 6.3.

Threshold Characteristics

A Josephson junction will switch from the superconducting to the resistive state if its critical current I_c is exceeded. For a magnetically controlled junction (or a SQUID), the control current I_{con} will decrease the critical current I_c, approximately as indicated in Fig. 6.7a. This dependence $I_c(I_{con})$ defines the "threshold characteristic" of the junction, as indicated in Fig. 6.7b, where $I_{c0} = I_c(0)$. Consider a junction initially biased with a current $I_b < I_{c0}$, with $I_{con} = 0$, which will be in the zero-voltage superconducting state. Now, if a control current I_{con} is applied to the junction, the operating point may cross the threshold into the resistive state, as indicated by an arrow in Fig. 6.7b. This will be valid for either a hysteretic or a nonhysteretic junction. The distinction between the two occurs when I_{con} is subsequently turned off. A nonhysteretic junction will again return to the zero-voltage state, whereas a hysteretic junction will remain in the

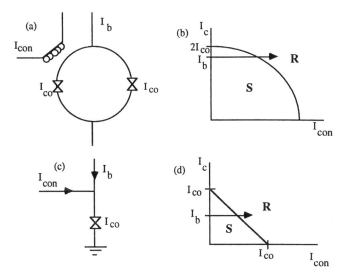

Figure 6.7. Critical current threshold curves for Josephson logic gates. (*a*) Magnetically controlled gate, where magnetic field from an inductively coupled control line reduces the critical current of the SQUID. (*b*) Threshold curve $I_c(I_{con})$ for (*a*). The arrow indicates the action of the control current to cross the threshold, thus switching the SQUID to the resistive state. (*c*) Current injection (or overdrive) gate, where a control current adds to the bias current I_b to exceed the critical current I_{c0}. (*d*) Efffective threshold curve $I_c(I_{con})$ for bias current in (*c*) with an arrow crossing the threshold similar to that in (*b*).

resistive state. In the latter case, the transition is not reversible; one must reduce I_b to zero to "reset" the junction in the $V = 0$ state.

The situation is virtually the same for a junction where the control current is directly injected (Fig. 6.7*c*). Here, the critical current is strictly a constant value I_{c0}, but from the point of view of the bias current input, the critical current takes the form $I_c = I_{c0} - I_{con}$. This forms a superconducting/resistive threshold curve in the same way as for the magnetically controlled gate (Fig. 6.7*d*). If the junction is initially operating in the superconducting state with $I_b < I_c$, the application of a control current can cause the operating point to cross the threshold, switching the junction to the resistive state. An underdamped junction will again remain in the resistive state until the total current $I_b + I_{con}$ is reduced to zero.

Logic Gates

Nothing that we have said is limited to a single control current. With either magnetic or injection coupling, we could easily have two control currents producing additive effects in depressing the critical current. This lends itself naturally to application to certain elementary logic gates, in particular AND and OR gates, each with two inputs and one output, with the control lines forming the inputs and the junction voltage the output. For an OR gate, either control current alone is sufficient to cross the threshold; for an AND gate, neither alone is sufficient, but the sum of the two can cross the threshold. And in both cases, there can be an output current that can be used in turn to control one or more subsequent gates. As we will discuss below, real AND and OR gates are somewhat more complex, with additional junctions and other circuit elements, but they are all based on these principles.

Consider, for example, the magnetically coupled OR gate in Fig. 6.8. This has two control lines, each inductively coupled to the SQUID loop. Either I_x or I_y (or both) is sufficient to depress the critical current below the bias current I_b. When this happens, the SQUID switches to the resistive state, and most of the bias current is diverted to the load resistance R_L, as the load-line analysis indicates. This is for an ideal hysteretic tunnel junction, and the output voltage for Nb would be close to the gap voltage $2\Delta/e \approx 3$ mV; the load resistance would typically be of order $R_N \sim 10\ \Omega$. Furthermore, this output current could be used as the control current for one or more gates in series, as suggested by the schematic of Fig. 6.8d. For the usual case where the control line is strongly coupled to the SQUID loop, the value of the induced circulating current in the

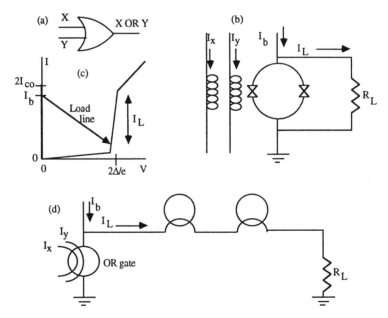

Figure 6.8. Basic OR gate in Josephson voltage-state logic. (*a*) Standard schematic symbol for OR gate with two inputs *X* and *Y*. (*b*) Circuit for basic magnetically controlled OR gate with two control currents I_x and I_y, sending output current I_L to a resistive load R_L. (*c*) *I–V* characteristic of SQUID with load line showing initial and final operating points corresponding to the 0 and 1 states. (*d*) Schematic of OR gate sending output to act as inputs for two other gates in series. The circle represents the SQUID; the curved line across the circle represents the inductively coupled control line.

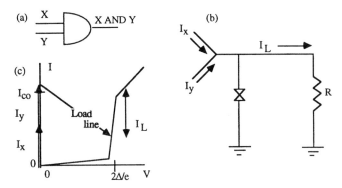

Figure 6.9. Basic AND gate in Josephson voltage-state logic. (*a*) Standard schematic symbol for AND gate with two inputs *X* and *Y*. (*b*) Circuit for basic current injection AND gate with two control currents I_x and I_y, sending output current I_L to a resistive load *R*. (*c*) *I–V* characteristic of load line similar to that for OR gate.

loop is close to that of the control current. This helps to explain why an output current that is of order I_{c0} is generally sufficient to switch other gates using magnetic control.

There is a similar situation with the injection coupled AND gate in Fig. 6.9. Here, there is no bias current, but the two control currents I_x and I_y together exceed the critical current I_{c0}. This causes the junction to switch to the resistive state, which in turn diverts most of the input current $\sim I_{c0}$ into the load. This, too, could be used to control other gates further down the line.

It is a well-known theorem of binary logic that any arbitrarily complex logical operation can be expressed in terms of the elementary operations AND, OR, and NOT. As we have shown, AND and OR gates can be obtained directly from the critical current threshold characteristic of Josephson junctions. Unfortunately, it is not so obvious how to make a NOT gate from a Josephson junction. This would require an input with $V > 0$ to produce an output with $V = 0$ and vice versa; this operation is also sometimes called INVERT. In fact, it is difficult to design a practical INVERT gate in this technology. A "timed-invert" gate has been developed but can be difficult to implement because it requires precise timing. An alternative approach that makes this unnecessary is "dual-rail logic," in which both a binary signal and its complement are carried in parallel through the entire set of logic operations. As illustrated in Fig. 6.10, constructing the dual side of the circuit requires transforming an OR gate into an AND gate, and vice versa. This is consistent with the standard theorems of logic that

$$\text{NOT}(A \text{ OR } B) = \text{NOT}(A) \text{ AND } \text{NOT}(B)$$
$$\text{NOT}(A \text{ AND } B) = \text{NOT}(A) \text{ OR } \text{NOT}(B)$$
(6.10)

Then, when the complementary bit is needed for an INVERT operation, it can be obtained from the dual side of the circuit without requiring an explicit NOT

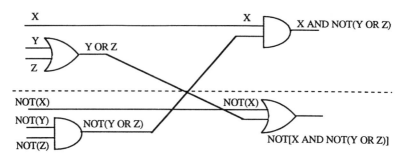

Figure 6.10. Illustration of dual-rail logic in which the NOT operation can be obtained without using an INVERT gate. The top half represents the operation X AND NOT (Y OR Z); the bottom half contains the complementary (or dual) logic circuits. The complementary operations include the transformations of AND gates into OR gates and vice versa. When a NOT operation is required, the signal is brought over from the complementary side of the circuit.

gate. With these understandings, dual-rail logic is straightforward to implement. It does require doubling the number of gates in the system, thus doubling the area required for the circuit, but nevertheless, this approach has been used in some Josephson logic systems.

Power Supply and Global Clock

In the discussion above, we have shown how a Josephson gate can be made to switch from the superconducting to the resistive state. However, given the hysteretic junctions used, such a gate must be reset to the superconducting state before it can be used again, by reducing the bias current to zero at every gate. Since these circuits are useful only if they can be operated rapidly and repetitively, this naturally suggests the use of a global clock system (Hasuo and Imamura, 1989), where the power is cycled on and off once each clock cycle. In a complex circuit with many junctions, this current bias could be distributed to the junctions in parallel, using, for example, a single voltage bias and a resistor network.

The ideal behavior of the bias current might take the periodic trapezoidal form shown in Fig. 6.11a. This includes a flat section on top (the "ON" interval) when the logic operations would occur, time intervals for ramped turn-on and turn-off, and an "OFF" time interval when $I = 0$. In some cases, this can be approximated by a sine wave with a dc offset. It is also possible to use a sine wave *without* a dc offset; in this case, the polarity of the ON bias alternates in successive clock cycles. Since the Josephson junction is a symmetric device, this alternating polarity is not a problem; all Josephson gates work in the same fashion if all the currents are reversed. In fact, one can use an external sine-wave current source with amplitude just above I_c, together with a series array of Josephson junctions (Fig. 6.11b), to generate something close to the ideal ac trapezoidal voltage source, but with alternating polarities (Fig. 6.11c). In effect,

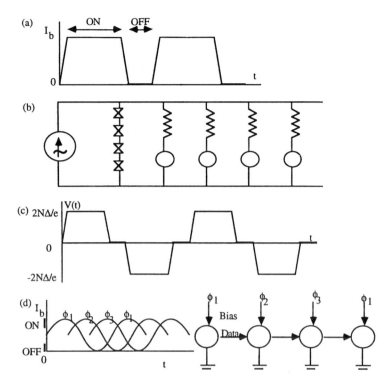

Figure 6.11. Power supply and clocking schemes for digital circuits based on hysteretic Josephson junctions, which must be reset once each period. (*a*) Unipolar current bias, alternating periodically between ON and OFF (reset) periods, with brief turn-on and turn-off ramps. (*b*) Schematic circuit for generating bipolar ac power supply and (*c*) resulting trapezoidal output voltage; the series array of Josephson junctions acts as a voltage regulator. (*d*) Three-phase unipolar power supply with 120° phase delays between sequential gates.

the sharp rise in the current at the gap voltage acts as a voltage regulator to limit the voltage amplitude in the ON state to $N\ 2\Delta/e$, where N is the number of junctions in series.

One implicit requirement of such a global biasing scheme is that the output value of a given gate must be stored during the OFF time. This requires a latch between each gate, where the latch accepts the data during the turn-off period and transmits it again during the next turn-on. The latch may also consist of Josephson junctions, for example based on persistent current in a SQUID loop. This can store a binary bit even in the absence of bias current. An alternative scheme that does not require the use of a latch involves a multiphase power source. This could involve, for example, three or four offset sinusoidal waves with the same frequency but with phase delays, so that the ON states of adjacent gates overlap but their OFF states are staggered in time (Fig. 6.11*d*).

For any of these powering/clocking schemes, one would like to run with as high a clock frequency as possible, consistent with correct operation of the gates. If the clock is run too fast, a standard problem is known as "punchthrough." This corresponds to the situation in which a junction does not properly reset to the superconducting 0 state during the OFF period. We can understand the basis for punchthrough using the rotating pendulum analog for a Josephson junction. As we pointed out earlier, an underdamped junction is equivalent to a rotating pendulum with low damping. If the drive torque (analogous to the bias current) is turned off, such a pendulum will continue to rotate for some time, followed by a long period of oscillation around the bottom (corresponding to Josephson "plasma oscillations") before it finally settles down. If the drive torque is turned on again before this settling process is complete, the pendulum may start rotating again. Similarly, if a bias current of either sign is turned on during the oscillation period, there is a significant possibility that the junction may switch prematurely into the resistive state. Furthermore, this punchthrough can also occur if the bias turn-on is too fast. For this reason, the clock speed of these voltage-state circuits is generally limited to several gigahertz, particularly if a low error rate is required for a large-scale digital system.

Memory Elements

Most digital systems require an array of memory elements that can store and retrieve binary data on the same time scale as the operation of the logic gates. (This might represent the fast cache memory of a computer system.) As we mentioned earlier, the lossless circulating current in a SQUID loop provides a natural basis for such a memory cell, and flux quantization provides a natural binary encoding scheme: 0 or 1 for zero or one flux quantum in the loop. In addition, a READ gate or "sense gate" is needed for fast readout and a WRITE gate to store the bit. Just as for logic circuits, these memory circuits can also be designed using hysteretic Josephson junctions (Wada, 1989).

The basic design of one type of Josephson memory element is shown in Fig. 6.12a. Here, the symmetrical superconducting loop is fed by a bias current in the center. A Josephson junction (or more typically, a SQUID) is located along one of the two parallel legs; this represents the WRITE gate. Let us assume that the junction is in the zero-voltage state and initially there is no circulating current in the loop, which of course is equivalent to zero flux quanta stored. If now we apply a bias current I_b, then the two legs present a similar inductance, the current splits evenly into the two legs, and the stored flux is still zero, assuming that the current $\frac{1}{2}I_b$ is still less than the critical current I_{c0} of the junction or SQUID. This corresponds to the 0 state of the memory cell.

There is also a control line magnetically coupled to the junction in this WRITE gate. If a control current depresses I_c sufficiently, then the junction is driven into the voltage state. The current will quickly redistribute to the other leg, but in the meantime, magnetic flux has crossed the junction. Furthermore, once this happens, the junction can now return to the zero-voltage state,

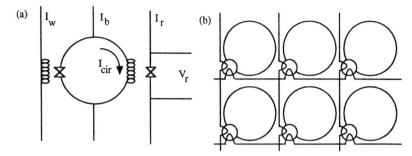

Figure 6.12. NDRO (nondestructive readout) Josephson memory cell, based on the presence (1) or absence (0) of trapped flux in storage loop. (*a*) Basic cell configuration with storage loop interrupted by Jospehson junction (or SQUID). The WRITE current I_w is magnetically coupled to the junction in the storage loop, causing the junction to enter momentarily into the resistive state, thus permitting flux to enter the loop. This is stored as circulating current I_{cir}, which is magnetically coupled to the output junction (or SQUID) on the READ line. (*b*) Array of NDRO memory cells showing the use of AND gates for the WRITE gates, permitting row and column addressing of memory cells.

trapping the flux inside. In a properly designed loop with inductance L, this flux $\Phi \sim (\frac{1}{2}L)I_b \sim \frac{1}{2}LI_c$ can be as small as a single flux quantum Φ_0, although several Φ_0 may be more typical. When the bias current is removed, conservation of flux in the loop requires that this flux be maintained indefinitely, with a circulating current $I_{\text{cir}} \sim \frac{1}{2}I_c$. This corresponds to a 1 stored in the loop.

This current I_{cir}, in turn, can be used as the control current for a READ gate, another SQUID magnetically coupled to the memory loop. The READ gate is normally biased with a readout current less than I_{c0} but more than I_c (in the presence of I_{cir}). If the memory cell is storing a 0, then $I_{\text{cir}} = 0$ and the READ gate will properly remain in the zero-voltage state. If however, a 1 is present, then $I_{\text{cir}} \neq 0$, and the READ gate will switch to the voltage state, reading out the content of the memory cell to another part of the circuit. The information in this cell can be read out repeatedly without affecting the stored flux, so that this circuit acts as a "nondestructive readout" cell, or NDRO.

One can easily design the WRITE gate as an AND gate with two control lines (see Fig. 6.12b). This makes it possible to consider a two-dimensional array of memory cells with control lines along the rows and columns. This permits unique addressing of each memory cell with a reduced number of control lines. For example, a 10 × 10 array could be addressed with 20 lines, instead of the 100 that would be required for a separate line to each one. A similar scheme could be used for the READ gates. Still, a possible disadvantage of this NDRO cell for large arrays is that each cell takes up quite a lot of area, requiring a main loop and two SQUID loops.

An alternative memory cell that is much more compact can be obtained using only a single two-junction SQUID but in a rather different (and somewhat subtle) mode. Consider a symmetrically biased SQUID with two identical

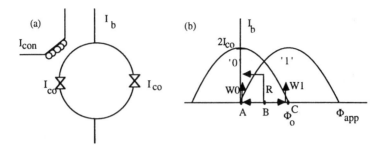

Figure 6.13. DRO (destructive readout) Josephson memory cell, based on presence (1) or absence (0) of a single flux quantum in two-junction SQUID loop. (*a*) Basic configuration with control line inductively coupled to SQUID loop consisting of two junctions with critical current I_{co} and a loop inductance $L = \Phi_0/2I_{co}$. A combination of bias and control currents I_b and I_{con} are used for both writing and reading. (*b*) Threshold curves as function of applied flux Φ_{app}, proportional to I_{con}. The curves for total flux $\Phi = 0$ and $\Phi = \Phi_0$ are shown together with current sequences for writing and reading. The WRITE sequences (W0, W1) make use of a vortex transition and keep the SQUID in the zero-voltage state. The destructive READ sequence (R) switches the SQUID into the resistive state if the stored flux is Φ_0. Points A, B, and C are described in the text.

junctions having critical current I_{c0} and a loop inductance L with $LI_{c0} = \frac{1}{2}\Phi_0$, as indicated in Fig. 6.13*a*. This loop inductance is *not* negligible compared to the junction inductances ($L_J \sim \Phi_0/2\pi I_{c0}$), so that the screening flux LI_{cir} is significant. The threshold characteristic here is now composed of two overlapping threshold curves for $\Phi = 0$ and $\Phi = \Phi_0$ ($n = 0$ and $n = 1$ flux quanta), as indicated in Fig. 6.13*b*. For the $\Phi = 0$ curve, the peak total current of $2I_{c0}$ corresponds to the two junctions in phase, with a circulating current $I_{cir} = 0$. The minimum of this curve corresponds to a circulating current $I_{cir} = I_{c0}$, the two junctions being 180° out of phase. Together, these correspond to an applied flux $\Phi_{ex} = \Phi_0$, half of which is being screened out by I_{cir}. But the fluxoid in going around the loop is still $\Phi = 0$. In contrast, consider the threshold curve for $n = 1$, which is shifted one flux quantum to the right, corresponding to a total fluxoid $\Phi = \Phi_0$. The peak of this curve again corresponds to $I_{cir} = 0$ and the two junctions in phase but with an enclosed flux of Φ_0.

If the operating point of the SQUID is below either of the two threshold curves, the SQUID may be in the zero-voltage state. In the overlap region, initial conditions will determine which configuration ($n = 0$ or $n = 1$) will be occupied. The bias and/or control currents may be changed in a way that causes the system to cross one of the threshold curves. If it crosses from either zero-voltage region to the resistive region, then the SQUID will clearly switch to the resistive state and remain there until the bias current is turned off (since these are still underdamped, hysteretic junctions). However, if the system crosses from one zero-voltage region to the other, the result is not quite so obvious. Will it transfer the flux Φ_0, corresponding to a single-flux-quantum voltage pulse, and then return to zero voltage? Or will it go irreversibly into the resistive state?

In general, any time a hysteretic junction goes instantaneously into the resistive state, it will remain there unless the bias current is quickly reduced to zero (or nearly zero). For a SQUID, the relevant current is the total current $I_{tot} = \frac{1}{2}I_b \pm I_{cir}$ passing through each of the junctions. In Fig. 6.13b, there are two points (A and C) where $I_{tot} = 0$ for $I_b = 0$ at the centers of $n = 0$ and $n = 1$ regimes. If an internal threshold-crossing leaves the system in (or near) one of these points, then the SQUID is automatically reset to the zero-voltage state. This is sometimes called a "vortex transition"; the voltage response is only the SFQ pulse corresponding to a single vortex crossing the junction. In contrast, an internal threshold-crossing higher up on the curve will leave the SQUID in the resistive state until it is reset by returning the bias current and control current to zero.

This vortex transition is an essential part of the WRITE operation of this compact memory cell. Consider an initial control current at point B with an applied flux of $\frac{1}{2}\Phi_0$, which could be in either the $n = 0$ or $n = 1$ zero-voltage states. To write a 1, an additional flux of $\frac{1}{2}\Phi_0$ is applied to move the operating point to C followed by a small bias current to switch to the $n = 1$ state. Then the bias current is removed followed by returning to point B. As long as the SQUID remains under the $n = 1$ threshold curve, the memory cell will retain the 1. To write a 0, the control flux is turned off, moving the system to point A, followed by an increase in bias current to switch to the $n = 0$ state.

In contrast to the WRITE operation, the READ operation requires crossing an internal threshold line outside of the vortex transition regime. For example, from point B, the bias current is first increased, followed by decreasing the control current to zero (opposite the sequence for writing a 0). If the SQUID was in the $n = 0$ state, then no threshold is crossed, and the SQUID remains in the zero-voltage state. On the other hand, if the SQUID was in the $n = 1$ state, then a "hard" threshold is crossed, driving the SQUID into the resistive state and giving a 1 voltage signal as the output. Of course, this readout process causes the stored flux to be lost. If necessary, the stored 1 can be rewritten after the SQUID is reset to the zero voltage by removing the bias current. A memory cell of this type is sometimes called a "destructive readout," or DRO, as opposed to the NDRO described in Fig. 6.12.

Practical Gate Families in Voltage-State Logic

The logic gates and memory cells described above are somewhat idealized examples. Those that have been developed and tested in real digital systems are somewhat more complex. An example is the "modified variable threshold logic" (MVTL) family of gates, which are still being applied to complex voltage-state logic circuits. The MVTL OR gate is shown in Fig. 6.14. This is a hybrid gate, including both magnetic coupling and current injection, which should be somewhat faster and more sensitive than either type by itself. The third junction is added to increase isolation between input and output. The resistor across the SQUID loop inductance is added in order to damp out LC oscillations that

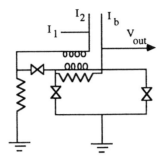

Figure 6.14. Schematic circuit of OR gate in MVTL (modified variable threshold logic) family of voltage-state logic. This consists of a control line magnetically coupled to a two-junction SQUID combined with current injection into the SQUID via a buffering junction. A damping resistor is placed across the coupling inductor to damp out LC resonant oscillations. The input currents are I_1 and I_2; the output is the voltage across the SQUID.

would otherwise slow the response and recovery of the gate. Gate delays of 2 ps/gate have been measured for 2-μm Nb junctions, although the clock speed of a microprocessor based on this family of junctions was still limited (due to the possibility of punchthrough and related effects) to about 1 GHz. Further discussion of these and other digital systems will be continued in Section 6.4 after a brief tutorial on SFQ logic circuits.

6.3 SINGLE-FLUX-QUANTUM LOGIC

In an attempt to take maximum advantage of the picosecond switching speed of Josephson junctions, several alternative approaches have been developed that make use of junctions with near-critical damping which do not exhibit hysteresis. These approaches are based on the transfer and control of single flux quanta (SFQ) Φ_0. As we have shown, a Josephson junction can act to regulate leakage of an SFQ, which is associated with a fast SFQ voltage pulse $\int V \, dt = \Phi_0$. This also corresponds to a rotation of the junction phase (or of the angle of the pendulum that represents it) by 360°. This rotation lends itself naturally to the development of binary logic circuits that can in principle be almost as fast as these pulses. The most successful such approach is known as RSFQ logic, for "rapid single flux quantum" (Likharev and Semenov, 1991). We will focus on the RSFQ approach, mentioning briefly some other alternatives at the end of the section.

Pulse Transmission

We have already considered one of the key components of SFQ logic, the discrete Josephson transmission line (JTL), an active line that permits the

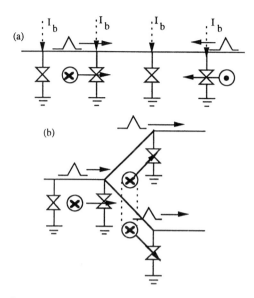

Figure 6.15. SFQ pulse transmission on active JTL. (*a*) Discrete JTL with parallel array of nonhysteretic junctions, each with critical current I_c and loop inductance $L = \Phi_0/2I_c$, biased with $I_b \sim 0.7 I_c$. This supports propagation of a fluxon to the right or an antifluxon to the left. (*b*) SFQ splitter with SFQ pulse on input line generating similar SFQ pulse on each of two output lines (DC current bias lines are present but not shown.)

propagation of an SFQ pulse (and accompanying fluxon) in either direction (Fig. 6.15a). As discussed briefly at the end of Section 5.4, this consists of a parallel array of identical nonhysteretic junctions, each biased near its critical current I_c, coupled by superconducting wires with inductance L given by $LI_c \sim \Phi_0$. If L is much smaller, then the fluxon is spread out over several junctions. If L is much larger, then one could have several flux quanta in a single loop. A typical value is $LI_C = \frac{1}{2}\Phi_0$, for which a fluxon propagates along the line but cannot rest quasi-statically in any of the loops. When an SFQ pulse passes a given junction, the phase of that junction rotates by 360°, corresponding to a flux of Φ_0 crossing the junction transversely. The junctions should be damped with $Q \sim 1$, which requires external resistive shunting for LTS junctions on the micron scale. Actually, the shunt resistor alone may yield a slightly larger Q since the line itself provides some additional damping. The current generation of HTS junctions are intrinsically overdamped, so that they are suitable without additional shunting.

Alternatively, a purely passive superconducting transmission line can sometimes be used for transmitting SFQ pulses with very low loss or dispersion. It should ideally have a characteristic impedance that matches the shunt resistances of the junctions (typically $\sim 2\,\Omega$). Both the active and the passive lines should be very fast, permitting pulse propagation at speeds of order 10^7 m/s (10 µm/ps) or greater. The nonlinear active line has the key advantage that it will automatically filter out low-level noise as well as regenerate the SFQ pulse with a standard shape. For these purposes, the flux quantum is often expressed in the units $\Phi_0 = 2.07$ mV-ps. For a typical shunted junction (see Section 6.1), the SFQ pulse as generated by such a junction has pulse height $\sim 2I_c R \sim 0.5$ mV and width ~ 4 ps. Note that this is both smaller in voltage and much shorter in time than the typical pulses obtained with voltage-state Josephson logic.

The JTL can also be split into two (or more) parallel lines (Fig. 6.15b), so that one SFQ pulse can give rise to many such pulses. It is also possible in principle to implement a similar "splitter" with a passive transmission line, but the matching condition at the intersection would be much more stringent. A mismatch in the passive splitter could bring about partial reflection of the pulse; in contrast, the active line can only propagate a complete SFQ pulse. Note that the SFQ splitter is actually symmetric among its three ports; an SFQ pulse incident from any one of the three would yield an SFQ pulse out along the other two lines. Note also that this JTL splitter might initially seem to conflict with "conservation of flux" in the system; how can one flux quantum split into two flux quanta? However, as we discussed earlier, a fluxon is not really a point particle; rather, it is a part of a closed flux line. From this point of view (Fig. 6.15b), the two fluxons after the split start out as part of the same flux line.

Pulse-Based Logic

In addition to merely transmitting the SFQ pulses, a set of logic gates are needed that will permit standard logic operations (OR, AND, NOT) to be performed on these very weak, fast pulses. Let us first consider a simple combiner geometry with two control lines injecting current directly into a Josephson junction (Fig. 6.16a), similar to the basic configuration of current-injection gates in voltage-state Josephson logic. Both inputs and outputs here would generally be connected to an appropriate (active or passive) transmission line. In addition, there is a low-frequency (dc) current bias line biased at close to I_c (typically 70–80%).

As we showed earlier, this defines a quasi-static threshold curve in which the effective critical current is depressed by the injected current (Fig. 6.16b). Similarly, the dc $I(V)$ curve and load line show a nonhysteretic characteristic (Fig. 6.16c). When an input pulse causes the junction to exceed I_c, it will switch to the resistive state but then return to zero voltage after the current pulse is removed. In the meantime, a pulse of voltage and current will be diverted to the load. However, neither the quasi-static threshold curve nor the dc load line analysis is strictly correct on the time scale of the junction response; they can only be used to obtain a qualitative guideline for the device performance.

To obtain a more accurate picture, we must consider the dynamics of the Josephson junction. It is convenient to do this in terms of the pendulum picture

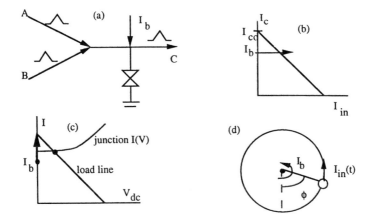

Figure 6.16. Basic concept of logic gate (OR or AND) using SFQ pulses (but see Fig. 6.19 for complete gates). (*a*) Combiner geometry, where input pulses on input lines *A* and *B* may generate output pulse on *C*. (*b*) Quasi-static threshold curve for critical current of junction in (*a*). (*c*) DC load line picture of switching of nonhysteretic Josephson junction. The diagrams in (*b*) and (*c*) provide only a qualitative picture of switching due to a fast current pulse. (*d*) Pendulum picture of junction switching due to fast current pulse. Rotation of the pendulum by 360° corresponds to generation of an SFQ voltage pulse on the output line.

of the junction, where initially the pendulum hangs statically at an angle from the bottom $\phi = \cos^{-1}(I_b/I_{c0})$, where I_b is the current bias (Fig. 6.16d). The total current bias corresponds to the torque on the pendulum. If a damped pendulum first reaches the top, then gravity will complete the rotation. The required torque actually decreases once the pendulum rises above the horizontal. First let us consider a single narrow current pulse that might be produced by a JTL with similar junctions. If the pulse current causes the total current to exceed I_c for a long enough time to push the pendulum up to an angle ϕ below the vertical, then the dc bias current I_b can do the rest of the work, even without any inertia. Given the damping, the pendulum will not continue to oscillate but will quickly return to the static angle ϕ at which it started. This rotation corresponds to an SFQ voltage pulse, that could then propagate down the output line. In contrast, if the current pulse is insufficient to cause a rotation, then the pendulum will rise but then fall back to its initial position. This corresponds to a smaller oscillating voltage with no net flux transfer.

Let us now consider the presence of two input lines *A* and *B* each of which may have either an SFQ pulse (binary 1) or no pulse (binary 0), both entering the junction at the same time. It is easy to see that we may choose junction and bias values so that either pulse is sufficient to trigger an SFQ output pulse; this is conceptually an OR gate. Alternatively, for slightly different parameters (such as reduced I_b/I_{c0}) we might require simultaneous currents from both inputs to get an output pulse; this would be the AND gate.

Buffering Junctions

Unfortunately, things are not this simple for several important reasons. First, although we have labeled these as input and output lines, transmission lines transmit SFQ pulses equally well in both directions. In fact, all three lines (A, B, and C) in Fig. 6.16a are fully in parallel, and this is the same geometry as the splitter of Fig. 6.15b. A triggering pulse sent to the junction on any one of the three lines will produce an output pulse on the other two. If we want the input lines to serve for input only, we need to provide a junction in series to act as a buffer on each input line. The resulting circuit is shown in Fig. 6.17a and is known as the "confluence buffer." Here, the bias current for the "SFQ generator junction" J_C goes partly to bias the buffer junctions J_A and J_B in the reverse direction. A voltage pulse traveling in the forward direction produces a current pulse in the same direction, which will subtract from the dc bias current and therefore not switch the junction. In contrast, a pulse traveling backward will add to the bias current, thus causing the junction to switch. This permits the fluxon to escape from that line by crossing the junction, thus blocking the further propagation of this SFQ pulse. In general, there will be a series buffer junction on all high-speed input lines, with a backward dc current bias. The dc current bias lines themselves need not carry pulses (in either direction), so that they can be separately filtered.

For example, let us assume that the parameters of this confluence buffer are chosen so that if an input pulse enters on line A or line B, an output pulse will leave on line C. Figure 6.17b shows the corresponding path of the fluxon entering from line A. It crosses junction J_C, splits so as to go down lines B and C, but escapes line B by crossing junction J_B. This circuit behaves like an OR gate; a pulse in either input yields an output pulse, with no reverse pulses on the input lines.

But what happens if a pulse enters on both input lines? The result depends on whether the input pulses overlap in time. If they are simultaneous, then there will

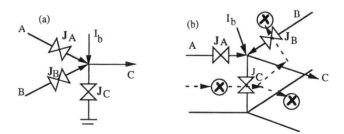

Figure 6.17. SFQ confluence buffer, an improved version of combiner circuit shown in Fig. 6.16. (a) Circuit geometry with buffer junctions J_A and J_B in series with input lines A and B to prevent propagation of SFQ pulse in reverse direction. (b) Illustration of fluxon path for confluence buffer with input SFQ pulse on line A. Note that the fluxon escapes across junction J_B rather than propagating out along input line B.

be a single SFQ output pulse. If they are staggered in time, then there will be a series of two SFQ output pulses on line C. This is still functioning somewhat as an OR gate, but a gate that obtains different outputs depending on when the two pulses arrive is generally unacceptable. A similar problem occurs for the prototype AND gate; if the pulses do not arrive simultaneously, then no output pulse will result. This illustrates the key problem with SFQ logic. The logic scheme is based on rapidly moving narrow pulses, but it is impractical to ensure that two pulses from different parts of the circuit arrive at the same time to within 1 ps. RSFQ logic deals with this problem by integrating one or more storage cells, also known as registers or latches, as part of each and every logic gate.

SFQ Memory Cell

This standard RSFQ latch is essentially the same two-junction SQUID that we have seen before in various forms, and stores its information in the quantized flux (and corresponding persistent current) trapped in the superconducting loop. Figure 6.18a shows the SQUID with a dc bias and two input lines, each with a series buffering junction, and one output. For $LI_c = 1.5\Phi_0$ and an asymmetric current bias as shown, the SQUID can exist in either of two bistable zero-voltage states, storing either zero or one fluxon with current distributions as indicated. Figure 6.18b shows the corresponding approximate threshold curves.

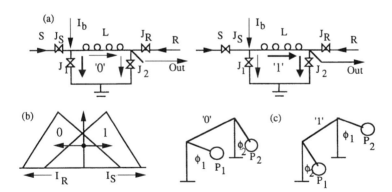

Figure 6.18. SFQ memory cell, also known as RS flip-flop and DRO register. (a) Circuit schematic of SQUID with loop inductance $L = 1.5\Phi_0/I_c$ showing static current distributions in 0 and 1 states, corresponding to the absence and the presence of a stored fluxon. (b) Approximate critical current threshold curves for RS flip-flop, showing zero-voltage regimes for the 0 and 1 states as a function of control currents on S and R lines. A current in the R line is generally opposite in effect to a similar current in the S line and is represented here as the negative half of the threshold plot. (c) Pendulum picture of RS flip-flop in 0 and 1 states (left and right, respectively), with pendula P_1 and P_2 connected by torsional spring. A counterclockwise rotation of pendulum P_1 takes 0 to 1; a similar rotation of P_2 releases the stored torque and restores the 0 state.

Let us assume that the SQUID is initially in the 0 state, with zero flux in the loop. If the current is biased below the critical current of the SQUID, most of the current will go into the left branch, since the loop inductance is large. A current pulse in the S input on the left will initially increase the current in the left junction J_1 above its critical current, causing the junction to rotate its phase. This corresponds to a fluxon crossing J_1 and entering the loop, and to crossing the threshold into the 1 state. This is similar to the vortex transitional memory cell described earlier (DRO, Fig. 6.13) except that the current through J_1 need not go to zero after switching; the damping will always permit it to reset properly to the zero-voltage state. The 1 configuration is equivalent to a fluxon stored in the loop; if I_b were reduced to zero, there would be a clockwise circulating current. Now, if there is a current pulse incident from the right "reset" line, this will initially increase the current in the right junction J_2, causing its phase to rotate and the trapped fluxon to escape thus returning the cell to the 0 state. This generates an SFQ pulse on the output line but not on the R line.

The behavior of this memory cell can also be viewed in terms of the pendulum picture (Fig. 6.18c). Here, the two pendula are coupled by a torsional spring that represents the inductor, as in Fig. 5.31c earlier. In the 0 state, most of the applied torque acts directly on pendulum P_1; only a small part of the torque is transferred to P_2, which rests at a smaller angle ϕ_2. An additional counterclockwise rotation of P_1 will transfer the torque to the torsional spring and switch the angles ϕ_1 and ϕ_2, which represents the 1 state. Now a rotation of pendulum P_2 will release the stored tension in the spring and revert the configuration back to the 0 state.

Note also that if the circuit is in the 0 state, then a readout pulse into the R-input will *not* trigger J_2 (since the effective bias current is rather small), but rather will trigger the right buffer junction J_R, which has a smaller critical current. Therefore, no output pulse will be generated. So if the SQUID is in the 0 state, a 0 signal (i.e., no SFQ pulse) will be generated, while if it is in the 1 state, a 1 output signal (an SFQ pulse) will be generated. In either case, the final state of the SQUID will be the 0 state. This process is effectively a "destructive readout" of the stored bit, so that this can be called a DRO register. Similarly, if a second SFQ pulse comes into the S line after the first, then the buffer junction J_s will switch rather than J_1, so that the loop remains in the 1 state. Because of this set of responses, this cell is also equivalent to a "set–reset flip flop" (or RS flip-flop) that alternates between the two states if the trigger signals are given in the proper sequence.

Finally, the RS flip-flop can also be viewed consistently from the fluxon picture. A fluxon entering from the left will be trapped in the loop until it is released by an antifluxon entering from the right. This reset procedure can be viewed as the antifluxon crossing junction J_2 and splitting. It then cancels the stored fluxon on the left and also propagates to the output line on the right as a fluxon. In contrast, if the loop is in the 0 state, then the antifluxon entering on R escapes across the buffer junction J_R, so that it need not cross J_2, and no fluxon propagates out along the output line.

Basic RSFQ Logic Gates

In addition to being a basic RSFQ memory cell, this RS flip-flop can also function as a latch for a logic gate. The data bit will enter the cell from the left and leave when a trigger signal (or timing signal) enters the right input. We can see how this works for the RSFQ OR gate shown in Fig. 6.19a. This is essentially the confluence buffer with its output connected to the RS flip-flop. Now, if two SFQ pulses enter inputs A and B at different times, only the first fluxon will be stored in the SQUID. The second fluxon will leave the line across the buffer junction J_s. Now, when the trigger pulse T is sent at the end, only a single SFQ output pulse is generated. This then functions as a more proper OR gate, in contrast to the simple confluence buffer. Note that of the six junctions that constitute this OR gate, four of them are buffer junctions and the other two form the latch. The OR output itself is really generated by the latch output junction J_2.

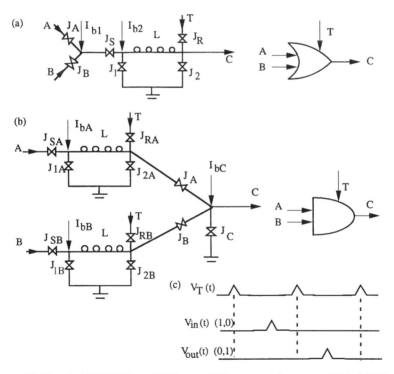

Figure 6.19. Practical RSFQ OR and AND gates (Likharev and Semenov, 1991). (*a*) Triggered OR gate with inputs A and B and trigger input T leading to output C. The symbol for a timed OR gate is also shown. (*b*) Triggered AND gate and symbol. (*c*) Transient voltages on trigger and I/O lines showing basic convention of timing in RSFQ logic; an SFQ pulse on an I/O (input or output) line during the time window between two trigger pulses represents a 1; the absence of such an SFQ pulse represents a 0.

Similarly, an improved AND gate can be designed that has an RS flip-flop before each input of a confluence buffer. If the input pulses A and B come in at different times, the latches serve to store them until they can be released simultaneously by the trigger pulse T. This in turn will trigger the (nearby) junction J_C, providing an SFQ pulse only when both inputs were 1. This gate has 11 junctions, where six junctions are input buffers, four constitute two latches, and only the last junction, J_C, generates the output.

Note also that for both AND and OR gates, there are multiple closed circulating loops (i.e., SQUIDs), which would be more evident if all the ground links were connected. This would give a much messier diagram; only the inductive loops that can store a flux quantum are shown. In fact, the other loops also include a series inductance, but one with $LI_c \sim 0.5\Phi_0$ (similar to those in the JTL), so that no static storage of flux is possible. The presence of all these closed superconducting loops makes these RSFQ circuits sensitive to the presence of stray magnetic fields and trapped flux. As for the similar situation with voltage-state logic devices, care must be taken in the use of magnetic shielding, the avoidance of any magnetized materials, and the design of "moats" to keep trapped flux away from the circuits.

Other RSFQ logic gates can also be designed using a similar approach. These include a NOT (or INVERT) gate, an exclusive-or (XOR) gate, and a variety of binary arithmetic gates. All contain one or more latches, together with a trigger input that resets the latch and outputs the data bit. In effect, the time between two consecutive trigger signals defines a "window" within which all inputs must arrive; this is the "basic convention" of RSFQ logic (Fig. 6.19c) (Likharev and Semenov, 1991). An SFQ pulse arriving during this period defines a 1; if no such pulse arrives on the given line, we have a 0. Furthermore, the triggered output signal can then be used as an input during the next window period. In this way, the timing associated with the trigger (or reset) pulses is fundamental to the logic design.

We indicated above that the RS flip-flop can also act as a DRO memory cell, where the "reset" line now carries the train of trigger pulses that define the time periods in which a new memory bit can enter the cell. This can be generalized to a linear array of memory cells, known as a shift register, as shown in Fig. 6.20. Here the bottom row of linked SQUIDs represents a series of binary bits, each cell (with loop inductance $L \sim 1.5\Phi_0/I_c$) contains either a single flux quantum (1) or no flux quantum (0). The upper row of SQUIDs do not store fluxons but rather form a Josephson transmission line that transmits the trigger/reset pulses entering from the right.

This shift register functions as a first-in, first-out (FIFO) memory. Each time a trigger pulse enters, it shifts the contents of each cell one cell to the right, starting with the rightmost cell. Then a new data bit can be inserted at the input on the left. Such a circuit can be very fast and compact, and similar circuits can be applied to the development of a variety of functions in digital arithmetic and signal processing.

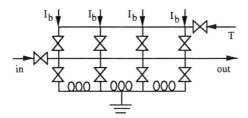

Figure 6.20. RSFQ shift register consisting of series array of DRO memory cells. The bottom row of cells stores the string of bits, which enter from the left. The top row of cells consists of a JTL that propagates the trigger pulse from the right and distributes it to each DRO cell. Each time a trigger pulse passes through the shift register, the contents of each cell move one cell to the right.

Finally, an important part of any superconducting digital system is its communication with the outside world, which generally operates at much lower frequencies. Two key circuits are therefore needed to generate and detect SFQ pulses using lower frequency signals. These are known as dc–SFQ and SFQ–dc converters. The simplest dc–SFQ converter is a single Josephson junction biased just above the critical current. This will generate a periodic train of SFQ pulses.

But in many cases, we would like to be able to produce just one SFQ pulse in a controlled fashion. The circuit in Fig. 6.21a is based on a current-biased dc SQUID with a loop inductance $L \sim \Phi_0/I_c$ and two bistable states, containing 0 and 1 fluxon. There are two current biases, one on the right that is fixed, and a second on the left (I_{in}) that can be "slowly" varied. The corresponding threshold curves as a function of the two biases are shown in Fig. 6.21b. As the current I_{in} is ramped up, the threshold between the 0 and 1 states is crossed, at which point junction J_1 switches, permitting a single fluxon to enter the loop across J_1 and an SFQ pulse (of opposite sign) to propagate down the output line. (Alternatively, one can say that J_1 generates a fluxon–antifluxon pair; the fluxon leaves to the right, while the antifluxon gets trapped in the loop.) When I_{in} is then decreased below current I_2, a threshold is crossed again, which now permits the trapped fluxon to leave the loop, this time across the other junction J_2. This does not generate an output pulse, but rather resets the circuit in preparation for another. So if an ac signal is sent to the input line, an SFQ pulse will be generated once per period, on the leading ramp of the wave.

There are several options to convert from an SFQ pulse to a dc voltage level. For example, if there is a repetitive train of SFQ pulses at frequency f, this can generate a Shapiro step at a dc voltage $V = \Phi_0 f$ in the dc $V(I)$ relation for a non-hysteretic junction. To detect a single SFQ pulse, one can simply use a hysteretic junction. A single SFQ pulse can trigger a transition to the resistive state with a standard output voltage ~ 3 mV. This must of course be reset by reducing the bias current to zero. Finally, an RSFQ-based SFQ–dc converter circuit has also

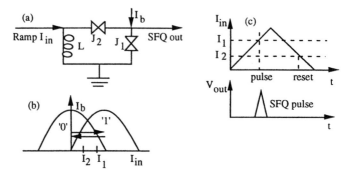

Figure 6.21. DC–SFQ converter. (a) Schematic circuit and current configuration in 0 and 1 states. (b) Approximate threshold curves; I_1 is the threshold current for the $0 \to 1$ transition, with I_2 for the return transition. (c) Input current ramp and output SFQ pulse, which occurs when I_{in} first exceeds I_1; another output pulse will not occur until after the cell is reset to 0 with $I_{in} < I_2$.

been developed by which a single SFQ pulse switches a nonhysteretic junction into the resistive state (with a typical voltage ~ 0.2 mV) but that can be easily reset using a second SFQ pulse.

Clock Generation and High Speed Testing

The RSFQ basic convention does not require that these trigger pulses be periodic, but RSFQ systems are generally implemented in this way, with the trigger pulses effectively providing the clock for a synchronous digital system. It would appear from Fig. 6.19c that the minimum clock period could approach just a few SFQ pulse widths. For a pulse width of ~ 4 ps, this suggests a clock frequency up to about 100 GHz, and indeed simple RSFQ systems have been shown to operate up to this range. More complex systems have operated up to at least 20 GHz, with higher speeds being developed. These very high speeds are the primary motivation for the further development of this technology.

There are several ways to generate the global clock for an RSFQ digital system using either an external source or one that is on the same chip with the superconducting circuit. For example, an external microwave signal can be used as the input signal to a dc–SFQ converter (Fig. 6.22a). This can be sent in from room temperature on a coaxial line for frequencies up to at least 20 GHz. This will generate an SFQ pulse once each period, and this pulse train can be distributed to the various gates via passive or active transmission lines, with splitters to enable sufficient parallel channels.

An on-chip clock source might consist of a single Josephson junction biased just above the critical current. If the current is very well controlled, this will generate a periodic pulse train at the Josephson frequency $f = V/\Phi_0$. A more practical alternative is a Josephson transmission line with its two ends connected to form a ring oscillator (Fig. 6.22b). If a single SFQ pulse is injected into this

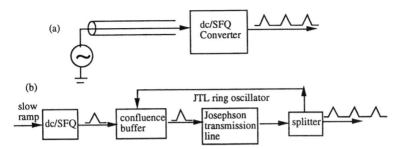

Figure 6.22. RSFQ clock generation for synchronous digital system. (*a*) Externally generated clock with microwave signal driving DC/SFQ converter; one SFQ pulse is generated each ac cycle. (*b*) Internally generated clock with SFQ pulse rotating around JTL ring oscillator. A slow ramp and dc/SFQ converter are used to insert a single SFQ pulse into the JTL ring. This pulse then rotates around and generates an output pulse at the splitter once each pass. (Current bias lines are present but not shown.)

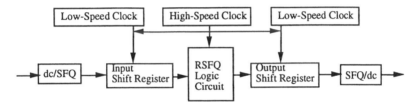

Figure 6.23. Block diagram of high-speed test circuit for RSFQ circuits. A logic circuit to be tested will have one or more input lines and one or more output lines; only one of each is indicated, and dc bias lines are present but not shown. An *N*-bit input sequence to be tested is first loaded into the input shift register at low speed using a dc/SFQ converter. These are then passed at high speed into the logic circuit, and the resulting output is passed to the output shift register. Finally, this output bit sequence is read out at low speed.

ring (using a standard RSFQ confluence buffer), then the pulse will continue to circulate indefinitely. If one unit of the JTL includes a splitter, then an SFQ pulse is coupled out of this ring once in each revolution of the pulse. The frequency is determined by the speed of the pulse around the ring, and one can monitor the frequency by measuring the dc average voltage on the JTL.

A key concern in distributing the clock signal is the "clock skew" associated with varying delays along different paths (Likharev and Semenov, 1991; Gaj et al., 1997). And, of course, there are similar delays in propagation of the various logic pulses themselves. As the clock frequency becomes ever higher these delays (typically of order 3 ps/stage of JTL and similar values for other cells) can lead to errors due to incorrect relative timing of the input and clock pulses. These delays can be at least partly compensated by careful design, but is quite likely that errors of this sort may serve to limit the ultimate speed of these circuits. In fact, many complex RSFQ circuits have been shown to function correctly up to clock frequencies of only ~ 20 GHz, although ideally they should continue to function several times faster.

An alternative approach to timing that may become necessary at the highest speeds involves giving up the concept of a global synchronous clock signal. Instead, each block of the circuit (or even each gate) could be "self-timed," with "send" and "acknowledge" signals triggering the exchange of data. The basic design of such an asynchronous RSFQ system has been explored, although it has not yet been fully implemented in complex circuits.

In contrast to the complexity of the clock distribution network, the power supply for the junctions is much simpler. In fact, each junction is biased with a fixed dc current of order I_c, generally in parallel. It is impractical and unnecessary to have a separate power supply for each junction, and the standard procedure is to have only one (or a few) external dc voltage lines, with an on-chip resistor network to distribute the currents appropriately among the various junctions. In practical RSFQ circuits, this resistor network dissipates far more power than that associated with switching of the junctions. Still, this is very small, of order 10–100 nW per junction, assuming $I_c \sim 100$ µA and the power distribution resistances ~ 1–10 Ω.

It is not a trivial matter to test the proper high-frequency operation of RSFQ logic circuits. Since these circuits are much faster than conventional room-temperature semiconductor circuits and operate at low voltage in a cryogenic environment, one cannot simply use standard semiconductor logic testers. In the most general case, it is necessary to determine the output of a given logic circuit for all possible combinations of input signals. One way to accomplish this (Kirichenko et al., 1997), using only low-speed input/output lines to the superconducting circuit, involves an N-stage Josephson shift register (Fig. 6.20) to store an initial arbitrary sequence of N input bits (one such shift register for each input line), running the logic circuit at high speed for N clock cycles, and storing the N output bits in another shift register (Fig. 6.23). The input shift register can be loaded from the outside at low speed using a dc/SFQ converter. Similarly, the output shift register can be read out at low speed using an SFQ/dc converter. This can be repeated as necessary using a variety of sequences and combinations of inputs.

Of course, this testing protocol requires two separate clock circuits, one for low speed and another for high speed, each containing exactly N SFQ pulses. The low-speed clock signal can be simply a series of "slow" pulses entered into a dc–SFQ converter; the exact speed does not matter. The high-speed clock circuit can be built around another N-stage SFQ shift register, this time loaded with a sequence of N 1's (again using another dc–SFQ converter). When these N SFQ pulses are released at high speed (as triggered by a high-speed clock such as those in Fig. 6.22), they form the high-speed clock sequence that causes the RSFQ logic circuit to be tested with the given input sequence. This may seem rather complicated, but it is only through tests such as these that the high-frequency limits of RSFQ logic circuits can be unambiguously determined. For current Nb or HTS junction technology, these limits tend to lie in the range of 20–30 GHz. At higher speeds, the circuits either start to produce a significant error rate or else stop operating entirely. Of course, it may be debatable whether it is the logic circuit or the test circuit that is failing, since they are based on the same technology.

Alternatives and Future Developments in SFQ Logic

Several other approaches have also been proposed for Josephson logic circuits based on single flux quanta and nonhysteretic junctions, and some prototypes have been demonstrated. One such approach is known as the "quantum flux parametron," or QFP (Harada, Hioe, and Goto, 1989). In contrast to the current injection that characterizes the RSFQ approach, the QFP requires magnetically coupled control lines, somewhat analogous to magnetically coupled logic gates in voltage-state logic. A QFP shift register with a data transfer rate in excess of 20 GHz has been fabricated and successfully demonstrated. Another alternative technology involves long Josephson junctions, for which the basic elements consist of fluxons traveling on continuous Josephson transmission lines, with splitters and combiners. These alternatives may have some advantages in certain regimes but are more likely to complement RSFQ circuits as these technologies develop rather than supplant them.

Most of the development of RSFQ circuits to date has been carried out using LTS Nb junctions resistively shunted to obtain $Q \sim 1$. As described earlier, the technology of such junctions is now well established, at least for junctions on the 1–5-µm scale. Junction reproducibility and uniformity are very important for RSFQ, as for all digital logic technologies. The technology of HTS Josephson junctions is still being advanced, but reproducible overdamped junctions with $I_c R \sim 0.1$–1 mV are starting to become available. These are ideal for application to RSFQ circuits (without the need for any extrinsic shunting), and indeed a number of such circuits have been fabricated and tested at speeds comparable to those for LTS RSFQ circuits. The key difference, though, is that they can operate up to much higher temperatures (up to 40 K and possibly higher). As this fabrication technology develops further, one can expect even faster speeds for such HTS circuits.

A related issue is that the scale of both LTS and HTS junctions is likely to move into the submicron range in the near future, in parallel with similar moves that have already taken place in the semiconductor world. This will permit much higher circuit densities and higher speeds, but in addition, as was pointed out earlier, this will tend to make all junctions intrinsically damped, both LTS and HTS. This is likely to reinforce the position of RSFQ as the leading candidate for the development of ultrahigh-speed, low-power Josephson digital circuits.

6.4 APPLICATIONS TO DIGITAL SYSTEMS

We have already discussed the ultrafast switching properties of Josephson junctions and how they may be combined to form logic gates that are much faster than those possible with conventional semiconductor technologies. However, it is a major step from the demonstration of performance in small circuits to application to large-scale digital integrated circuits and systems. In recent years, quite a number of potential applications have been proposed, and several of these have been demonstrated in advanced prototypes. It remains to be seen which of these applications will develop into real commercial products.

The key problem is that the semiconducting competition continues to advance rapidly, both in integration scale and clock speed. The current state-of-the-art Si VLSI chip (very large scale integrated circuit) contains of order a million transistors and operates at up to 1 GHz in clock frequency. The challenge for superconducting circuits is not merely to match current Si circuit performance, but rather to greatly surpass the performance that may be anticipated for the next 5–10 years, a formidable task indeed for any new technology. The application with the greatest potential payoff is a general-purpose computer, but this is also the most ambitious. It is more likely that superconducting digital circuits will first achieve acceptance for much smaller circuits that put a special premium on the ultrahigh speed and low power that are the hallmarks of Josephson logic. These may include fast switching networks, analog-to-digital (A/D) converters, and digital filters. In particular, digital signal processing for cryogenic sensors (for SQUID magnetometers or radiation detectors) may provide a natural opportunity for the application of medium-scale superconducting logic circuits. We will review recent achievements in these areas in the section below, with the understanding that this field will continue to evolve rapidly in the future.

Computer Systems

The versatile digital computer practically defines the modern age. Although computer architecture has become increasingly complex, the primary elements of any computing system are the central processing unit, or CPU, and the memory array (Fig. 6.24). The CPU carries out the arithmetic and logic operations at a characteristic clock rate and stores intermediate and final results in the memory. When all the components of the CPU are integrated on a single chip, as is common in most of today's computers, it is known as a microprocessor. The memory is typically organized in a hierarchy, where the fastest "cache" memory (which may sometimes be located on the microprocessor chip) is accessible on a time comparable to the clock time. There is also typically a larger "main memory" that can be somewhat slower.

For an ultrafast Josephson computer, it is probably necessary that both the CPU and the fast cache memory be superconducting and located in close proximity to each other. Even a minimal demonstration computer such as

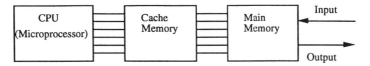

Figure 6.24. Highly simplified schematic diagram of generic computer. For a superconducting computer, the CPU and the fast cache memory are integrated superconducting circuits; the larger main memory might be semiconducting, possibly also cryogenic.

a 4-bit microprocessor with 4 kbits cache memory requires at least a medium-scale integration with thousands of gates (and several Josephson junctions per gate) for both the microprocessor and the memory array. Such a level of integration, in turn, requires a highly reliable and reproducible technology for the fabrication of Josephson junctions. When IBM initiated its ambitious program to develop a Josephson computer in the 1970's, the fabrication technology for these devices barely existed. A materials technology based on Pb-alloy Josephson junctions was developed, but despite major efforts, these junctions turned out to be insufficiently reliable for large-scale integrated circuits. IBM terminated its development program in 1983, and other U.S. corporations soon followed.

In the mid-1980s, the focus of efforts toward the Josephson computer shifted to Japan, where the Ministry of International Trade and Industry (MITI) coordinated the efforts of several Japanese corporations and laboratories (Fujitsu, Electrotechnical Laboratory, Hitachi, NEC). The materials technology had shifted to $Nb/AlO_x/Nb$ trilayer junctions, which proved to be much more robust and reliable and provided a firm basis for the fabrication of medium-scale integration of Josephson junctions on the 2–5-µm scale. The device technology was still based on voltage-state Josephson logic using hysteretic junctions. These Japanese efforts led by the late 1980s to the successful demonstration of several prototype Josephson microprocessors and memories based on voltage-state logic, with clock rates of order 1 GHz and 10,000 junctions or more. This was indeed significantly faster than competing technologies, with a much lower power level. Parameters of some representative systems are given in Table 6.1.

For example, the group at Fujitsu (Hasuo, 1993) designed and fabricated a Josephson microprocessor using a basic computer architecture modeled after an old standard 4-bit Si microprocessor, the AMD2901, which had also been implemented in GaAs. This contained some 5000 Josephson junctions with a 2.5-µm scale and fit on a 5-mm substrate. The superconducting Nb circuit operated at a clock speed of 770 MHz, as compared with 20 MHz for a comparable bipolar Si circuit and 70 MHz for GaAs, fabricated with similar device linewidths but operating at room temperature. Furthermore, the power dissipation of the superconducting circuit was only 5 mW, as compared with 1.4 W for Si and 2.2 W for GaAs. So for similar fabrication technologies, the Josephson implementation is far superior. In terms of Josephson memory arrays, the same

Table 6.1. Selected Voltage-State Josephson Microprocessors and Memory Arrays

Circuit	Gates	Junctions	Clock Rate (MHz)	Access Time (ps)	Power (mW)
4-bit microprocessor	1800	5000	800	—	5
64 × 64 RAM	4000	14,000	—	600	19
Microprocessor + ROM	3000	24,000	1100	—	6

Source: From Hasuo, 1993.

group fabricated a 4-kbit random-access memory (RAM), which contained some 14,000 junctions. This exhibited 600 ps memory access time and dissipated only 19 mW. Finally, the same microprocessor was fabricated using reduced-size junctions (1.5 µm) and integrated together with a multiplier and a 8-kbit read-only memory (ROM) on the same chip. This totaled some 24,000 junctions and ran at a slightly faster 1.1 GHz.

These prototypes confirmed that a Josephson computer was possible but also illustrated the major hurdles that would lie ahead. Integration on this scale was possible only with limited yield, due in part to continuing parameter variations among the junctions and also in part to effects such as trapped flux that are unique to Josephson junctions and SQUIDs. Furthermore, a 4-bit computer was already quite small and obsolete by this time, but it would be difficult to increase the integration scale and decrease the junction size (for, say, a 32-bit microprocessor) without a major improvement in fabrication technology. Finally, a 1-GHz clock speed was not sufficiently high given the continued rapid advance in Si computer technology. Again, a decision was made not to continue to pursue this path.

Despite this, the interest in superconducting computers was maintained by two major breakthroughs in the late 1980s. First, the discovery of high-temperature superconductors brought renewed attention to the field, and second, and more importantly for digital systems, the development of RSFQ logic promised dramatically higher clock speeds of 30 GHz and over 100 GHz for smaller junctions. These speeds are well in excess of what can be expected in the near future from conventional Si technology. On the other hand, current Nb junction technology may limit the integration scale of RSFQ logic to the same degree as for voltage-state logic. At the present time, HTS junction technology is even less mature. For these reasons, fabrication of a full-scale RSFQ microprocessor (Bunyk et al., 1997) would be a formidable task and has not yet been demonstrated.

Another related problem with Josephson computing is the inability (thus far) to make fully functional large-capacity memory arrays; the limit thus far has been 4 kbits. For this reason, a variety of proposals have been made for hybrids of Josephson logic with advanced Si memory arrays. It is worth noting in this regard that properly designed Si-CMOS memory cells work quite well at cryogenic temperatures, even at 4 K. Furthermore, they can perform substantially faster than at room temperature, with subnanosecond access times possible. Although hybrid prototypes have not yet been fully tested, this may provide at least a partial solution to the Josephson memory limitation (Van Duzer, 1997).

Switching Networks

Given the evident difficulty of competing with a mature silicon technology in general-purpose computers, several more specialized digital applications are being investigated for superconducting circuits composed of up to several thousand Josephson junctions. One class of such applications consists of high-speed switching networks. Such a network would be required, for example, for routing a signal in a high-speed communications system, or alternatively between processors in a parallel computing system (Van Duzer, 1997). In both cases, the signals are binary bits at a very high rate.

For example, in an optical fiber communication system, the standard data rate is typically either 2.5 Gbits per second (Gbps) or 10 Gbps, with even higher rates expected in the future. These are of course optical pulses, but they can be converted to voltage pulses using a fast semiconductor photodetector and back into optical pulses using a laser diode. The pulses are typically grouped in "packets," where each packet has a specific destination. Such a packet is likely to go through several switching networks before it ends up at its final destination. We can think of each such network as consisting of a two-dimensional $N \times N$ array of fast switches that direct a particular packet of bits from one of the N input lines to one of the N output lines (Fig. 6.25). This is sometimes referred to as a "crossbar" or "crosspoint" architecture. The address information is typically contained in the data stream itself, so that these packets should be "self-routing."

The current generation of fast switching networks is typically based on arrays of fast GaAs switches. These are very fast, but not quite fast enough, so that in most cases a series of electrical pulses must be sent sequentially into a number of parallel channels in order to slow down the rate sufficiently. Of course, these must be brought back together again at the end in the proper sequence, which requires careful timing. In addition, the total number of switches is greatly increased, which further increases the power consumption of the network.

In contrast, fast superconducting SFQ switches are clearly fast enough for the direct serial bit rate, so that a relatively simple $N \times N$ switching network may be sufficient. The performance of a prototype 4×4 switch operating at 8 Gbps was recently demonstrated, where the switches were superconducting Nb SFQ circuits operating at 4 K. It is likely that improvements can lead to substantial increases both in the number of elements in the array and in the speed.

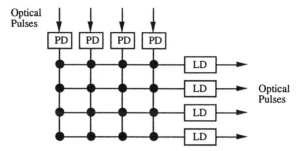

Figure 6.25. Conceptual diagram of optical fiber communication switching network. Fast packets of optical pulses at a rate of ~10 Gbits/s enter on N input fibers. Each packet is converted to a series of fast electrical pulses with a semiconductor photodetector (PD) and then routed to one of N output lines via a fast switch at the intersection (this is known as a "crossbar" switch design). A laser diode (LD) converts the electrical pulses back to optical pulses, which are sent out again on optical fibers. In a recent prototype, the switching network consisted of superconducting SFQ circuits; in some cases, the optoelectronic elements may also operate at cryogenic temperatures.

Analog Sampling and Interfacing

Another potential application of fast superconducting switches involves sampling and digital representation of fast analog signals. This is not strictly a digital systems application, but most signals originate in the analog world, and analog-to-digital (A/D) and digital-to-analog (D/A) converters are essential parts of the interface between the analog and digital worlds. Furthermore, it is important that these be carried out with as great precision and speed as possible, and prototype superconducting circuits are starting to address these issues.

In the laboratory, the first question about a fast analog signal is generally whether its time dependence $V(t)$ can be displayed on an oscilloscope. If the signal is a single-shot (non-repetitive) waveform on the picosecond time scale, this is very difficult, since it is much faster than standard semiconducting electronic circuits. The fastest standard oscilloscopes have effective bandwidths of order 1 GHz. One can do considerably better with a periodic signal or one that can be made periodic with a repetitive trigger. Here, a measurement can repetitively sample a specific part of a fast waveform on a somewhat slower time scale. The central components of such a "sampling oscilloscope," as shown in Fig. 6.26, are a fast pulse generator and a fast "comparator," both of which can be easily implemented using Josephson circuits (Hamilton et al., 1989). The arrival of the pulse selects the time to be sampled, and the comparator determines the value of the waveform at that time. The time resolution is essentially determined by the pulsewidth, which as we have seen can be of order 4 ps or less for an SFQ pulse. The Josephson junction or SQUID is a natural threshold detector for current and forms the basis for a current comparator. In essence, when the bias current I_b, signal current I_s, and pulse current I_p combine together to exceed the critical current, the Josephson comparator will generate an output

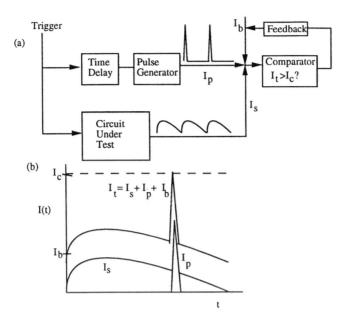

Figure 6.26. Basic concepts of Josephson sampling oscilloscope for fast repetitive signals. (a) Block diagram in which dc bias current I_b is added to signal and pulse currents I_s and I_p. The variable delay permits the pulse to be swept across the signal. The pulse generator and comparator (and in some cases the delay circuit as well) can be easily constructed using Josephson junctions. (b) Time dependence of currents entering comparator circuit. The bias current is adjusted (using external feedback) to maintain the peak comparator current at the threshold current I_c.

voltage. This can be implemented either with SFQ or with voltage-state logic. For a repetitive signal, I_b can be readjusted until I_c is just reached (so that the output is triggered half the time); in that case, the signal current is given by

$$I_s = I_c - I_p - I_b \qquad (6.11)$$

where all of the quantities on the right are known. This adjustment of I_b can be achieved using (non-cryogenic) feedback techniques. The time delay can then be electronically swept to span the entire period of the waveform; this may also be superconducting. Several versions of fast sampling circuits of this type have been used to measure repetitive signals up to ~ 100 GHz, typically for on-chip sampling of other fast superconducting circuits. In addition, a commercial 70-GHz Josephson sampling oscilloscope has been marketed by Hypres, Inc. for measuring repetitive signals at room temperature. The effective bandwidth was limited by the transmission line structure from 300 K to 4 K, rather than by the intrinsic speed of the superconducting circuit.

More generally, in evaluating a fast signal, we want not only a display of its $V(t)$ but also its precise measurement. This is generally accomplished by digitizing the signal. For binary representation of an analog quantity with an N-bit number, there is a magnitude associated with the least significant bit (LSB) and similarly with the most significant bit (MSB); this establishes the dynamic range. For example, if there are 10 bits and the LSB represents 1 µV, then the MSB represents $2^9 = 512$ µV, and the largest quantity represented (with a sequence of ten 1's) is 1023 µV. (If negative values are allowed, then one of the bits is needed to represent the sign.) In general, we would like to maximize both the sampling rate and the resolution (number of bits), but there is normally a trade-off to be made (Lee and Peterson, 1989). The standard equation relating bandwidth and resolution is given by

$$t_q = \frac{1}{\pi 2^N f_{max}} \qquad (6.12)$$

Here t_q is the "quantization time" or "aperture time" associated with assigning a particular binary bit to the digital representation of the signal, f_{max} is the maximum input bandwidth, and N is the effective number of bits (which need not always be an integer). The aperture time is essentially limited by the characteristic switching time of the basic logic gates. For example, if $N = 10$ and $f_{max} = 300$ MHz, this corresponds to $t_q \sim 1$ ps. If the measurement actually requires 10 ps, then we must be satisfied with a reduced bandwidth of 30 MHz, or alternatively with a reduced accuracy to about 7 bits.

Conventional A/D converters are of course based on semiconductor circuits. There are a number of different circuit designs to achieve the desired digitization of an analog signal, depending in part on whether the premium is on precision or on speed. The key advantage of superconducting A/D converters is that the aperture time should be significantly shorter than that for semiconductors, permitting in principle both greater precision and higher speed. Recent superconducting prototypes have indeed matched or exceeded the best semiconducting A/D converters, with further improvement anticipated. Examples of such Josephson A/D converters using several different design approaches are shown in Fig. 6.27, with recent performance listed in Table 6.2. Both voltage-state and SFQ logic designs have been proposed, although SFQ designs are somewhat faster and are now predominating (Likharev, 1990; Przybysz, 1993). All approaches include one or more simple comparators (or quantizers), essentially threshold detectors based on junctions or SQUIDs, that determine whether the analog signal is larger or smaller than some fixed value. The approaches differ in the number of such comparators that are being processed in parallel.

A fully parallel approach for N bits of resolution requires $2^N - 1$ comparators (Lee and Peterson, 1989). For example, for $N = 10$ bits, there are 1024 distinct analog ranges and 1023 values separating these, with a comparator for each of them. The key advantage of this "flash" approach is its high speed; the entire digitization can be accomplished in a single clock period. The major disadvantage is that this requires quite a large number of precision-matched gates and junctions for a reasonable number of bits, which is a serious problem at the

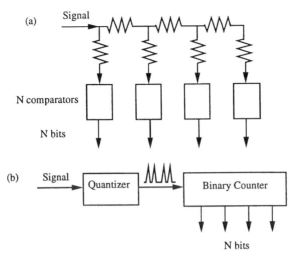

Figure 6.27. Some alternative approaches to superconducting A/D converters. (*a*) Flash-periodic A/D, with *N* parallel comparators generating *N* bits. (*b*) Counting-type A/D, with *N* bits generated by binary counter.

Table 6.2. Selected Types of Superconducting Nb A/D Converters and Their Typical Performance

Type of A/D	Number of Bits	Bandwidth
Parallel-serial (flash-periodic)	3	10 GHz
Serial (counting)	12	6 MHz
Oversampling (sigma–delta feedback)	16	4 MHz (simulated)

Source: From Van Duzer, 1997.

current stage in Josephson fabrication technology. Fortunately, an alternative approach that is also very fast requires only N comparators, one for each bit (e.g., 10 comparators instead of 1023 for $N = 10$). This method is sometimes known as flash-periodic, or bit-parallel, or parallel-serial. It is based on using the periodic threshold characteristic of a SQUID as the comparator.

A completely different approach to A/D converters is based on using only a single input comparator, known in this context as a quantizer. This is again essentially a junction or SQUID, but here a given voltage generates a series of SFQ pulses (based on the Josephson relation), which can then be counted up to determine the voltage. This is known as a serial or "counting" A/D and is much slower due to both the multiple pulse generation and the counting process. It is, however, possible to obtain a rather high precision with a fairly simple circuit. Furthermore, it is possible to obtain additional bits of resolution (at the expense of some bandwidth) with additional signal processing.

One can think of this in analogy to analog measurement using a SQUID magnetometer. As was discussed earlier in Section 5.3, a SQUID measures flux in units of the flux quantum Φ_0, and it can certainly count multiple flux quanta. But a well-designed SQUID system can also measure a small fraction of a flux quantum. Furthermore, this is normally operated in a feedback mode, where a null detector permits linear operation with both high resolution and a large dynamic range (many flux quanta). In an analogous way, the pulse output of the quantizer can be averaged to obtain additional bits of resolution by a process known in the digital domain as "oversampling." One type of oversampling can be combined with feedback (using a simple D/A converter) to obtain a much higher dynamic range, as with the case of the SQUID. This is commonly used for semiconductor A/D converters, where it is known as "sigma–delta" conversion. Efforts along these lines have been proceeding for superconducting circuits as well, although the circuits are somewhat more complex than for the simple counting A/D (Przybysz, 1993). Finally, a SQUID magnetometer itself can be operated with on-chip digital feedback; this will be discussed below in the context of integrated cryogenic sensors.

A D/A converter presents similar considerations to those for an A/D converter; there is again a trade-off between speed and resolution (numbers of bits). There has not been quite as much development of Josephson D/A converters, but several recent prototypes have been demonstrated. These include an 8-bit direct digital frequency synthesizer (Silver, 1997) designed using voltage-state logic that can generate rapidly varying sine waves up to 4 GHz in frequency. A quite different kind of D/A converter is a Josephson array voltage standard with a digitally programmable voltage. Given a stable precise external source of microwaves, such a system will be able to generate an arbitrary voltage waveform (in the 1-V range) to extremely high precision.

Digital Signal Processing and Cryogenic Sensors

Once a signal is translated into digital form using an A/D converter, then it is possible to use digital logic circuits to carry out further processing, opening up a wide range of possibilities. These digital signal processing (DSP) circuits may include fast digital filters, correlators, fast Fourier transforms (FFT), and other programmable functions. Unlike a general-purpose computer, these computations can generally be achieved using only a relatively small read-only memory (ROM), which is feasible even with present-day Nb fabrication technology. One potential class of DSP applications is to microwave communications and radar, which are pushing the frequency limits of conventional semiconducting technology. As the demand for higher bandwidth increases, high-speed DSP offers the possibility of more efficient utilization of these new channels. Furthermore, superconducting DSP circuits are extremely low in power dissipation, which may make them particularly appealing for some applications such as those on satellites.

However, superconducting circuits are generally perceived to have one serious disadvantage: they require cooling down to cryogenic temperatures. The most advanced fabrication technology is for Nb circuits, which generally require cooling to less than about 5 K for optimum operation. This causes two serious problems for many digital circuit applications. First, as discussed in Appendix C, closed-cycle refrigeration to these temperatures typically requires at least 1000 times more power delivered (and dissipated) at room temperature than the cooling power available at 4 K. Second, and equally important, an ultrahigh bandwidth transmission structure between room-temperature semiconductor circuits and cryogenic (and low-voltage) superconducting circuits is difficult to achieve. These reasons may help to explain the lack of current applications for superconducting digital systems, despite the superior high-speed performance.

On the other hand, there are a set of potential applications that already require cryogenic cooling for one or more sensors or other detectors. These are generally critical components in advanced scientific instruments, which seek to obtain the highest possible performance. These components include not only superconducting sensors (to be discussed further in Chapter 7) but also some low-noise semiconducting sensors for detection of photons and particles. If the signals from these cold sensors can also be processed or pre-processed at low temperatures, this can drastically reduce the amount of room-temperature instrumentation and interfacing needed to take full advantage of these devices. These systems will probably provide the first real implementation of superconducting DSP technology.

Probably the foremost superconducting sensor is the SQUID for detection of ultrasmall magnetic fields. As was discussed earlier in Chapter 5, virtually all SQUIDs are operated in a feedback mode, where the phase-sensitive detection and amplification take place at room temperature. All of this processing could in principle be carried out using superconducting circuits on the same chip as the SQUID sensor itself. Several demonstrations of this have been carried out, making use of modern DSP approaches to provide the feedback in the digital domain. The magnetic signal itself is then available for digital processing or output, so that such a digital SQUID can be regarded as a high-performance A/D converter. Although this technology is still evolving, the low-noise performance of such digital SQUIDs (Yuh and Rylov, 1995) is now approaching that of commercial analog SQUID systems and will likely soon surpass them. Furthermore, arrays of SQUID sensors are being developed for biomedical applications (such as for magnetic source imaging of the brain), and integrated digital SQUIDs are ideally configured for superconducting digital preprocessing of the array of data.

There are also a variety of cryogenic detectors and detector arrays that are used in astronomy and high-energy physics, often (but not always) at liquid helium temperature (4 K). For example, reconstruction of a collision in particle physics requires an accurate measurement of the relative timing of many events on the subnanosecond time scale. Each of many cooled (semiconducting) detectors will generate a fast electrical pulse when a given particle is incident. This pulse can be transmitted to a nearby fast SFQ digital counter to provide accurate timing of the event. A recent prototype 14-bit time-to-digital converter using RSFQ logic and Nb junctions demonstrated 50 ps resolution and dissipated less than 1 mW (Mukhanov and Rylov, 1997).

Cryogenic detectors are used for astronomical applications across the electromagnetic spectrum, from radio frequencies through infrared and ultraviolet to x-rays. Some of these are superconducting, but advanced semiconducting detectors are also generally cooled to reduce thermal noise. For imaging applications, a two-dimensional array of detectors is typically placed at the focal plane of a telescope. Each detector provides an electrical signal that contributes to a given pixel of the image. Substantial improvements in performance and reduced power can be obtained if the signals are digitized and processed by cryogenic electronics at the focal plane before transmitting the partly processed data to room-temperature electronics. In some cases, a combination of both cryogenic semiconducting and superconducting electronics may be appropriate. A recent prototype (Silver, 1997) demonstrated this using a 128×128 semiconducting infrared focal plane array coupled to a GaAs multiplexer and a superconducting NbN A/D converter (12 bits at 1 MHz bandwidth), all operating at 10 K (Fig. 6.28).

As the above example indicates, superconducting digital circuits are not necessarily limited to operating at 4–5 K if an alternative higher temperature material is available. In most cases, the limiting factor is the reliability of the junction fabrication technology. As the number of junctions increases, this becomes increasingly more critical. Niobium junction technology is the most mature, and indeed circuits with several thousand junctions can now be made that perform a variety of digital switching and processing functions but must operate at around 4 K. The NbN junction technology is also rapidly maturing and is now capable of medium-scale integrated circuits with junctions on the 3–5-μm scale. Although YBCO junction technology is still not quite as mature, there have also been several complete YBCO digital circuits based on SFQ logic and operating at 40 K and above, including simple A/D converters and basic RSFQ gates (Van Duzer, 1997). As this technology improves, there will certainly be more complex YBCO digital circuits fabricated and tested. The higher

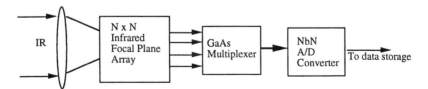

Figure 6.28. Block diagram of infrared focal plane array with semiconducting and superconducting electronics on focal plane at 10 K. A prototype using a 12-bit NbN A/D Converter was fabricated and successfully tested.

operating temperature may not necessarily lead to better device performance than Nb at 4 K, but it may help to develop real-world applications for ultrafast superconducting digital circuits.

SUMMARY

- For a JJ with $Q > 1$, a current pulse will cause the junction to switch from $V = 0$ to $V \sim 1$ mV, and to remain in the voltage state until the bias current is reduced (hysteresis). These two voltage levels form the basis of voltage-state logic.
- For $Q \lesssim 1$, a current pulse will cause the junction to generate a single-flux-quantum (SFQ) voltage pulse before returning to $V = 0$. The presence or absence of an SFQ pulse forms the basis of SFQ logic.
- This can be viewed in analogy to a rotating pendulum. A static pendulum corresponds to $V = 0$, a rotating underdamped pendulum to the voltage state $V > 0$, and a single damped rotation to an SFQ pulse.
- The temporal response of a JJ in the CRSJ model is limited by the slower of two characteristic times—$\tau_1 = RC$ and $\tau_2 = L_J/R$—and by $Q = \sqrt{\tau_1/\tau_2}$. For underdamped junctions with $Q > 1$, τ_1 is slower; for overdamped junctions with $Q < 1$, τ_2 is slower.
- For ideal LTS tunnel junctions on the micron scale with $I_c \sim 100$ μA, $Q \gg 1$ and the switching speed (~ 10–100 ps) is limited by C. Here Q can be reduced to 1 for faster switching (to ~ 1 ps) by either adding a resistive shunt or decreasing the junction area to the submicron range.
- Current HTS junctions have internal resistive shunting with $Q \ll 1$ typical, but switching times ~ 1 ps are feasible.
- Switching energies of Josephson logic gates are $< 10^{-18}$ J, much smaller than anticipated for semiconductor logic.
- Voltage-state logic gates are based on threshold characteristics of Josephson junctions or SQUIDs. Current from one or more control lines causes the gate to cross the threshold from $V = 0$ to the voltage state; this forms the basis for OR and AND gates, among others.
- Voltage-state logic requires a global power supply that periodically reduces the bias current to zero, resetting the gates to $V = 0$ and permitting another logic step to be processed.
- Memory cells in voltage-state logic are based on flux stored in SQUID loops, with write and read gates.
- A model 4-bit microprocessor with 1 GHz clock rate and a 4-kbit RAM with subnanosecond access time, with thousands of junctions each, have been demonstrated using Nb technology.
- The leading SFQ logic family, RSFQ logic is based on picosecond SFQ pulses propagating on discrete Josephson transmission lines, with splitters,

combiners, and buffering stages to direct pulses. All RSFQ logic gates require internal latches and triggering pulses to achieve picosecond timing.
- Complex RSFQ logic circuits have been demonstrated using LTS Nb technology at clock speeds above 20 GHz, with higher speeds anticipated. Some HTS RSFQ circuits have also been demonstrated at comparable speeds.
- Developing applications for ultrafast digital superconducting systems include switching networks, A/D and D/A converters, and digital signal processing for cryogenic sensors.

REFERENCES

W. Anacker, Ed., "Josephson Computer Technology," special issue of *IBM J. Res. Devel.* **24**, 105–264 (March 1980).

P. Bunyk, A. Y. Kidiyarova-Shevchenko, and P. Litskevitch, "RSFQ Microprocessor: New Design Approaches," *IEEE Trans. Appl. Supercond.* **7**, 2697 (1997).

K. Gaj, E. G. Friedman, and M. J. Feldman, "Timing of Multi-GHz RSFQ Digital Circuits," *J. VLSI Signal Process.* **16**, 247 (1997).

C. A. Hamilton, D. G. McDonald, J. E. Sauvageau, and S. R. Whitely, "Standards and High-Speed Instrumentation," *Proc. IEEE* **77**, 1224 (1989).

Y. Harada, W. Hioe, and E. Goto, "Flux Transfer Devices," *Proc. IEEE* **77**, 1280 (1989).

S. Hasuo, "Josephson Microprocessors," in *The New Superconducting Electronics*, Ed., H. Weinstock (Kluwer, Dordrecht, 1993).

S. Hasuo and T. Imamura, "Digital Logic Circuits," *Proc. IEEE* **77**, 1177 (1989).

H. Hayakawa, "Computing," in *Superconducting Devices*, Ed., S. T. Ruggiero (Academic, New York, 1990).

A. F. Kirichenko, O. A. Mukhanov, and A. I. Ryzhikh, "Advanced On-Chip Test Technology for RSFQ Circuits," *IEEE Trans. Appl. Supercond.* **7**, 3438 (1997).

G. S. Lee and D. A. Peterson, "Superconductive A/D Converters," *Proc. IEEE* **77**, 1264 (1989).

K. K. Likharev and V. K. Semenov, "RSFQ Logic/Memory Family: A New Josephson Junction Technology for Sub-THz Clock Frequency Digital Systems," *IEEE Trans. Appl. Supercond.* **1**, 3 (1991).

K. K. Likharev, "Progress and Prospects of Superconductor Electronics," *Supercond. Sci. Technol.* **3**, 325 (1990).

O. A. Mukhanov and S. V. Rylov, "Time-to-Digital Converters based on RSFQ Digital Counters," *IEEE Trans. Appl. Supercond.* **7**, 2669 (1997).

J. X. Przybysz, "Josephson Analog-to-Digital Converters," in *The New Superconducting Electronics*, Ed. H. Weinstock (Kluwer, Dordrecht, 1993).

A. H. Silver, "Superconductivity in Electronics," *IEEE Trans. Appl. Supercond.* **7**, 69 (1997).

Y. Taur, Y. J. Mii, D. J. Frank, H. S. Wong, D. A. Buchanan, S. J. Wind, S. A. Rishton, G. A. Sai-Halasz, and E. J. Nowak, "CMOS Scaling into the 21st Century: 0.1 μm and Beyond," *IBM J. Res. Devel.* **39**, 245 (1995).

T. Van Duzer, "Superconductor Electronics, 1986–1996," *IEEE Trans. Appl. Supercond.* **7**, 98 (1997).

Y. Wada, "Josephson Memory Technology," *Proc. IEEE* **77**, 1194 (1989).

P. F. Yuh and S. V. Rylov, "An Experimental Digital SQUID with Large Dynamic Range and Low Noise," *IEEE Trans. Appl. Supercond.* **5**, 2129 (1995).

PROBLEMS

6.1. NbN junctions for digital circuits. NbN is being developed for digital circuit applications at ~ 10 K. Assume that the energy gap is $2\Delta = 5$ meV, that $\lambda = 300$ nm, that the insulator consists of 1 nm of MgO with $\varepsilon_r = 10$, and that the critical current density is 1000 A/cm^2 for $T \ll T_c$.

(a) For a junction with area $A = 25$ μm^2, determine I_c, L_J, λ_J, C, and R_n.

(b) Determine τ_1, τ_2, τ_0, and Q for this junction. Which is the limiting characteristic time?

(c) What shunt resistance would be necessary to achieve $Q = 1$ for RSFQ applications? What would be the limiting time in that case?

(d) If we make the junction smaller while maintaining I_c, for what junction size would $Q = 1$ without the need for external shunting? (Assume that the change in the insulator thickness can be neglected in this estimate.) What is the limiting characteristic time here?

6.2. Subgap resistance and characteristic times. In the estimation of characteristic times in Section 6.1, we generally assumed that the relevant resistance is R_n, even for unshunted junctions with a large subgap quasiparticle resistance. For a high-quality Nb junction, for example, the subgap resistance is generally at least a factor of 10 above R_n at 4 K, with a larger ratio at lower T. In a transient switching process, this would have the greatest effect in the resetting time, when the pendulum may undergo low-amplitude oscillations around its equilibrium position.

(a) What effect would the subgap resistance have on a typical unshunted high-Q junction? Would this be expected to limit the maximum clock rate of a voltage-state logic circuit or introduce errors?

(b) What effect does a large R_{sg} have on the dynamics of a typical externally shunted junction and on digital circuit applications?

(c) Now consider a very small junction that achieves $Q \sim 1$ without external shunting. What effect would R_{sg} be expected to have on the SFQ response of such a junction? Would this be expected to limit the maximum clock rate of an RSFQ logic circuit or introduce errors?

6.3. Transient response of underdamped junction in SPICE. The transient junction responses shown in Fig. 6.4 for an underdamped Josephson junction were obtained numerically using SPICE. Appendix B discusses several versions of SPICE and other circuit simulation programs that include the Josephson junction as well as a way to incorporate the JJ into standard SPICE. In this problem you are to reproduce these simulations using a version of SPICE or equivalent.

(a) For a high-Q junction with $I_c = 100$ μA, $R_n = 20$ Ω, and $C = 0.5$ pF, calculate and plot the transient voltage response to a supercritical 150-μA current pulse, 10 ps wide, of a junction prebiased at 70 μA. Identify the relevant characteristic times in the problem and label on the plot.

(b) Repeat for a junction with a subgap resistance $R_{sg} = 100$ Ω for $V < 2\Delta/e = 2.8$ mV. Estimate the switching time into the voltage state and the resetting time back into the zero-voltage state.

6.4. Critically damped junction in SPICE. Reproduce the SPICE simulations for a critically damped junction, as in Fig. 6.5 (see Appendix B for further details on SPICE simulations).

(a) For a junction with $Q \approx 1$, having $I_c = 100$ μA, $R = 2.6$ Ω, and $C = 0.5$ pF, calculate and plot the transient voltage and phase response to a supercritical 350 μA pulse, 2 ps wide, of a junction prebiased at 70 μA.

(b) Do the same for a smaller pulse with height of 200 μA, still above the critical current. Does the output correspond to an SFQ pulse? What is the net phase change?

(c) Calculate the dc $V(I)$ relation for these junction parameters. This may be achieved point by point or by ramping the bias current very slowly. The dc voltage must be averaged over many Josephson periods for an accurate result.

6.5. Voltage-state DRO memory cell. Consider the destructive readout (DRO) memory cell in Fig. 6.13.

(a) For the SQUID loop initially in the 0 state, a 1 is first written and then read out, followed by resetting the cell to its initial condition. Sketch the sequences of pulses needed for the control and bias currents and the corresponding voltage output $V(t)$. Include the SFQ voltage pulse that occurs during this process.

(b) Do the same thing for the SQUID loop initially in the 1 state, with a 0 written to the cell, followed by readout and resetting.

6.6. RSFQ OR and AND gates. Consider the RSFQ OR and AND gates as shown schematically in Fig. 6.19.

(a) For the OR gate of Fig. 6.19a, assume that on both inputs A and B, a 1 arrives at the same time. Shortly after this, a pulse on the trigger (T) line releases the output on C. Sketch the sequence of SFQ voltage pulses on each of the lines (A, B, C, and T) as well as the transient SFQ voltages across each of the six junctions.

(b) Do the same for the case of 1 pulses arriving on A and B at slightly different times. Does this give only a single output pulse at C, as required? In particular, does the extra fluxon escape, and across which junction?

(c) For the AND gate of Fig. 6.19b, assume that a 1 arrives on A and a 0 at B. By following the voltage pulses across the relevant junctions, show that no output is generated on C after the trigger pulse is released. What happens to the fluxon that approaches junction J_c?

(d) Now consider the AND gate with a 1 on both A and B. Show that a pulse is released at the output following a trigger input to both registers.

6.7. High-speed testing of RSFQ circuit. In Fig. 6.23, a generic approach to high-speed testing of RSFQ circuits is illustrated. Show how this could be applied to the simple example of an AND gate.

(a) How many input and output shift registers would be needed, and how many cells per register, to test all possible combinations of inputs?

(b) Generate the input sequences and corresponding output sequences of 1's and 0's.

(c) Assume that the high-speed clock runs at a frequency of 50 GHz and that the pulses are ~ 5 ps wide. During the high-speed testing of the circuit, show the expected voltage pulses on each input and output line of the AND gate, including the trigger input from the high-speed clock. Label the axes with appropriate units.

6.8. Josephson A/D converter. For a given analog-to-digital converter technology, the trade-off between the bandwidth of the analog signal and the resolution of the binary representation is given by Eq. (6.12).

(a) If the aperture time $t_q = 2$ ps, determine the effective number of bits for a 500-MHz signal. If the maximum input range of the signal is 0–0.1 mV, what is the size of the least significant bit? The most significant bit?

(b) What is the maximum frequency that can be represented with 10 bits accuracy?

(c) If the input quantizer is simply a Josephson junction in Fig. 6.27c, how fast must the binary counter operate to cover the entire input range? Does this seem possible using RSFQ circuits?

7

SUPERCONDUCTING RADIATION DETECTORS

As a final class of developing superconducting electronic applications, we will consider in this chapter how superconducting thin films and tunnel junctions may be used for the sensitive detection of radiation across the electromagnetic spectrum. We have already discussed in Chapter 5 the low-frequency detection of magnetic flux using SQUIDs, so we will not deal with that here. But superconducting devices are also used in sensitive detection of microwaves for radio astronomy, and new applications in detectors for infrared imaging and for x-ray astronomy are rapidly developing. These systems depend on a variety of different characteristics, but a common theme is that they take advantage of a sharp threshold in some property of the superconductor. For example, the resistance $R(T)$ jumps sharply at T_c, and similarly for the voltage at the critical current I_c, the current through a tunnel junction at the gap voltage $2\Delta/e$, and the magnetization of a type I superconductor at the critical field H_c. If the superconductor is biased just below the threshold and an rf current or an x-ray photon acts to push the device over the threshold, then we have the basis for a sensitive detector. As we have seen, superconductors can also make fast switches, so that these detectors can also be very fast. These two aspects of detection, sensitivity and speed, are often complementary, but in many cases superconductors offer an unrivaled combination of both. For these reasons, despite the need for cryogenic cooling, superconducting devices are increasingly being used in high-performance detection and imaging systems.

Electromagnetic detectors can be classified into three major categories, largely associated with the frequency range for which they are used (Table 7.1). In the low-frequency range, typically up through microwave frequencies, a "modulation detector" is fast enough to follow the ac current or voltage directly, and individual photons can be neglected. In high-frequency range, generally from the

Table 7.1. Classes of Electromagnetic Detectors and Superconducting Examples

Detector Class	Range	f (Hz)	λ (µm)	Superconducting Example
Modulation	Radio, microwave	$<10^{12}$	>1000	SIS mixer
Thermal	Infrared	10^{11}–10^{15}	1–1000	YBCO bolometer
Photon	Visible, UV, x-rays	$>10^{14}$	<1	Tunnel junction photon detector

visible through the x-ray range, a "photon detector" is based on the response due to the absorption of single photons of energy $\mathscr{E} = hf$. For a low enough photon flux, individual photons can be measured. In the intermediate regime, essentially covering the infrared, photons are not yet large enough to detect individually, but the frequency is too high for direct ac measurement, so the detector corresponds to the average effect. In this range, the energies of many small photons (or high-frequency currents) are sufficient to produce local heating, and "thermal detectors" are used. Superconducting devices have contributed substantially to all three regimes, which are discussed sequentially in the sections below.

7.1 MODULATION DETECTORS AND MIXERS

Conventional analog radio technology makes use of nonlinear devices to modulate a signal onto a sinusoidal carrier wave at the transmitter and to demodulate it at the receiver. These operations are conventionally carried out using semiconductor technology up to frequencies of order 10 GHz or higher. Superconducting tunnel junctions offer the possibility to extend similar techniques to create highly sensitive receivers up to considerably higher frequencies. We will address several such devices, focusing on the very successful SIS quasiparticle mixer. But first, we will review the basic principles of more standard modulation detectors in order to show how superconducting devices can offer improved performance.

Direct and Heterodyne Detection

Consider first a simplified dc $I(V)$ relation for an idealized diode, with $I = 0$ for $V < 0$ and $I = V/R$ for $V > 0$ (Fig. 7.1). The time-dependent $I(V)$ relation should be the same, provided that V does not vary too fast, compared to the characteristic response time of the diode. Assume that the voltage consists of the sum of a dc and a sinusoidal ac component at frequency ω_s:

$$V(t) = V_0 + V_s \cos(\omega_s t) \qquad (7.1)$$

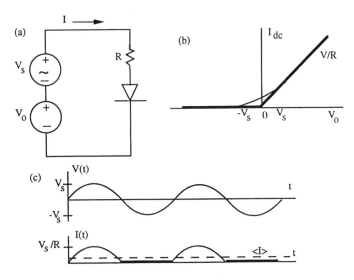

Figure 7.1. Rectifying behavior of simplified diode. (*a*) Schematic circuit. (*b*) Piecewise linear approximation of diode dc $I(V)$, with rounded time average in presence of ac voltage. (*c*) $V(t)$ and $I(t)$ for dc bias $V_0 = 0$.

where V_0 is the voltage bias and V_s is the amplitude of the signal we are trying to detect. For a purely linear element, of course, the current response would also have just a dc component and one at ω_s, but the diode is not a purely linear element. In particular, at and around $V = 0$, it is highly nonlinear. As illustrated in Fig. 7.1c, for $|V_0| < V_s$, the ac current will be partly rectified by the diode, giving a dc-average current $\langle I(t) \rangle$ greater than the intrinsic dc $I(V)$. If we fix V_s and sweep the dc voltage V_0, this rectified current rounds out the sharp corner and is dependent only on the ac amplitude, not on its frequency. The effective current response $\delta I = \langle I \rangle - I(V_0)$ is thus the output signal for this diode detector.

This piecewise linear approximation of a diode characteristic is somewhat oversimplified, particularly in its treatment of the nonlinearity. A better approximation for a semiconducting diode has the standard thermal exponential factor in the form

$$I = I_0 \left[\exp\left(\frac{eV}{kT}\right) - 1 \right] \qquad (7.2)$$

This has a more gradual curvature near the origin, with a characteristic voltage width $\Delta V \sim kT/e$ (~ 25 mV at room temperature) and is everywhere differentiable. If we now consider the same dc + ac voltage bias of Eq. (7.1) and assume that $V_s \ll \Delta V$, we can expand the current in a Taylor series expansion, up to the first nonlinear term:

$$I(t) \approx I(V_0) + I'(V_0)(V - V_0) + \tfrac{1}{2}I''(V_0)(V - V_0)^2$$
$$= I(V_0) + I'V_s\cos(\omega_s t) + \tfrac{1}{2}I''V_s^2\cos^2(\omega_s t)$$
$$= I(V_0) + \tfrac{1}{4}I''V_s^2 + I'V_s\cos(\omega_s t) + \tfrac{1}{4}I''V_s^2\cos(2\omega_s t) \quad (7.3)$$

where $I' = dI/dV = I_0 e/kT \exp(eV_0/kT)$ and $I'' = d^2I/dV^2 = I_0(e/kT)^2 \exp(eV_0/kT)$ are the first and second derivatives of Eq. (7.2) at $V = V_0$. Here, the dc output signal is $\delta I = \tfrac{1}{4}I''V_s^2$. Note that this output is proportional to the curvature I'' of the dc $I(V)$ relation and also to the square of the voltage amplitude, V_s^2, both of which are true more generally. This is sometimes called a "square-law" detector or a "direct detector." The diode is acting as a mixer that is mixing (or multiplying) the input signal with itself, thus giving second-order contributions at the sum and difference frequencies, namely dc and $2\omega_s$. The ac power absorbed is also proportional to V_s^2:

$$P = \langle V_s\cos(\omega_s t)I'V_s\cos(\omega_s t)\rangle = \tfrac{1}{2}I'V_s^2 \quad (7.4)$$

so that the dc output is directly proportional to the input power. With that in mind, it is conventional to define the responsivity

$$R = \frac{\delta I}{P} = \frac{I''}{2I'} = \frac{e}{2kT} = \frac{1}{2\Delta V} \quad (7.5)$$

which has units of amperes per watt or reciprocal volts. Note that this rectification is essentially independent of frequency, so that the power in a very wide bandwidth can be detected. We normally would like as large a responsivity as possible, so that we can obtain a large output signal from a small input signal power. This requires an $I(V)$ relation with a sharp nonlinearity at the operating voltage.

There is an alternative mode of operation of the same diode that can yield a much larger output for the same weak input. Rather than mixing the weak signal at ω_s with itself, we mix it with a strong "local oscillator" (or "pump") at a second frequency ω_p that is close to ω_s. This generates a response at the reduced difference frequency $\omega_s - \omega_p$, which is known as the "intermediate frequency" ω_{if} or sometimes the "beat frequency." Indeed, this is the standard process for detecting a weak rf signal by demodulating the signal from the rf carrier using "heterodyne detection." We can see this for the example given above, where the voltage bias now includes both a dc and an ac part (Fig. 7.2):

$$V(t) = V_0 + V_s\cos(\omega_s t) + V_p\cos(\omega_s t) \quad (7.6)$$

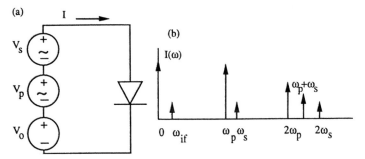

Figure 7.2. Heterodyne mixer with diode biased with local oscillator (pump) at frequency ω_p in addition to dc voltage V_0 and signal at ω_s. (a) Schematic circuit. (b) Spectrum of current response. The intermediate frequency $\omega_{if} = \omega_s - \omega_p$ is normally amplified and the rest filtered out.

If we again take the Taylor series expansion around the operating point V_0, we obtain

$$I(t) = I(V_0) + I'[V_s\cos(\omega_s t) + V_p\cos(\omega_p t)]$$
$$+ (\tfrac{1}{2}I'')[V_s\cos(\omega_s t) + V_p\cos(\omega_p t)]^2 \quad (7.7)$$

The final term gives a dc contribution as well as ones at $2\omega_s$ and $2\omega_p$ and in addition at $\omega_{if} = \omega_s - \omega_p$ and $\omega_s + \omega_p$. Focusing on the component at ω_{if}, we have

$$\delta I_{if}(t) = (\tfrac{1}{2}I'') V_s V_p \cos(\omega_{if} t) \quad (7.8)$$

This is similar to the output signal from the direct detector, in that both are proportional to the second derivative $I''(V_0)$. However, note that the amplitude here is proportional to $V_s V_p$, which is $\gg V_s^2$ if the amplitude of the pump $V_p \gg V_s$. As long as we can provide a stable pump of known frequency and amplitude, heterodyne detection should provide for a much more sensitive detector. Furthermore, spectral information about the weak input signal can also be inferred, unlike the case of direct detection for which this information is lost. On the other hand, if the total power in a wide frequency band is desired, then direct detection as described earlier may be preferred.

Equation (7.8) suggests that the *if* signal can become arbitrary large if the pump amplitude is large enough. Actually, this is not true; in fact, the *if* output power is always less than the input signal power ω_s. A standard theorem of classical mixer theory proves that the "power conversion gain" G from ω_s to ω_{if} cannot exceed $\tfrac{1}{2}$. Still, a mixer performs a valuable function in converting the signal from a very high frequency, at which linear amplification may be difficult or unavailable, to a lower frequency where it can be properly amplified and filtered.

Conventional nonlinear elements for high-frequency microwave receivers are generally based on Schottky diodes, in which a metallic electrode makes a nonohmic contact to a semiconductor such as GaAs. In some cases, the Schottky diode may be cooled to cryogenic temperatures to decrease the effective thermal width associated with the nonlinearity, in order to enhance the performance for either a direct detector or a heterodyne detector. Of course, if one is willing to cool down to 4 K, then LTS detectors start to become competitive. Although the results above were derived with a standard semiconductor nonlinear element in mind, these same principles will continue to apply for a superconducting detector.

Superconducting Tunnel Junction Detector

Consider the dc I–V curves for an unshunted hysteretic Josephson junction (Fig. 7.3). (We will address a resistively shunted junction later.) The ideal superconductor–insulator–superconductor (SIS) tunnel junction exhibits two branches in the characteristic: the zero-voltage supercurrent and the quasiparticle current. Ignoring the supercurrent for the moment, the quasiparticle current for $T \ll T_c$ is very small until the gap voltage $V = 2\Delta/e$, when it rises sharply. In theory it is actually discontinuous; in practice the rise can occur over a voltage of less than 100 µV, very small indeed compared to the ≈ 25 mV for the diode at room temperature. It is this sharp nonlinearity in the quasiparticle current that is key to the superior performance as a nonlinear element.

But there is implicitly another important requirement: that this ideal quasiparticle dc $I(V)$ is also the instantaneous $I(V)$ for a rapidly varying voltage. Several effects might limit the frequency response of this quasiparticle current, including the tunneling traversal time, the energy gap, the junction capacitance, and the ac Josephson current. First, the intrinsic response time of the tunneling currents through a few nanometers of insulator is sub-picosecond and does not limit the performance of the device. More relevant is the superconducting energy gap; above the gap frequency $2\Delta/h$ (~ 700 GHz for Nb), the superconducting electrodes behave as normal metals without an energy gap. A very important practical consideration is that these junctions have significant shunt capacitance, which tends to shunt the signal at high frequencies. As we will discuss below, this capacitance can be partly compensated by special tuning structures, but it is also important that the junctions be very small (submicron) in order to minimize the capacitance. Finally, it is important to take into account the effect of the ac Josephson current. When the junction is biased near the energy gap, there are Josephson oscillations at $f = 4\Delta/h$ (~ 1.4 THz for Nb). These may be largely shunted by the shunt capacitance, but if not, they can act to make the mixing more "noisy." In practice, it is possible to suppress the ac Josephson current by application of a magnetic field in the plane of the junction, which does not significantly affect the quasiparticle current.

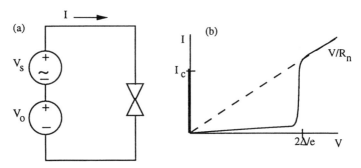

Figure 7.3. Hysteretic superconducting tunnel junction with dc and rf voltage bias. (*a*) Circuit schematic. (*b*) DC $I(V)$ showing sharp rise in quasiparticle tunneling current for $V > 2\Delta/e$, which is the nonlinearity upon which SIS detection is based.

Quantum Detection

For a purely classical detector, the amplitude of the input voltage signal V_s can be arbitrarily small. However, quantum effects become important when the amplitude of the energy per electron eV_s becomes comparable to the photon energy hf. We can see this in the energy band picture of SIS tunneling (Fig. 7.4). The sharp current rise at the gap voltage at $V = 2\Delta/e$ occurs when the peak in the filled states of the lower band first lines up with the peak in the empty states of the upper band. Let us now assume that there is a high-frequency ac field across the junction. If the lower band is below this threshold by $\Delta\mathscr{E} = hf$, then absorption of a single photon can serve to assist the electron in reaching the threshold energy and then tunneling across to the other side. This corresponds to a current step at $V = (2\Delta - hf)/e$, effectively a reduced replica of the main current rise at the gap edge.

Furthermore, multi-photon events are also possible for higher amplitudes, giving rise to a series of "photon-assisted tunneling" steps, each separated by a voltage width hf/e. These include a set of steps on the positive side of the energy gap that correspond to tunneling with "stimulated emission" of one or more photons, instead of absorption. For a large-amplitude ac signal, the series of voltage steps near the energy gap will approximate the rounded dc average that one obtains from the classical theory for reduced photon energy. Of course, if the rise at the energy gap were more gradual, the series of discrete steps would be smeared out. It is important to note that these discrete steps are not necessarily unique to a superconducting detector; what is special here is that the sharpness of the gap structure made these visible.

It is also interesting to note that these photon-assisted steps near the gap edge are present under the same conditions that also give rise to Shapiro steps due to the Josephson current, separated by $hf/2e$ (with the extra factor of 2 because these carriers are Cooper pairs), which are essentially reduced replicas of the zero-voltage supercurrent. The full theoretical formula for these photon-assisted

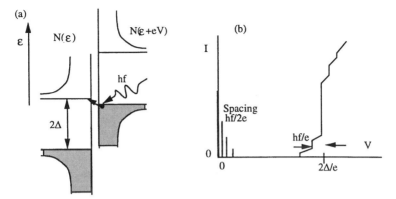

Figure 7.4. Photon-assisted tunneling in SIS. (*a*) Energy diagram of tunneling between two superconducting electrodes. Absorption of a photon in lower band permits tunneling across to other side at energy *hf* below threshold of energy gap. (*b*) Corresponding *I*(*V*) showing first and second photon-assisted tunneling steps on each side of gap, each separated by voltage *hf*/*e*. Also shown is the (reduced) Josephson current together with several Shapiro steps, each separated by *hf*/2*e*.

steps (Tinkham, 1996, p. 213) is also somewhat similar to that for Shapiro steps:

$$\langle I(V) \rangle = \sum J_n^2 \left(\frac{eV_S}{hf_s} \right) I_0 \left(V + \frac{nhf}{e} \right) \qquad (7.9)$$

where the sum is over all integers n, J_n is the nth-order Bessel function, and $I_0(V)$ is the dc quasiparticle current in the absence of radiation. However, note that the width of these steps is proportional to J_n^2, rather than to J_n as for the Shapiro steps [Eq. (5.31)].

As a direct detector of a weak signal in the quantum limit, only the single-photon steps would be visible, since the multi-photon events are higher order. (This is also consistent with the leading terms in the Bessel functions for small argument.) The junction would be biased at a voltage hf below the energy gap and the step height measured. In the ideal case, one electron would cross the junction for each photon absorbed. If there are N photons/sec absorbed, this corresponds to an average current $\delta I = Ne$ and an average power $P = Nhf$ for a responsivity of

$$R = \frac{\delta I}{P} = \frac{e}{hf} \qquad (7.10)$$

This quantum-limited responsivity is similar to the classical responsivity in Eq. (7.5), in that it is inversely proportional to hf/e, the step width in the presence of the radiation. However, for a given frequency, this limits the responsivity even if the dc I'' is infinitely large. This represents the quantum limit of the responsivity for any direct detector.

To put in some specific numbers, for $f = 100$ GHz, $hf/e = 400$ μV (a significant fraction of the gap voltage ~ 3 mV for Nb), much greater than the width of the gap structure in a typical high-quality LTS tunnel junction. As the frequency increases toward $2\Delta/h \sim 700$ GHz (for Nb), the location of the photon-assisted quasiparticle current step moves toward zero voltage, causing possible interference with the Shapiro step and the supercurrent. However, even higher frequencies are possible (despite resistive losses in the superconducting electrodes), with the quasiparticle step moving to negative voltage and overlapping the negative step from the other side moving to positive voltage. This suggests an effective upper limit of order $4\Delta/h \sim 1.5$ THz for Nb.

The photon energy is extremely small in the 100-GHz range (less that 1 meV), and it is impractical to consider observing or counting single photons. But the SIS in the quantum limit shares some aspects with photon detectors that operate at much higher frequencies. In particular, the quantum-limited responsivity for a photoconductor takes the same form as Eq. (7.10), corresponding to one electron per absorbed photon. Furthermore, unlike the case of the classical square-law detector, this quantum response does permit spectral resolution, in that the threshold for the step is a function of the photon energy.

SIS Quasiparticle Mixer

The quasiparticle current can also be used as the basis for a heterodyne mixer that again can approach the quantum limit. The pump or local oscillator is large enough to produce substantial photon-assisted steps, as shown in Fig. 7.4, and the voltage is normally biased in the middle of the $n = 1$ step. The signal is normally too weak to produce a significant change in the dc characteristics. The intermediate frequency is typically of order 1 GHz, at which there are high-quality low-noise amplifiers. The theoretical analysis of quantum mixing (Tucker and Feldman, 1985) is complicated, but a key result is that there can actually be net conversion gain under ideal circumstances, that is, $G > 1$, unlike the classical case where there is always conversion loss. Another important figure of merit for a mixer is the effective noise temperature $T_N = P_N/Bk$, where P_N is the noise power and B is the bandwidth of the measurement. Quantum fluctuations place a lower limit $T_N \sim hf/k$, and some SIS mixers have come close to this limit (Richards and Hu, 1989). For example, $T_N < 10$ K has been obtained at $f = 100$ GHz.

As mentioned above, the capacitance of a real tunnel junction can place a severe constraint on device performance. We can obtain some estimates of the requirements for typical junction sizes by considering the capacitance in more detail. To avoid capacitive shunting at the signal frequency f_s generally requires $\omega_s R_n C \leq 1$, where R_n is the normal-state resistance of the junction. Also, one would prefer $R_n \sim 100$ Ω, so that the junction can be efficiently matched to typical waveguide and antenna coupling structures. Therefore, for a signal frequency $f_s \sim 100$ GHz, we need a capacitance $C < 15$ fF. For typical $Nb/AlO_x/Nb$ junctions, the capacitance per unit area (as we estimated earlier in

Chapter 5) is about $50 \, \text{fF}/\mu\text{m}^2$, which requires a junction having area $A < 0.3 \, \mu\text{m}^2$. A junction of this size is difficult to make reliably but is generally necessary to obtain a sensitive detector for this frequency range. (Incidentally, Schottky diode detectors for this frequency range have similar dimensional constraints, also because of shunt capacitance.) Furthermore, taking $I_c R_n = \pi \Delta / 2e \sim 2 \, \text{mV}$ for Nb, this gives $I_c \sim 20 \, \mu\text{A}$, which for a 0.3-$\mu\text{m}^2$ junction corresponds to a current density $J_c \sim 7000 \, \text{A/cm}^2$, somewhat higher than the typical current densities for present-generation digital Josephson circuits. This is still a hysteretic junction, however, since the characteristic inductance is $L_J = \Phi_0 / 2\pi I_c \sim 15 \, \text{pH}$, $\omega_0 = 1/(L_J C)^{1/2} \sim 2 \times 10^{12}$, and $Q = \omega_0 R_n C \sim 3$, or $\beta_c = Q^2 \sim 9$. A final consideration is the effect of the ac Josephson current, which generally reduces performance by producing additional frequency mixing at undesired frequencies. But at the typical operating point near the gap voltage (3 mV), the Josephson frequency is $\sim 1 \, \text{THz}$, which is largely shunted out by the capacitance.

For even higher frequencies, it becomes even more difficult to design a junction that is not limited by capacitance. If we wanted to make a detector optimized for a frequency of 300 GHz, this would require reducing the capacitance by a further factor of three, to obtain a junction of area $< 0.1 \, \mu\text{m}^2$. But this would no longer be an ideal hysteretic junction, and in operation the Josephson current would no longer be shunted out. A magnetic field corresponding to one flux quantum in the junction could then be used to reduce the ac Josephson current. Another way to get rid of the Josephson current entirely is by replacing one of the electrodes with a normal metal. However, the nonlinearity at $V = \Delta/e$ for the SIN (superconductor–insulator–normal tunnel junction) is not quite as sharp as the corresponding structure for the SIS. An alternative approach that has been more successful at high frequencies is to accept the capacitance of the SIS but to cancel out its effect using an on-chip inductance $L = 1/\omega_s^2 C \sim 10 \, \text{pH}$, essentially constituting a resonant input filter (Fig. 7.5a). This would limit the bandwidth of the detector to $1/Q$ of the resonator, but a practical compromise can often be obtained.

Another important consideration for the design of a high-frequency receiver is the coupling structure for the high-frequency radiation (both the signal and the local oscillator). The junction size is much smaller than the free-space radiation at 100 GHz (3 mm at 100 GHz), so that a carefully designed waveguide or antenna structure is necessary. (The *if* signal can be coupled out using standard transmission lines.) Waveguides are intrinsically relatively narrow band and become more difficult to work with as the wavelength moves into the submillimeter range. In this regime a thin-film antenna may be preferable, and several configurations have been investigated (Wengler, 1992), including the bow-tie and the spiral antenna, both of which are very broadband. These antennas can also be fabricated lithographically out of the same low-loss superconducting thin film (such as Nb) as the junctions, so that they may be easily integrated together. The radiation may be focused onto such an antenna quasi-optically using a lens made out of an appropriate material (such as quartz or Teflon) that

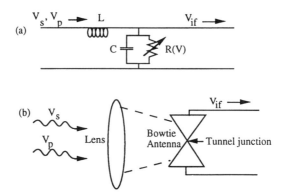

Figure 7.5. RF coupling schemes for SIS mixers. (*a*) A waveguide may be used to couple in rf signal and local oscillator and a transmission line to couple out the *if* signal. An integrated inductance is used to cancel out the effect of the capacitance. (*b*) Quasi-optical coupling through a lens to a wide-band bow-tie antenna, with the SIS tunnel junction in the center.

is transparent in this range. Most receiver systems for radiation at 500 GHz and above use this sort of quasi-optical coupling (Fig. 7.5*b*).

A related issue is providing the high-frequency pump power needed for efficient operation of the SIS mixer. This must be a stable, narrow-linewidth oscillator in the frequency range 100 GHz–1 THz and beyond. The required power is not very large, typically on the microwatt scale at the junction. One approach is to use a lower frequency oscillator (such as a semiconductor Gunn diode at ~ 100 GHz) and pass that through an appropriate nonlinear crystal that produces a multiplied signal. This process is not very efficient but can provide sufficient power at several times the input frequency (as high as about $n = 5$). An alternative approach is to use a superconducting local oscillator. As discussed earlier in Chapter 5, a Josephson junction forms an ideal voltage-controlled oscillator with a frequency of $2e/h = 483$ GHz/mV. A single RSJ is too noisy, but a long junction or a junction array is likely to be sufficient for such purposes. The power output is small but is tunable over an appropriate range, is well matched to the detector junction, and can be readily integrated on the same chip as the detector. Several projects to design and build such an integrated SIS mixer and local oscillator are in progress.

At present, the greatest interest in SIS mixer receivers is for radio astronomy in the millimeter and submillimeter range. In fact, a number of radio telescopes internationally have adapted such Nb SIS receivers covering the 100–500-GHz range, with even higher frequencies being developed. The primary interest in this range is in spectroscopy and imaging of cold nebulas, where the intensity is low, and any improvement in sensitivity is strongly desired. Narrow spectral lines are typically associated with rotational states of various molecular gases, so that the narrow-band capabilities of SIS mixers are in demand.

Josephson Radiation Detectors

Virtually all quasiparticle tunneling radiation detectors are based on conventional low-temperature superconductors, primarily Nb, Pb, and NbN. This is largely due to the fact that HTS tunnel junctions with similarly ideal (hysteretic) quasiparticle characteristics have not yet been fabricated. But as we discussed earlier, HTS junctions that behave like RSJs can now routinely be fabricated. Would it be possible to obtain high-performance modulation-type radiation detectors with these junctions?

In fact, mixers and direct detectors based on RSJs predated the development of LTS quasiparticle devices. The first practical LTS Josephson junctions, used for SQUIDs and other applications, were generally point contacts that were similar to RSJs in behavior. Such Josephson junctions certainly exhibit strongly nonlinear characteristics that can also be used for mixing. Unfortunately, the performance of all of these Josephson devices as radiation detectors is generally not very sensitive, and they were rapidly superceded by the developing SIS devices.

This poor performance initially seems rather surprising, in that shunted junctions would seem to form a natural basis for radiation detection. Consider an RSJ with a dc current bias I_0 and an rf signal current $I_s \sin(\omega_s t)$ (see Fig. 7.6). First, they are not capacitively limited and should show nonlinear effects up to at least the characteristic frequency $R/L_J = 2eI_cR/h$, which can approach 500 GHz for a high-quality LTS junction. Second, an rf current of amplitude I_s should cause a reduction in the effective critical current by the same amount, which would seem to make for a sensitive broadband direct detector. Third, in the presence of a weak rf current at frequency f_s (and in the absence of noise), a Shapiro step forms at voltage $V = hf_s/2e$ of approximate width I_sI_cR/V. This would seem to make for a sensitive narrow-band detector. Finally, the ac Josephson current could in principle act as an internal local oscillator for a mixer; indeed, this is the basis for the Shapiro steps.

Unfortunately, things are not quite this simple. Of greatest significance is the fact that the dc $V(I)$ relation is not the same as the high-frequency relation.

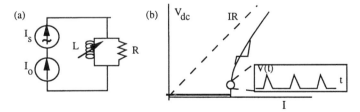

Figure 7.6. The RSJ as modulation detector. (a) Circuit schematic with dc current bias and rf current signal, and RSJ represented by linear resistance in parallel with nonlinear inductance. (b) DC $V(I)$ for RSJ showing the Shapiro step that is caused by the rf current. The inset shows the instantaneous $V(t)$ at a bias point just above I_c.

Unlike the nonlinear quasiparticle resistance, the RSJ consists of a linear resistance in parallel with a highly nonlinear inductor. For current I just above the critical current I_c, the instantaneous voltage consists of a series of sharp, narrow SFQ pulses with spectral content up to $\sim R/L_J$. If we modulate the current bias, we are modulating the frequency of these pulses. Given the asymmetry, this does indeed increase the dc average voltage, but it also generates lots of high harmonics, not merely the second harmonic as in the quasiparticle case. The presence of these high-order harmonics acts to increase the effective noise level of the device. Furthermore, as we have discussed, the internal Josephson oscillation generally does not have a sufficiently narrow linewidth to act as a good local oscillator (LO) for a mixer. But if we use an external LO for the Josephson mixer, the effective noise temperature T_N is far higher than the quantum limit. Typical values for an HTS RSJ mixer at 300 GHz are $T_N \sim 1000$ K (Shimakage et al., 1997), a factor of 100 larger than for the optimized SIS quasiparticle mixers. There may be some room for improvement here, but it is unlikely that HTS Josephson mixers will displace quasiparticle mixers, or even cooled Schottky diodes, for high-frequency astronomical applications.

On the other hand, if it becomes possible to fabricate true HTS tunnel junctions exhibiting a sharp rise in the quasiparticle current at a voltage $2\Delta/e \sim 30$ mV, this would be expected to have a major impact on terahertz mixer technology. For these hypothetical junctions, if capacitance limitations could be overcome (which is perhaps even more difficult), then we might expect to be able to extend modulation techniques including heterodyne detection to 15 THz, well into the infrared range (20 µm wavelength), as well as to move operation to somewhat higher temperatures.

In the absence of such revolutionary HTS tunnel junctions, there is a completely different kind of superconducting mixer, the hot-electron bolometric mixer, that is developing for applications in the terahertz range. This mixer is based on a quasi-thermal detection mechanism, so we will postpone its discussion until the next section.

7.2 THERMAL AND QUASI-THERMAL DETECTORS

As the frequency increases, it eventually becomes difficult for a detector to follow the electromagnetic waveform directly. In this regime, which generally includes the entire infrared range, most detectors measure only the average power deposited by the radiation. The simplest way to do this is with a thermometer that measures the local temperature rise δT in the detector. This implicitly assumes thermal equilbrium, but even in strictly nonequilibrium systems, an effective temperature rise can often be assigned. We refer to these as thermal and quasi-thermal detectors. These are by no means limited to superconductors, but superconducting elements tend to make some of the most sensitive thermometers for such systems. The most common device of this type is the superconducting bolometer, which uses the sharp dependence of the resistive transition near T_c as the thermometer. LTS thin-film bolometers have been developed for sensitive infrared detectors and recently for heterodyne mixers in the terahertz range. With the advent of HTS materials, there has been renewed interest in the development of a bolometer that can operate at liquid nitrogen temperatures and above.

Thermal Properties of Solids

But before describing how superconductors may be used to make such thermal detectors, let us first review some basic thermal properties of solids, including superconductors. The two basic thermal parameters of any material are its thermal conductivity κ and heat capacity (per unit volume) c. Both of these are related to the thermal current density J_t by the standard relations

$$J_t = -\kappa \nabla T \qquad \nabla \cdot J_t = -c \frac{\partial T}{\partial t} \qquad (7.11)$$

The first equation defines the flow of thermal current down a temperature gradient, and the second describes how a net inflow of thermal current causes a local increase in temperature. These equations are directly analogous to similar equations for the flow of electrical current, where the temperature T itself is analogous to the electrical potential or voltage V. The thermal conductivity is analogous to the electrical conductivity, and the heat capacity is equivalent to an electrical capacitance distributed through the volume. This can be represented in one dimension by the transmission line model of Fig. 7.7, with a distributed series resistance and shunt capacitance to ground. Alternatively, it can be described by the single differential equation

$$\frac{\partial T}{\partial t} = D \nabla^2 T \qquad (7.12)$$

where $D = \kappa/c$ is the thermal diffusivity or diffusion constant, with units of m^2/s. This makes it clear that rather than propagating with a single velocity, a heat pulse will spread diffusively, out to a distance $x \sim (D\tau)^{1/2}$ in a time τ. For example, if $D \sim 1$ cm^2/s, then the characteristic heating or cooling time is 1 s on the centimeter scale and 1 µs on the 10-µm scale.

On the microscopic level, heat is carried by two distinct species, the phonons (or lattice vibrations) and the electronic carriers (free electrons and/or holes). For an electrical insulator without carriers there is only the phonon contribution. A conductor has parallel contributions from both, so that

$$\kappa = \kappa_l + \kappa_e \qquad c = c_l + c_e \qquad (7.13)$$

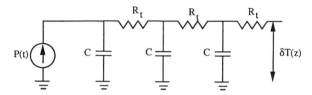

Figure 7.7. Transmission line representation of one-dimensional heat flow in a solid where temperature T is the analog of voltage. The current source represents a source of heat, the distributed resistance per unit length is $R_t = 1/\kappa A$, and the distributed capacitance per unit length is $C = cA$, where A is the cross-sectional area, κ the thermal conductivity, and c the heat capacity per unit volume.

where the subscripts l and e are the lattice and electronic contributions, respectively. All of these quantities depend strongly on temperature, particularly in the cryogenic domain, where all thermal quantities decrease to zero as T goes to zero. Typical dependences (following the Debye model) are (Kittel, 1996)

$$c_l \approx \begin{cases} 3N_a k_B & T > \Theta \\ 200 N_a k_B \left(\dfrac{T}{\Theta}\right)^3 & T \ll \Theta \end{cases} \qquad (7.14)$$

where N_a is the atomic density and Θ is the Debye temperature, typically ~ 300 K, and

$$c_e \approx 5 N_e k_B \frac{T}{T_F} \qquad T < T_F \qquad (7.15)$$

where N_e is the density of free electrons and $T_F \sim 10{,}000$ K is the Fermi temperature. The thermal conductivity contributions are typically proportional to the corresponding heat capacities, since the heat transported is proportional to the heat carried by the relevant species of particle. For a good metal at low temperatures, both are dominated by the electronic contribution. For higher T, κ_e is still likely to be dominant, but c_l may well be the larger contributor. The detailed analysis is complex and sample dependent, since c is a volume parameter that is largely independent of the sample microstructure, whereas κ is a transport coefficient that is strongly dependent on sample purity. For example, a single crystal generally has a much higher thermal conductivity than a granular composite of the same material.

There is an important universal (but approximate) relationship, known as the Wiedemann–Franz law, between κ_e and the electrical conductivity σ of a metal:

$$\kappa_e = LT\sigma = \frac{\pi^2 k_B^2}{3e^2} T\sigma \qquad (7.16)$$

where σ is the electrical conductivity and $L = \pi^2 k_B^2/3e^2$ is known as the Lorenz constant. It is important to note, however, that this does *not* apply to superconductors. In fact, Cooper pairs in a superconductor cannot store or transport heat at all (being essentially in the quantum ground state), so that both κ_e and c_e approach zero for $T \ll T_c$. In contrast, closer to T_c, when most of the electrons are unpaired quasiparticles, both κ_e and c_e are similar to the values for $T > T_c$.

Consider, for example, a typical temperature dependence of κ and c for the HTS material YBCO, shown in Fig. 7.8 (Uher, 1990). These temperature dependences are dominated by lattice contributions rather than electronic values over most of this range. Note that κ actually increases below T_c due to the increased mobility of phonons when collisions with thermal quasiparticles are reduced; there is no scattering off Cooper pairs. There is a well-known anomaly in c_e just below T_c, but it is not visible on this scale, since c_l is dominant. Note that the value of $D \sim 10^{-6}$ m²/s near T_c. This corresponds to heat spreading over a distance ~ 1 μm in 1 μs.

In thermal transport at low temperatures, the thermal conductance across the boundary between two materials can be at least as important as the conductance within each of them. In particular, in thermal transport between a film and an electrically insulating substrate, the heat must be carried via phonons. If one thinks of phonons as acoustic waves in the two materials, then one would expect to find a substantial reflection from the interface if the two materials have different "acoustic impedances." This is the basis for the thermal boundary resistance or "Kapitza resistance" associated with such an interface, typically going as $R_t \sim 1/T^3$ at low temperatures. This thermal boundary resistance usually dominates the thermal properties of a film–substrate combination at low temperatures. Typical values are $\sim 10^{-4}$ m²K/W at $T = 4$ K and 10^{-7} m²K/W at $T = 100$ K (Nahum et al., 1991).

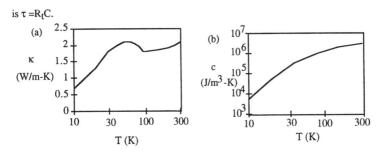

Figure 7.8. Temperature dependence of thermal conductivity κ (a) and heat capacity c (b) for typical sample of the high-temperature superconductor YBCO. Both are dominated by the lattice contribution over most of this range. Note the anomaly in κ below T_c, where the freezeout of the quasiparticles permits the phonons to conduct heat more effectively.

Thermal Time Constants

The simplest thermal detector is one in which the incoming radiation heats up a single lumped thermal element, with a single thermal link to the thermal bath. This can be represented by the simple lumped RC network shown in Fig. 7.9, where the voltage to ground is analogous to the temperature rise. The capacitance C represents the heat capacity of the detector element (in units of joules per kelvin), and the resistance R_t the thermal resistance (in kelvins per watt) between the element and the thermal bath (equivalent to the electrical ground). The current source represents the heat flux P (in watts) absorbed in the element. In steady state, the temperature rise is $\delta T = PR_t$, and the response time is $\tau = R_t C$.

So to make a sensitive detector, we would like to make R_t large, but this tends to make τ large, slowing the response and reducing the effective bandwidth $1/R_t C$ of the detector. To obtain the greatest sensitivity while maintaining a reasonably fast response time, it is necessary to reduce the thermal mass of the detector (i.e., the heat capacity C) to a minimum. One approach to achieve this is to fabricate the detector element on a very thin substrate (such as a membrane) that is only weakly connected to the rest of the system. The remaining thermal link would probably be due to the electrical leads.

We can estimate the approximate response time and responsivity for a YBCO microbolometer by considering values from Fig. 7.8. Taking a thin YBCO film ~ 100 nm thick and 10 μm² in area, we have $C \sim 10^{-11}$ J/K at 90 K (neglecting a contribution from the substrate). If the thermal link consists of four YBCO legs with length 10 μm and area 1 μm × 100 nm, we have $R_t \sim 10^7$ K/W, corresponding to $\tau \sim 100$ μs. This demonstrates that a small bolometer element can be

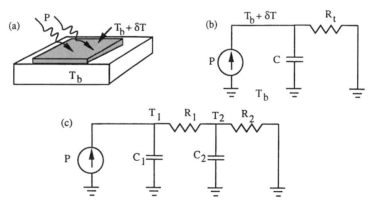

Figure 7.9. Schematic of thermal radiation detector. (*a*) Simplified picture of incident radiation with power P heating up detector element above bath temperature T_b. (*b*) Thermal equivalent lumped-element circuit with heat capacity C and thermal resistance R_t between detector and thermal bath. (*c*) Two-stage thermal equivalent circuit with heat deposited in thin-film detector passing through substrate to thermal bath.

relatively fast. Note that in viewing the detector element as a lumped capacitor, we are implicitly assuming that the characteristic heating time for the element itself ($\sim x^2/D$) is faster than this value of τ. For an element 3 μm across and $D \sim 10^{-6}$ m^2/s, the internal time is ~ 10 μs, validating the assumption. But this also makes clear why a fast thermal element must be small along the direction of heat flow.

It is worth noting that any real thermal system is likely to be more complicated than this simple one-stage RC network. For example, if the detector element is a thin film on a substrate, there is typically a thermal barrier between the film and the substrate, particularly at low temperatures. This might then correspond to the two-stage network shown in Fig. 7.9c, where the transient dynamics of the system depend on the relative magnitudes of the two time constants $\tau_1 = R_1 C_1$ and $\tau_2 = R_2 C_2$. The film–substrate stage is typically very fast (microseconds or less), but it can sometimes be important in describing fast transients. In particular, if the radiation comes in the form of a very fast pulse, then the thermal response of the detector can exhibit two (or more) characteristic response times.

Bolometers

The most critical element for any thermal radiation detector is the thermometer used to read out the temperature rise. A number of different thermometers are used for thermal detectors at room temperature, including a pyroelectric detector (which measures capacitance), a thermocouple (thermal emf), and Golay cell (gas pressure changes). However, for more sensitive detectors, it is common to cool down to cryogenic temperatures, and the most common cryogenic detector, the bolometer (Richards, 1994), is based on a temperature-dependent electrical resistance $R(T)$.

The most important parameter for a bolometer is the thermal coefficient of resistance of the sensing element $\alpha = (dR/dT)/R$. A superconductor near its transition is particularly attractive in this regard due to its sharp, narrow transition (Fig. 7.10). For an ideal superconducting thin film, the width ΔT of the resistive transition near T_c can be very narrow indeed, with ΔT typically ~ 1–10 mK for LTS and 100 mK for HTS. The peak value of α at T_p in the middle of the transition is $\alpha_p \sim 1/\Delta T \sim 10$–$1000$/K. If the bath temperature $T_b < T_c$, then a small electric heater on the film can be used to maintain the film at the peak value of the thermal response. If the superconductor is then biased with a fixed (dc or ac) current, then for radiational heating $\delta T \ll \Delta T$, the voltage change will be $\delta V = I\,\delta R \sim IR\,\delta T/\Delta T$. The magnitude of the bias current I should be small enough to avoid excessive self-heating or broadening of the transition but large enough for a substantial voltage. Of course, the signal saturates if δT approaches $\frac{1}{2}\Delta T$. The dynamic range can be increased if the temperature bias is maintained at the optimum location by feedback.

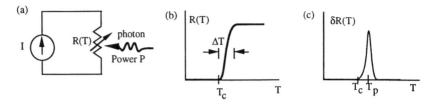

Figure 7.10. Superconducting thin film as bolometer. (*a*) Electrical schematic of current-biased strip with incident radiation. (*b*) $R(T)$ near superconducting transition. (*c*) Typical photoresponse $\delta R \approx (\partial R/\partial T)\,\delta T$ as function of bath temperature showing peak at T_p.

An important figure of merit for a bolometer is its voltage responsivity

$$R_v = \frac{\delta V}{P} = I \frac{dR}{dT} R_t \qquad (7.17)$$

where P is again the absorbed radiated power. Clearly, one would like all three factors to be as large as possible. To take a specific example, consider the YBCO thin-film element described above. If we assume that it has $dR/dT \sim 10\,\Omega/K$ (corresponding to $\Delta T \sim 0.1$ K and $R \sim 1\,\Omega$) and choose $I = 1$ mA, then we obtain a respectable responsivity $R_v \sim 1000$ V/W (or more to the point for sensitive detection, $R_v \sim 1$ mV/μW). But note that the Joule heating would be relatively large: $I^2 R \sim 1$ μW, so that the temperature rise due to self-heating is ~ 10 K! This illustrates the difficult trade-offs that must be made in the optimization of a bolometer of this type.

A superconductor is not unique in acting as a sensitive low-temperature thermometer, and as in many areas, its main competition is semiconducting. If we take the approximate resistance of an intrinsic semiconductor with energy gap E_g as $R \sim \exp(-E_g/kT)$, then $\alpha = d\ln(R)/dT \approx -E_g/kT^2$. For a typical $E_g \sim 1$ eV, this gives $\alpha \sim -600$ at 4 K, comparable to that for a superconductor (the sign does not really matter). Furthermore, the temperature does not have to be biased in a narrow transition region, which may be a significant advantage. Still, there has been significant interest in recent years in developing a bolometer based on a HTS thin film that can operate around 90 K, where semiconductors may not hold a clear advantage. The performance at 90 K would not be as high as for a detector at 4 K, but the cooling requirements would be substantially less.

We have not mentioned the wavelength of the incident radiation, since a good bolometer responds over a broad range of wavelengths. All that is necessary is a means to absorb the radiation, converting it to heat. This can sometimes occur in the superconductor itself, or alternatively in a separate layer that absorbs strongly across the frequency range of interest (a "composite bolometer"). It is notable in this regard that HTS cuprates such as YBCO are black and transparent, with an optical penetration depth ~ 100 nm in the visible and near infrared. Bolometers are typically used from microwaves through the infrared. For shorter wavelengths (higher photon energies), single-photon detectors (described later) tend to be more common.

An important quantity for any radiation detector is the minimum radiated power that it can detect. In an optimized system, this is limited by the background noise in the system, often expressed as a noise voltage V_n in $V/\sqrt{Hz}$, since the noise is generally proportional to the square root of the effective bandwidth. This leads to the definition of "noise-equivalent power" $NEP = V_n/R_v$ (in $W/\sqrt{Hz}$) as an important figure of merit; clearly, a small NEP is preferred. A closely related parameter is the "detectivity" D^*, defined as $D^* = \sqrt{A}/NEP$, normally quoted in units of $cm\sqrt{Hz}/W$, where A is the effective area of the detector to the incident radiation. For many conventional infrared detectors, the NEP increases with $\sqrt{A}$, so that D^* provides a better measure of intrinsic performance. Since background noise is often thermal noise that decreases at low temperatures, NEP and D^* for optimized bolometers generally improve as the temperature is reduced, so that an LTS detector at 9 K is generally better than an HTS detector at 90 K. Still, values of $D^* \sim 10^{10}$ have recently been demonstrated for HTS microbolometers, significantly better than room-temperature detectors (with $D^* \sim 10^8$). Typical values of D^* for a variety of types of infrared detectors are shown in Fig. 7.11, including some cooled semiconducting photon detectors (Richards, 1994).

A possible disadvantage of the resistive readout is that it requires biasing the superconductor in the middle of its superconducting transition. Furthermore, self-heating limits the amount of bias current that can be used, effectively limiting the voltage. Finally, the nonzero resistance gives rise to Johnson noise, limiting its sensitivity. The question arises whether a superconductor can act as a thermometer in its low-temperature $R = 0$ state. One such temperature-dependent property is the kinetic inductance L_k, which is proportional to the

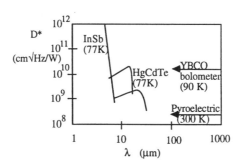

Figure 7.11. Typical values of detectivity D^* for different classes of infrared detectors as a function of wavelength (Richards, 1994). Thermal detectors are essentially independent of wavelength, in contrast to photon detectors. Both room-temperature and cryogenic detectors (including semiconducting InSb and HgCdTe with two different band gaps) are indicated. YBCO bolometers at 90 K are superior to alternatives for long infrared wavelengths.

penetration depth $\lambda(T)$ (or λ^2 for a very thin film). A figure of merit of such a detector is $(1/L)(dL/dT)$, which diverges as $\sim 1/(T_c - T)$ close to T_c. While this may not be particularly large at low T, this may be compensated by the reduced noise level, which permits greater amplification. This kinetic inductance must be read out using an ac (rather than a dc) current, but otherwise this can provide for a bolometer in a similar way as the resistive readout. The impedance of such a device will typically be very small, but it should be well matched to a low-noise SQUID amplifier. A complete system of this type has not yet been optimized, but preliminary noise measurements suggest the posssibility of a much greater sensitivity at cryogenic temperatures (below ~ 4 K) than the resistive bolometer (Sauvageau and McDonald, 1989).

An alternative configuration is to use an SIS tunnel junction rather than a single thin film as the sensitive thermometer. There are three key junction parameters that are strongly dependent on temperature: the critical current I_c, the energy gap 2Δ (or gap voltage $V_g = 2\Delta/e$), and the subgap quasiparticle resistance R_{sg}. A detector based on $I_c(T)$ can consist of a nonhysteretic junction biased just above the critical current; the sensitivity parameter $(1/V)(dV/dT)$ should be $\approx 1/(T_c - T)$. A similar sensitivity would arise for a hysteretic junction biased near the gap voltage. The subgap resistance increases exponentially at low T, very much like a semiconductor with a small energy gap $E_g = 2\Delta$, yielding an effective $\alpha = (1/R)(dR/dT) \sim -2\Delta/kT^2$. Although thermal detectors based on these characteristics have been demonstrated, they typically do not perform quite as well as either transition-edge bolometers or semiconductor bolometers (with the much larger value of E_g). On the other hand, the subgap resistance of a superconducting tunnel junction provides the basis for a very high-resolution *photon* detector, as we will discuss further in the final section.

Nonequilibrium Heating

In the preceding analysis of thermal detectors, we have implicitly made one critical assumption, namely that the detector element is in thermal equilibrium at a temperature T. Although this is normally assumed, it need not always be the case. Thermal equilibrium is maintained by the random interactions of a large number of electrons and phonons and determines the distributions of their energies. Electron–phonon interactions occur via inelastic electron–phonon scattering events, whereby an electron changes its energy and momentum accompanied by emission or absorption of a single phonon. This phonon, in turn, may go on to interact with another electron. (This inelastic scattering should be distinguished from the more frequent elastic scattering of electrons off impurities, which changes the momentum of an electron but not its energy and does not involve a phonon). The inelastic scattering rate is determined by a number of factors, including temperature, energy, and the band structure of the given material. However, there is a characteristic average inelastic scattering time τ_{ep} that is typically ~ 20 ps at 10 K and decreases as strongly as $1/T^3$ at low temperatures. For times much shorter than τ_{ep}, the conduction electrons are typically not in thermal equilibrium with the phonons.

Consider how the electromagnetic energy is initially deposited in a material. For relatively low frequencies, one has $I^2 R$ Joule heating, which couples energy directly into the electronic carriers. For much higher frequencies, a photon is typically absorbed by a single electron in the valence band, creating an electron–hole pair (Fig. 7.12a). These high-energy excitations quickly lead (via electron–electron interactions on the 1-ps scale) to a larger number of lower energy excitations, depending on the initial photon energy. In both cases, energy is initially deposited into the electron system, and to a first approximation we may assume that the resulting electrons exhibit an increased effective temperature T_e, even if they do not initially share this energy with the phonons. This is what we mean by "nonequilibrium heating." Over a slightly longer time scale, this excess energy is transferred to the phonons (at temperature T_p) and from there out of the element.

The nonequilibrium heating that we have been describing applies to a normal metal, but the situation is very similar in a superconductor (Kadin and Goldman 1986; Tinkham, 1996, Chap. 11). The energy of an absorbed photon is typically distributed among a larger number of lower energy quasiparticles, thus effectively raising their temperature to a slightly elevated T_e. These excess heated quasiparticles, in turn, act to depress the superconducting energy gap to a self-consistent value $\Delta(T_e)$. Under certain special circumstances, it is possible to obtain a distribution of quasiparticles that is somewhat nonthermal, but in most practical cases, an effective temperature T_e is still a good first approximation. Close to T_c, the effective relaxation time should still be similar to τ_{ep}. However, for every low temperatures $T_e \ll T_c$, the relaxation process requires that two quasiparticles recombine to form a Cooper pair, giving off the excess energy in a phonon. This also scales with τ_{ep} but can be much slower since it

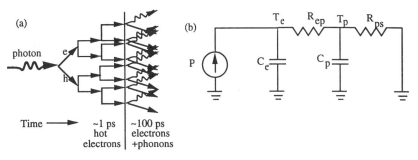

Figure 7.12. Nonequilibrium electron heating in a thin superconducting film. (*a*) Scattering diagram of electron heating process with cascade of thermal electrons at effective temperature T_e followed later by emission of phonons. (*b*) Schematic lumped thermal circuit representing electron cooling, first to the phonons in the film and then to the phonons of the substrate.

requires two oppositely directed quasiparticles to "find each other." This latter process is analogous to electron–hole recombination in a photoexcited semiconductor, which can also sometimes be quite slow.

There is one distinct type of nonequilibrium distribution in a superconductor that is not quasi-thermal. This is referred to as "charge imbalance" and involves a net excess charge in the quasiparticles, which otherwise have an equal number of electrons and holes. This charge is compensated by the Cooper pairs to maintain overall electrical neutrality. Such a charge imbalance can be produced by quasiparticle charge injection into a superconductor but is not normally obtained by absorption of radiation, so we will not deal with this further.

This sequential process of nonequilibrium electron cooling may be represented using a two-stage thermal RC network, as in Fig. 7.12b, very similar to the one in Fig. 7.9c. In this case, the heat is injected into the electrons (with heat capacity C_e), transferred to the phonons via electron–phonon scattering (through an effective thermal resistance $R_{ep} = \tau_{ep}/C_e$), and then transferred to phonons in the thermal bath by the usual mechanisms. The only distinctive aspect of this nonequilibrium relaxation process is that it occurs throughout the volume of the material, rather than just at its extremities, as for the normal cooling process. For this reason, nonequilibrium heating of electrons is most significant for an ultrathin film.

Consider, for example, a 10-nm film of NbN at 10 K deposited on a polished Si wafer. Such a thin film is semitransparent, but its high sheet resistance (up to $\sim 1000\ \Omega$) enables it to absorb a significant fraction of incident radiation. In terms of the thermal equivalent circuit of Fig. 7.12b, the nonequilibrium thermal resistance R_{ep} may be larger than the thermal resistance to the substrate R_s, so that the nonequilibrium effects are indeed dominant. For typical parameters at 10 K, estimated values are $\tau_{ep} \sim 20$ ps and R_{ep} and R_s are both $\sim 10^{-6}$ K-m^2/W (Johnson, Herr, and Kadin, 1996). Similar fast hot-electron effects would also be expected if such a film were operated well below T_c as a kinetic-inductance bolometer.

An alternative variety of hot-electron bolometer makes use of submicron transverse dimensions (rather than an ultrathin film) to ensure rapid cooling (Bumble and LeDuc, 1997). In this configuration, there is a thin, narrow superconducting constriction connecting two thicker films in thermal equilibrium. The constriction is driven into its resistive transition by a large current density, producing electron heating in the constriction. An appropriate antenna structure can be used to direct high-frequency currents into the same constriction (since the device is much smaller than wavelengths in the infrared), where they can cause a detectable temperature rise δT.

A key characteristic of both types of hot-electron bolometers is that they are not capacitively limited in the same way as SIS tunnel junction detectors. These are small devices, but the device length is comparable to its width. We may estimate a capacitance $C \sim \varepsilon_0 (1\ \mu\text{m}) \sim 10$ fF and a typical resistance $R \sim 1\ \Omega$ for a time constant $RC \sim 10$ fs or a cutoff frequency of 100 THz! So if we can couple in the radiation effectively, these devices should make for very broadband detectors through the infrared.

Bolometric Mixer

We have implicitly been assuming that a bolometer detects the average power absorbed by the element. However, for most fast bolometers, such as the hot-electron bolometers discussed above, the resistive element itself also absorbs the radiation. At least for relatively low frequencies (before single-photon effects start to become dominant), this absorption occurs via direct I^2R Joule heating of the high-frequency currents. This permits a sufficiently fast bolometer to be used as a mixer in a similar manner to those discussed earlier for modulation detectors. Recent experiments have suggested that superconducting hot-electron bolometers may permit sensitive heterodyne detection techniques to be extended further into the terahertz range (Bumble and LeDuc, 1997; Semenov et al., 1997).

Consider a resistive element biased with an rf local oscillator (pump) at frequency ω_p and a weaker signal at frequency ω_s, where we assume that ω_p and ω_s are relatively close. The total current is then given by

$$I(t) = I_p \cos(\omega_p t) + I_s \cos(\omega_s t) \tag{7.18}$$

The Joule heating of the device then produces an instaneous power

$$P(t) = I^2 R = R[\tfrac{1}{2}(I_p^2 + I_s^2) + I_p I_s \cos(\omega_p - \omega_s)t + I_p I_s \cos(\omega_p + \omega_s)t \\ + (\tfrac{1}{2} I_p^2)\cos(2\omega_p t) + (\tfrac{1}{2} I_s^2)\cos(2\omega_s t)] \tag{7.19}$$

The dc component gives rise to the usual bolometric temperature rise $\langle \delta T \rangle = R_t \langle P \rangle$, but the components at other frequencies can also cause temperature modulations at the appropriate frequency $\delta T(\omega)$, which can be read out of the device as resistance modulations. The temperature components at very high frequencies will be shorted out by the heat capacity (see Fig. 7.12b), but the component at the much lower intermediate frequency $\omega_{if} = \omega_s - \omega_p$ may remain. For example, if $\omega_s = 1001$ GHz and $\omega_p = 1000$ GHz, then $\omega_{if} = 1$ GHz. A hot-electron bolometer with characteristic time $\tau \sim 100$ ps is far too slow to respond directly to the terahertz signals but will give a strong mixed response of amplitude $\delta T_{if} = R_t I_p I_s R$ at the beat frequency. In fact, it has recently been shown that a device of this type will produce a mixed response to two ~ 100-THz signals due to stabilized semiconductor lasers provided only that ω_{if} was in the gigahertz range (Lindgren et al., 1994). Further development of superconducting hot-electron bolometers is continuing, and they may become practical for sensitive astronomical receivers in the terahertz range, just beyond the limits of SIS quasiparticle mixers.

7.3 SINGLE-PHOTON AND PARTICLE DETECTORS

There is another class of radiation detectors that depends fundamentally on the quantum nature of the radiation. A photon is absorbed, giving rise to an electrical pulse in the detector. One may either count these pulses (the photon-counting mode) or average over them to determine the total rate (photon-integrating mode). Furthermore, these detectors normally exhibit some dependence on photon energy, so that a determination of the radiation spectrum may be possible. We have focused in this chapter on detection of electromagnetic radiation, but detection of high-energy particles occurs by a similar quantum process, so we will also discuss some approaches to superconducting particle detection.

Semiconductor Photon Detectors

Before addressing some of the newer superconducting photon detectors, let us first review some of the basic properties of semiconductor photon detectors. The key parameter in a semiconductor is the energy gap $\mathscr{E}_g$, typically ~ 1 eV. If the semiconductor is undoped, there will be few electrons in the conduction band or holes in the valence band, so that the material is essentially an insulator, particularly at low temperatures. A photon with energy $hf > \mathscr{E}_g$ will then pass through the semiconductor without absorption. If a photon with energy $hf > \mathscr{E}_g$ is absorbed, this creates an electron and a hole (we are implicitly assuming a direct-bandgap semiconductor), which quickly relax down to the gap edge. For the time being, let us assume that the photon energy is only a bit more than the energy gap, so there is insufficient energy to create additional excitations. The electron and hole could recombine (with the additional energy released again as a photon), but this could take a relatively long time, since their density (and those of other excitations) is small. In the meantime, we clearly have a nonequilibrium situation, since we are not simply heating up the semiconductor.

If an electric field is now applied across the semiconductor, the electron will go in one direction and the hole in the other, both corresponding to a net current parallel to the applied field (Fig. 7.13a). If the electron and the hole both reach their respective electrodes before recombining, these charges can be collected (let us assume ohmic contacts). This provides the basis for a "photoconductive" detector. A "photovoltaic" detector is similar, except that there is a built-in electric field as in the depletion region of a junction diode and no external field need be applied. In either case, this process corresponds to a net transfer of one electron charge between the electrodes, or equivalently a current pulse that integrates in time to the same charge e. If every such photon gives rise to a single electron pulse, then the integrated photoresponse corresponds to a quantum-limited current responsivity of e/hf, which has units of amperes per watt. (Recall that this is the same limiting responsivity as for photon-assisted tunneling in an SIS mixer in Section 7.1.)

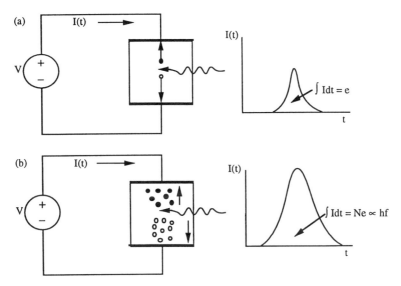

Figure 7.13. Schematic of semiconducting photodetector. (*a*) Voltage-biased photoconductor after absorption of single photon with energy $hf \gtrsim \mathcal{E}_g$. The excited electron (solid circle) and hole (open circle) move in opposite directions toward the appropriate electrode, leading to a current pulse of integrated charge *e*. (*b*) Similar photoconductor after absorption of high energy (x-ray) photon with $hf \gg \mathcal{E}_g$, leading to many excited electrons and holes and a much larger current pulse. In this regime, the collected charge is proportional to the photon energy and so can be used for x-ray spectroscopy.

A real semiconductor photodetector will also exhibit some "dark current," random current pulses even in the absence of absorbed photons, due to thermal excitation of electron–hole pairs. This, of course, is noise that will limit the ultimate sensitivity of the photodetector. For this reason, the most sensitive semiconductor photodetectors are typically cooled to cryogenic temperatures, particularly ones for the infrared that require a small-bandgap semiconductor.

But what happens if the photon energy is much larger than the energy gap? The initial absorption process is likely to be similar, creating an electron–hole pair, but each of these excitations can quickly relax (via electron–electron interactions) with the creation of additional electrons and holes. If a photon energy $hf = \mathcal{E}_g$ creates two excitations, one might expect that a photon energy $hf = N\mathcal{E}_g$ might create up to $2N$ excitations, equally divided between electrons and holes. This is indeed an upper limit, and a more typical number, taking into account statistical variations of the energy distribution, is closer to $1.7N$ (the extra energy gets distributed to low-energy phonons). If all of these excitations can be collected before recombining, this will give a much larger current pulse for each absorbed photon (Fig. 7.13*b*).

Furthermore, the energy of the absorbed photon is directly proportional to the size of the pulse (i.e., the integrated charge), so that the energy distribution of the photon flux may be determined. This provides the basis for "energy-dispersive x-ray spectroscopy" (EDX) using a semiconductor photodetector. With photon energies $hf \sim$ keV and a gap energy $\mathscr{E}_g \sim 1$ eV, this provides for a rather high discrimination of photon energies in the x-ray range. The energy resolution is not quite as good as the energy gap ($\Delta\mathscr{E} \approx \mathscr{E}_g\sqrt{N}$), but systems of this type are now standard instrumentation on electron microscopes, where they are used to identify and image chemical elements by the characteristic x-ray fluorescence of atoms bombarded by the electron beam.

Superconducting Tunnel Junction Photodetectors

In certain respects, a superconductor is similar to a semiconductor with a very small gap, ~ 1 meV instead of ~ 1 eV. Of course, there are obvious differences; a semiconductor at $T = 0$ is an insulator, not a superconductor. And radiation below the energy gap is reflected from a perfect superconductor. But as we have pointed out, the quasiparticle characteristics of a superconducting tunnel junction (SIS) for $V < 2\Delta/e$ act very much like a semiconductor. In both cases, the low-temperature resistance is large and exhibits thermally activated conduction. For this reason, a superconducting tunnel junction can form the basis for a sensitive photon detector, with higher resolution down to lower energies than is possible for semiconductor detectors.

These devices are still being refined, but the basic principle is indicated in Fig. 7.14. A superconducting tunnel junction is cooled down to very low temperatures $T \ll T_c$ and biased with a subgap voltage. (A magnetic field may be applied to suppress the Josephson current.) An absorbed photon creates a number of electron and hole excitations in one of the superconducting electrodes. Before these can recombine, they tunnel through the barrier, giving rise to a current pulse of integrated amplitude proportional to the energy of the absorbed photon.

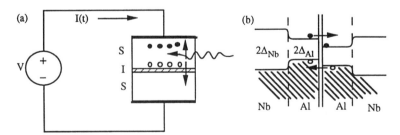

Figure 7.14. Schematic of superconducting tunnel junction photodetector. (*a*) Excess quasiparticles following absorption of high-energy photon. Dark circles represent electrons and open circles holes. (*b*) Approximate band diagram of Nb/Al/AlO$_x$/Al/Nb heterostructure tunnel junction showing trapped quasiparticles near the tunnel barrier.

Prototype devices based on this principle have been fabricated and have been shown to provide single-photon detection all the way from x-rays, through ultraviolet and visible, to the near infrared, all with good energy resolution, at rates that may approach 10 kHz (Verhoeve et al., 1997). There are a number of design details that have been used to enhance the rate of quasiparticle tunneling relative to recombination. Typical junctions consist of a heterostructure of Nb and Al on either side of the insulating (Al_2O_3) barrier (Fig. 7.14b). At the low operating temperature ($T \sim 0.4$ K), both the Nb and the Al are superconducting, but the Al layer has a much smaller energy gap. Quasiparticles created in the Nb diffuse into the Al and then relax down to the gap edge in Al. Then, they are trapped in the thin Al layer near the tunnel barrier, which tends to increase the tunneling rate. Devices of this sort are being developed for integrated arrays that may be used for high-sensitivity astronomical imaging and spectroscopy.

Thermal Photon Detectors

It is also possible under certain circumstances to use a thermal (or quasi-thermal) detector as a photon detector. This requires a device that is so sensitive (and so small) that a single photon can raise the temperature by a measurable amount. Then, if the photon flux is slow enough relative to the thermal time of the detector, each photon absorption will create an electrical pulse of magnitude that is proportional to the energy of the photon. In this mode, one typically refers to a "microcalorimeter" rather than a bolometer.

For example, consider the YBCO bolometer discussed earlier, with heat capacity $C \sim 10^{-11}$ J/K and a thermal time constant ~ 100 μs. For an x-ray photon with $hf \sim 1$ keV $\sim 10^{-16}$ J, we have $\delta T \sim 10$ μK with a 10-kHz measurement bandwidth. This seems rather small and may be marginal for such a bolometer at 90 K. However, similar LTS microcalorimeters at 4 K and below should have the requisite sensitivity even for photons with somewhat lower energies and again are being actively pursued for some astronomical applications.

Another recent proposal for a thermally based photon detector relies on the formation of a localized nonequilibrium "hotspot" after absorption of a photon (Kadin and Johnson, 1996). Consider an ultrathin superconducting film similar to that discussed earlier in the context of the hot-electron bolometer. Photon absorption in the film will lead to a cascade of lower energy electronic excitations that may be described as nonequilibrium hot electrons at a temperature above that of the phonons in the film. However, for a low flux of photons, these hot electrons will not be uniformly distributed across the film but rather will spread out from the initial absorption site with a finite diffusion constant D_e before cooling down by phonon emission in a time $\sim \tau_{ep}$. Thus, a localized circular hotspot of radius $\Lambda \sim \sqrt{D\tau_{ep}}$ will be present on this time scale, with increased effective electron temperature δT_e that is proportional to the photon energy (see Fig. 7.15). Estimates indicate that this temperature rise can be quite

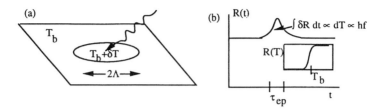

Figure 7.15. Photon-induced hotspot in ultrathin film. (*a*) Localized nonequilibrium hotspot with excited quasiparticles diffusing out to a distance $\sim \Lambda$ in a time τ_{ep} heated above the ambient temperature by δT. (*b*) For an increasing $R(T)$ near T_c (inset), each hotspot causes a resistance pulse of height proportional to δT and width $\sim \tau_{ep}$.

substantial. For example, for an ultrathin granular film with normal-state sheet resistance $\sim 4000\,\Omega$ and $\tau_{ep} \sim 50$ ps at 10 K, an infrared photon with energy $hf \sim 1$ eV will cause local heating with $\delta T \sim 5$ K over a transverse radius $\Lambda \sim 70$ nm.

If a film of this type is current-biased in the middle of the resistive transition, then the hotspot will have an increased sheet resistance, which will temporarily increase the resistance of the overall film, thus leading to a fast resistance pulse. But in addition to counting photons, the pulse height should also be approximately proportional to the photon energy, permitting spectroscopy as well. Essentially, this is acting as a very fast microcalorimeter, with a characteristic thermal time $\tau_{ep} \sim 100$ ps. This hotspot photodetector has not yet been verified experimentally, in part since it requires extremely fast, low-level (sub-millivolt) electronics. On the other hand, it is likely to be well matched to ultrafast superconducting digital electronics, so that a complete integrated system may be within the realm of possibility. Preliminary estimates have indicated that such a fast photon detector may be feasible for photons in the infrared range, with wavelength up to ~ 10 μm.

Fluxonic Photodetectors

Similar nonequilibrium hotspots would also be expected to form in a film held well below the critical temperature T_c. One might initially not expect such a hotspot to create a voltage in the film, since even if the spot were heated above T_c, the surrounding region with $R = 0$ would short it out. However, a small hotspot with a depressed value of the energy gap can act to nucleate a vortex–antivortex pair in the film (Kadin et al., 1990). If the transport current is large enough, the Lorentz force can pull apart this vortex pair, sending the vortex in one direction and the antivortex in the other, thus inducing a voltage. Indeed, the current distribution around a localized hotspot is very similar to that around a vortex pair oriented properly for separation (see Fig. 7.16). Assuming that each vortex can reach the edge of the film without being trapped

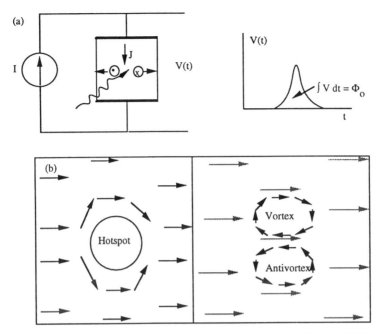

Figure 7.16. Photon-assisted vortex pair creation in a superconducting thin film. (*a*) Schematic circuit with separation of photon-produced vortex–antivortex pair and associated SFQ voltage pulse. This photofluxonic response is the electrical dual to the photoconductive response in a semiconductor (Fig. 7.13*a*). (*b*) Equivalent current configurations for localized hotspot (left) and nucleating vortex–antivortex pair (right) in presence of large transport current.

or recombined, this corresponds to a net transfer of one flux quantum Φ_0 across the width of the film. This, in turn, is equivalent to an SFQ voltage pulse.

This description may sound familiar; indeed, it is precisely the electromagnetic dual of the photoconductive response in a semiconductor (compare Fig. 7.16*a* with Fig. 7.13*a*). Here, we are suggesting that each absorbed photon may give rise to a voltage pulse of integrated amplitude Φ_0 in the presence of a current. This photoresistive (or "photofluxonic") response corresponds to an average quantum-limited voltage responsivity of Φ_0/hf, directly analogous to the limiting current responsivity of e/hf for a semiconductor photoconductor. Such a detector could be very sensitive and very fast; for 1-eV infrared photons, $\Phi_0/hf \sim 10{,}000$ V/W, and the "collection time" is given by the vortex transversal time, estimated to be <1 ns for a thin granular film on the micrometer scale. Despite the appealing nature of this picture, this proposed mechanism is probably oversimplified and has not yet been verified in this form. However, some thin granular LTS films (in particular, NbN and $BaPb_{0.7}Bi_{0.3}O_3$) have exhibited fast photoresponses up to $\sim 10{,}000$ V/W that could not be attributed to heating; fluxonic photodetection remains a likely candidate to explain these observations.

Particle Detectors

As we have seen, the photon is in many respects a point particle, which deposits all of its energy at one point in a detector. The same types of superconducting detectors that can be used to count single photons can also be adapted to measure other particles that deposit energy into a device (Booth and Goldie, 1996). These include superconducting tunnel junctions and bolometric microcalorimeters, for example. In addition, several types of other novel particle detectors have been proposed for certain unusual types of particles, which may or may not even exist.

One novel type of particle detector is based on an array of small type I superconducting grains in an applied magnetic field. Each grain is much larger than the magnetic penetration depth, so that most of the grain is in the Meissner state with the magnetic field excluded. When a particle deposits its energy in one of these grains, the grain heats up into the normal state, permitting magnetic flux to enter and giving an output signal on a SQUID coupled to the grain. Preliminary tests with x-ray photons using spherical particles of Pb, Sn, or In ~ 10 μm across have indicated that such an array can actually operate in the metastable "superheated" superconducting state with H slightly above H_c. The incident particle deposits enough energy (~ 1 keV) to cause the grain to switch irreversibly into the normal state, making detection easier. Such a superheated superconducting granular detector has been suggested in a search for hypothetical "weakly interacting massive particles," or WIMPs.

Finally, let us mention a completely different kind of particle detector for another hypothetical fundamental particle. This particle is the magnetic monopole, which would be the quantum of magnetic charge in the same way as the electron is the quantum of electric charge; it would correspond to a net source of magnetic field lines. Now, as far as we know, such magnetic monopoles do not exist in nature; all magnetic fields are magnetic dipoles produced by moving electric currents (even spins on the atomic level), and magnetic field lines form closed loops. If magnetic monopoles existed, then the fundamental magnetic flux equation would have to be modified to read $\int B \cdot dS = Q_m$, the total magnetic charge (in units of flux) inside a closed surface. Despite the lack of experimental evidence, some elementary particle theories have postulated the existence of such magnetic charge, quantized in units of twice the flux quantum $2\Phi_0 = h/e$. There might be a relatively small number of these magnetic monopoles moving through space and possibly interacting only very weakly with matter through their magnetic field.

A few researchers (Cabrera, 1982) have attempted to search experimentally for such a magnetic monopole using a superconducting detector. Consider what happens if such a particle passes through the center of a closed superconducting loop that initially contains zero flux (see Fig. 7.17). As the monopole approaches the plane of the loop, the applied flux increases from 0 to Φ_0 (i.e, half the total flux of the monopole). This will be opposed by a reverse flux $LI = \Phi_0$ in the loop, to maintain flux conservation. As the monopole moves through the loop

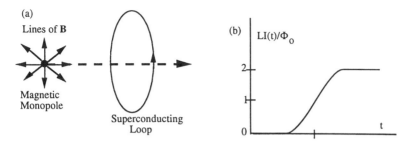

Figure 7.17. Detection of hypothetical particle—magnetic monopole—passing through closed superconducting loop. (*a*) Magnetic field configuration of magnetic monopole, which acts as a source of magnetic flux $2\Phi_0$, approaching center of loop, which initially contains no magnetic flux. The induced current I can be measured by coupling to a SQUID. (*b*) Approximate time-dependence of induced current I in loop, where the induced flux is $LI = \Phi_0$ for the monopole in the center of the loop and $2\Phi_0$ after it has passed through.

and away from it, the other half of the monopole flux causes the induced current to increase further to $I = 2\Phi_0/L$, corresponding to a net flux of $2\Phi_0$ trapped in the loop indefinitely. (Contrast this with a magnetic dipole approaching the loop; the circulating current first increases, but then decreases back to zero after the magnet passes through the other side.) If the particle passes through quickly, the trapped flux $n\Phi_0$ in the loop suddenly jumps from $n = 0$ to $n = 2$, without any flux cutting the wire itself. This would seem to be a unique indication of such a magnetic monopole, following from the fundamental properties of a superconductor. This trapped flux (or the circulating current in the loop) could be measured with a SQUID magnetometer.

Remarkably, one experiment (Cabrera, 1982) actually did detect one such sudden flux jump in a detector based on a single superconducting loop. This created a great deal of excitement among researchers worldwide, and several other groups attempted to reproduce these results. In these follow-up experiments, multiple loops were used in order to check for coincidence, that is, the monopole passing through more than one loop in sequence. Unfortunately, with these more stringent checks, no further events were detected. The general conclusion was that the initial event was perhaps some sort of experimental artifact, perhaps associated with the release of a trapped fluxon in the niobium wire constituting the loop. So it appears that Maxwell's equations in their standard form are safe, at least for the present.

This chapter has dealt with a number of distinctly different approaches toward the application of superconducting devices to sensitive radiation detection; some may be more practical than others. As they continue to develop in the future, it is likely that integrated superconducting systems will be developed that couple superconducting detector arrays with advanced on-chip superconducting digital signal processsing. Although few of these superconducting devices have yet reached the market, this combination offers potential performance that goes well beyond what can be achieved with more conventional technologies.

SUMMARY

- Three major categories of superconducting radiation detectors are modulation detectors and mixers (for rf and microwaves), thermal detectors (for infrared), and photon detectors (for visible through x-ray).
- The sharp rise in tunneling current at $V = 2\Delta/e$ in an SIS tunnel junction permits quantum limited detection and mixing up to at least ~ 500 GHz, with application to radio astronomy.
- A superconducting transition-edge bolometer uses the resistive transition at T_c as a sensitive thermometer to measure optical power absorbed in a thermally isolated detector. Using HTS YBCO may provide a sensitive infrared photodetector at 90 K.
- Nonequilibrium hot-electron effects can be used to design a very fast superconducting microbolometer that may even be used for terahertz mixing.
- An SIS tunnel junction acts as a small-gap semiconductor for $V < 2\Delta/e$; absorption of a high-energy photon creates many quasiparticle excitations and a current pulse. This is the basis for spectrally sensitive photon-counting detectors from x-rays through visible radiation.
- A microcalorimeter is a bolometer in which the thermal response from a single absorbed photon can be measured and counted, and functions as a spectrally sensitive photon detector.
- A fluxonic photodetector uses photoexcitation of vortices in a superconducting thin film in direct analogy to excitation of electrons in a semiconductor; this may account for some large infrared photoresponses in granular superconducting films.
- Superconducting particle detectors are similar in certain respects to photon detectors; they count each particle passing into the detector.

REFERENCES

N. E. Booth and D. J. Goldie, "Superconducting Particle Detectors," *Supercond. Sci. Technol.* **9**, 493 (1996).

B. Bumble and H. G. LeDuc, "Fabrication of a Diffusion Cooled Superconducting Hot Electron Bolometer for THz Mixing Applications," *IEEE Trans. Appl. Supercond.* **7**, 3560 (1997).

B. Cabrera, "First Results from a Superconductive Detector for Moving Magnetic Monopoles," *Phys. Rev. Lett.* **48**, 1378 (1982).

M. W. Johnson, A. M. Herr, and A. M. Kadin, "Bolometric and Nonbolometric Infrared Photoresponses in Ultrathin Superconducting NbN Films," *J. Appl. Phys.* **79**, 7069 (1996).

A. M. Kadin and A. M. Goldman, "Dynamical Effects in Nonequilibrium Superconductors," Chap. 7 in *Nonequilibrium Superconductivity*, Ed. D. N. Langenberg (North-Holland, Amsterdam, 1986).

A. M. Kadin and M. W. Johnson, "Nonequilibrium Photon-Induced Hotspot; A New Mechanism for Photodetection in Ultrathin Metallic Films," *Appl. Phys. Lett.* **69**, 3938 (1996).

A. M. Kadin, M. Leung, A. D. Smith, and J. M. Murduck, "Photofluxonic Detection; A New Mechanism for Infrared Detection in Superconducting Thin Films," *Appl. Phys. Lett.* **57**, 2847 (1990).

C. Kittel, *Introduction to Solid State Physics*, 7th ed. (Wiley, New York, 1996).

M. Lindgren, M. A. Zorin, V. Trifonov, M. Danerud, B. S. Karasik, G. N. Gol'tsman, and E. M. Gershenzon, "Optical Mixing in a Patterned $YBa_2Cu_3O_7$ Thin Film," *Appl. Phys. Lett.* **65**, 3398 (1994).

M. Nahum, S. Verghese, P. L. Richards, and K. Char, "Thermal Boundary Resistance for $YBa_2Cu_3O_7$ Films," *Appl. Phys. Lett.* **59**, 2034 (1991).

P. L. Richards, "Bolometers for Infrared and Millimeter Waves," *J. Appl. Phys.* **76**, 1 (1994).

P. L. Richards and Q. Hu, "Superconducting Components for Infrared and Millimeter-Wave Receivers," *Proc. IEEE* **77**, 1233 (1989).

J. E. Sauvageau and D. G. McDonald, "Superconducting Kinetic Inductance Bolometer," *IEEE Trans. Magn.* **25**, 1331 (1989).

A. D. Semenov, Y. P. Gousev, K. F. Renk, B. M. Voronov, G. N. Gol'tsman, E. M. Gersehnzon, G. W. Schwaab, and R. Feinäugle, "Noise Characteristics of a NbN Hot-Electron Mixer at 2.5 THz," *IEEE Trans. Appl. Supercond.* **7**, 3572 (1997).

H. Shimakage, Y. Uzawa, M. Tonouchi, and Z. Wang, "Noise Temperature Measurement of YBCO Josephson Mixers in mm and Sub-mm Waves," *IEEE Trans. Appl. Supercond.* **7**, 2595 (1997).

M. Tinkham, *Introduction to Superconductivity*, 2nd ed., Section 6.7, Chap. 11 (McGraw-Hill, New York, 1996).

J. R. Tucker and M. J. Feldman, "Quantum Detection at Millimeter Wavelengths," *Rev. Mod. Phys.* **57**, 1055 (1985).

C. Uher, "Thermal Conductivity of High T_c Superconductors," *J. Superconductivity* **3**, 389 (1990).

P. Verhoeve, N. Rando, A. Peacock, A. van Dordrecht, A. Poelaert, and D. J. Goldie, "Superconducting Tunnel Junctions as Photon Counting Detectors in the Infrared to the Ultraviolet," *IEEE Trans. Appl. Supercond.* **7**, 3359 (1997).

M. J. Wengler, "Submillimeter-Wave Detection with Superconducting Tunnel Diodes," *Proc. IEEE* **80**, 1810 (1992).

PROBLEMS

7.1. SIS detector.

(a) Plot the ideal SIS characteristic $I(V)$ for a Nb junction ($2\Delta = 3$ meV) with $I_c = 200$ µA at low T. Include both positive and negative voltage and the zero-voltage current.

(b) An rf voltage of 2 mV amplitude at 500 GHz is incident on the junction. Plot the changes in the $I(V)$, including both the quasiparticle and Shapiro steps. (A plot of Bessel functions is shown in Fig. 5.11.) Assume that capacitance is *not* a limitation.

(c) Do the same thing for a frequency of 1 THz.

(d) If this junction is 1 µm across and the magnetic penetration depth is 90 nm, estimate the magnetic field that would be needed to substantially suppress the Josephson current. How does this compare to the critical field(s) of Nb?

7.2. SIS mixer. Model an SIS mixer as a nonlinear resistance in parallel with an ideal Josephson element and a capacitance. Assume that this is Nb with $2\Delta = 3$ meV, $C = 10$ fF, and $R_n = 100\ \Omega$, that the signal frequency is 200 GHz, and that the intermediate frequency is 1 GHz.

(a) Determine the frequency of the local oscillator, and estimate roughly the magnitude of the optimum LO voltage amplitude and power.

(b) If the mixer is operated in the middle of the $n = 1$ step just below the energy gap, estimate the Josephson frequency, the critical current, and the rf voltage at this frequency, ignoring the capacitor. Compare to (a).

(c) Now include the effect of C, and estimate the reduced rf voltage across the junction at f_J. Also estimate the reduction in the signal current coupled to the nonlinear resistor.

(d) Estimate the inductance needed to tune out the capacitance at the signal and intermediate frequencies. Could this be produced using superconducting lines on the micron scale?

7.3. YBCO microbolometer. Consider a YBCO thin-film element that is mounted on a very thin substrate and connected to its thermal environment primarily via four thin-film YBCO leads. Assume that the film is 100 nm thick with $T_c = 90$ K and a transition width $\Delta T = 0.1$ K, and a resistivity above T_c of 100 µΩ-cm. The active region is a square with area 10 µm², and each lead is 1 µm wide by 10 µm long. This is to be used as a bolometer near T_c.

(a) Estimate the thermal diffusivity of YBCO and the time needed for thermal equilibration across the active region.

(b) Estimate the heat capacity C of the active region and the thermal conductance along each lead. Determine the characteristic heating time for the element. Is a lumped-element model a reasonable approximation?

(c) Estimate the film resistance and the maximum thermal coefficient of resistance at T_c. If this can be biased by a current of 300 µA and the temperature adjusted to remain in the middle of the resistive transition, what is the voltage responsivity R_v of this bolometer?

(d) If the noise level in the device is 3 nV/$\sqrt{\text{Hz}}$, what is the minimum power that can be measured? This is the noise-equivalent power (NEP). Estimate the corresponding detectivity D^* for this detector.

(e) What is the maximum detectable power if the bolometer is to remain in the resistive transition? What dynamic range does this imply?

7.4. Hot-electron bolometric mixer. In a hot-electron bolometer, the electrons in the thin-film element are heated by the absorbed radiation to a higher effective temperature T_e than that of the lattice (or phonons) T_p. Assume that the electron–phonon scattering time $\tau_{ep} = 20$ ps, that the thermal conductance between the film and the substrate is 10 W/cm² K, and that the contributions to heat capacity (per unit volume) are $c_e = 2$ mJ/cm³ K and $c_p = 7$ mJ/cm³ K. Also assume that the electrical resistivity of the film is 300 µΩ-cm, with $T_c = 10$ K. These parameters are typical of thin NbN.

(a) For what film thickness d will the thermal resistance between the electrons and the phonons R_{ep} be comparable to that between the phonons and the substrate R_{ps}?

(b) For this thickness, estimate the sheet resistance of the film for $T > T_c$ and the temperature coefficient of resistance if the transition width $\Delta T = 0.1$ K.

(c) Estimate the temperature rises δT_e and δT_p for a steady-state absorbed power of 1 pW/µm². What would be the fractional change in resistance for a film biased in the middle of the resistive transition?

(d) Now consider the operation of this device as a mixer for the same signal power as above. If the pump power is 100 nW/µm², estimate the fractional amplitude of the resistance modulation and compare to the above result.

7.5. Fluxonic photodetection. A novel form of superconducting photoresponse assumes that each absorbed photon creates a vortex–antivortex pair that can be pulled apart and collected at the two edges of the film in a large transport current. (This is dual to creation of an electron–hole pair in a semiconductor, which can be collected in an applied voltage.)

(a) Explain why this is equivalent to the transfer of one flux quantum across the entire film. Based on this assumption, show that the average voltage responsivity is $R_v = \Phi_0/hf = 1/2ef$. If some of the vortices recombine before reaching the edge of the film, what does this do to the responsivity?

(b) Estimate the vortex velocity in a thin film with resistivity ρ_n, superconducting coherence length ξ, and current density J. Use values $\xi = 30$ nm, $J = 10^6$ A/cm², and $\rho_n = 200$ µΩ-cm. If the film is 3 µm wide, what is a characteristic response time?

(c) Assume that a vortex and antivortex travelling toward each other will recombine. If the scale for this is the vortex core size ξ, estimate the maximum rate of photon absorption before vortex recombination becomes dominant.

7.6. Superconducting tunnel junction photon detector. The subgap conductance of an ideal SIS tunnel junction becomes exponentially small at low T. The absorption of a single high energy photon can create a large number $N \approx \mathcal{E}/1.7\Delta$ of quasiparticle excitations at the gap edge. These can be collected in a small voltage, giving rise to a current pulse of amplitude proportional to the photon energy $\mathcal{E}$.

(a) For a Nb junction with $2\Delta \approx 3$ meV and an x-ray photon with $\mathcal{E} \approx 1$ keV, estimate the number of electrons N created, and the integrated charge in a current pulse from a single absorbed photon. If the collection time is ~ 10 μs, estimate the peak current.

(b) The statistical variation in the number of collected electrons N is approximately $\sqrt{N}$. Using this approximation, estimate the spectral resolution for the 1-keV x-ray photon. How would this compare to the expected spectral resolution using a semiconductor detector with a 1-eV gap?

(c) For a high-quality Nb/Al junction, the leakage current in the relevant regime is ~ 0.1 pA/μm². For a junction with area 100 μm², estimate the number of electrons collected in the 10-μs collection time in the absence of incident photons. Use this to estimate roughly a lower limit on the energy of a detectable photon.

EPILOGUE: FUTURE PROSPECTS FOR SUPERCONDUCTING CIRCUITS

In the preceding chapters, we have explained the basic principles of superconductors and superconducting circuits and introduced some of the many possible applications. With this as perspective, we may be in a position to assess the further development of superconducting devices in the near future.

First of all, the field of superconductivity and its applications go back long before the recent discovery of high-temperature superconductors (HTSs). Although there has been much attention worldwide on the new HTS materials, there has also been considerable research and development on more conventional low-temperature superconducting (LTS) circuits and systems, going back some 40 years or more. From the point of view of practical applications, many of the LTS devices have already established a small but growing market. For example, Nb–Ti superconducting magnets are standard components of magnetic resonance imaging (MRI) systems for medical diagnostics. Similarly, Nb integrated circuits now form the basis for the International Standard Volt (via the inverse ac Josephson effect), and Nb SQUIDs provide state-of-the-art detection of weak magnetic fields. These devices are essential for their applications; a more conventional technology would simply not have worked.

This brings up a key point. There are many cases where a superconductor is better than the competition, but this is not enough for market penetration. This observations is not unique to superconductors; it is true for any substantially new technology. If a superconducting device is essential to perform a necessary function, then it will be adopted despite the need for cryogenic cooling. These devices will be introduced first in specialized (expensive) high-performance equipment such as advanced medical and scientific instrumentation. As such LTS devices become more common, the cryogenic cooling technology itself (including closed-cycle cryocoolers) will be improved, encouraging the wider application of superconducting devices. This has already happened to some degree, and this trend will continue into the future.

In the digital realm, silicon has continued its relentless increase in both speed and integration density, making it difficult for any alternative technology to gain a foothold. Partly for this reason, some researchers have chosen to explore future computer architectures where silicon is unlikely to be competitive. One such arena is for a projected massively parallel computer that could achieve 10^{15} floating-point operations per sec, known as the "petaflops computer." In one implementation of such a computer (Gao et al., 1996), 10,000 ultrafast LTS RSFQ microprocessors would be closely integrated together in a single small module. The heating alone would make this impossible for conventional semiconductor circuits; it would dissipate some megawatts of power. In contrast, the superconducting equivalent is projected to dissipate a modest 300 W. There are innumerable problems in getting the components of such a project to work properly, but this is an effort at developing the "superconducting supercomputer" of the future.

Another computer architecture that is still speculative at best is the quantum computer. A quantum computer takes advantage of some of the subtle properties of coherent quantum systems to perform certain kinds of massively parallel computations, but *without* requiring a massively parallel circuit (DiVincenzo, 1995). Such a quantum computer could clearly *not* be obtained using conventional silicon circuits. Several different approaches to achieving quantum computations are being investigated, but none has been demonstrated beyond just a few bits (known as quantum bits or "qubits"). It has recently been proposed that such qubits may be implemented using LTS SQUIDs and ultrafast Josephson digital circuits at low temperatures (Bocko, Herr, and Feldman, 1997). This would require a higher level of quantum coherence than we normally deal with in superconductors; in particular, the Josephson junctions could not exhibit any resistive dissipation on the time scale of the computations. It is not yet clear whether this is even possible, but if so, it might revolutionize the entire field of computing.

Where does all this place HTS circuits and devices? HTS materials are far more difficult to produce in a controlled way than most LTS materials, and the technology is in a relatively primitive state. What is needed for HTS technology are essentially the same things that have proved so valuable for LTS devices: a reliable high-current superconducting wire and a reliable Josephson tunneling junction. Further, HTS devices must now compete with established LTS devices as well as with noncryogenic technology. For these reasons, despite some promising potential markets (such as filters for cellular telephones), it is probably unlikely that HTS devices will develop a major market in the next decade. But this does not mean that research and development in these materials should be halted. On the contrary, present-day LTS applications are the direct consequence of several decades of R&D, and the same thing will likely take place for HTS materials in the decades to come.

In the late 1980s, there was a common expectation that the maximum superconducting critical temperature would continue to rise sharply and that superconductivity at room temperature was just around the corner. That has not come to pass, although one cannot yet rule out the possibility. However, it is important to understand that room-temperature superconductivity (RTS) would not necessarily be the solution to all our problems. If we can speculate on the properties of an ideal RTS material, it would have $T_c \sim 600$ K, so that it could operate at room temperature at about one-half of T_c, in order that its superconductivity would be "at full strength." However, since the superconducting coherence length tends to scale inversely with T_c, we would expect $\xi \sim 2$ Å or less. Even if the material were isotropic, this would likely make it very difficult to conduct large currents across grain boundaries, which is the key problem with the HTS materials. Furthermore the small coherence length and high temperature would tend to make thermal vortex fluctuations a serious problem, even more so than for the current HTS. It would be rather ironic if the RTS were discovered only to find that it cannot be used at room temperature!

It has been said, only half in jest, that superconductivity is the technology of the future and always will be. There is something to this; certainly many overly optimistic predictions of widespread superconducting applications (e.g., computers or levitating trains) have not been borne out. But in several areas, superconducting circuits are starting to make a serious impact, and as the twenty-first century proceeds, superconductivity may finally become a key enabling technology of the present *and* the future.

REFERENCES

M. F. Bocko, A. M. Herr, and M. J. Feldman, "Prospects for Quantum Coherent Computation Using Superconducting Electronics," *IEEE Trans. Appl. Supercond.* **7**, 3638 (1997).

G. Gao, K. K. Likharev, P. C. Messina, and T. L. Sterling, "Hybrid Technology Multithreaded Architecture," in *Frontiers '96*, Proc. 6th Symp. on Frontiers of Massively Parallel Computing (IEEE Computer Society, Los Alamitos, California, 1996).

D. P. DiVincenzo, "Quantum Computation," *Science* **270**, 255 (1995).

BIBLIOGRAPHY

There is truly a massive literature on the subject of superconductivity and related topics. In addition to the specific references at the end of each chapter, a partial list of useful general references is given below.

Textbooks and Monographs on Superconductivity and Josephson Effects

G. Burns, *High-Temperature Superconductivity, An Introduction* (Academic, New York, 1992).

M. Cyrot and D. Pavuna, *Introduction to Superconductivity and High-T_c Materials* (World Scientific, Singapore, 1992).

J. C. Gallop, *SQUIDs, the Josephson Effects and Superconducting Electronics* (Adam Hilger, Bristol, 1991).

K. K. Likharev, *Dynamics of Josephson Junctions and Circuits* (Gordon and Breach, New York, 1986).

T. P. Orlando and K. A. Delin, *Foundations of Applied Superconductivity* (Addison-Wesley, Reading, MA, 1991).

T. P. Sheahen, *Introduction to High-Temperature Superconductivity* (Plenum, New York, 1994).

M. Tinkham, *Introduction to Superconductivity*, 2nd ed. (McGraw-Hill, New York, 1996).

T. Van Duzer and C. W. Turner, *Principles of Superconductive Devices and Circuits* (Elsevier North-Holland, New York, 1981).

Edited Collections of Review Articles

S. T. Ruggiero and D. A. Rudman, Eds., *Superconducting Devices* (Academic, New York, 1990).

T. Van Duzer and C. E. Taylor, Eds., *Proc. IEEE* (Special Issue on Superconductivity) **77**, 1107 (1989).

H. Weinstock and M. Nisenoff, Eds., *Superconducting Electronics* (Springer-Verlag, Berlin, 1989).

H. Weinstock and R. W. Ralston, Eds., *The New Superconducting Electronics* (Kluwer, Dordrecht, 1993).

Proceedings of the Biennial Applied Superconductivity Conferences

IEEE Transactions on Applied Superconductivity, vol. 7, pp. 62–377 (June 1997).

IEEE Transactions on Applied Superconductivity, vol. 5, pp. 61–3434 (June 1995).

IEEE Transactions on Applied Superconductivity, vol. 3, pp. 51–2998 (March 1993).

IEEE Transactions on Magnetics, vol. 27, pp. 814–3405 (March 1991).

APPENDIX A
TRANSMISSION LINES AND ELECTROMAGNETIC WAVES

Standard Model for Transmission Line

The standard model of waves on a transmission line is used repeatedly in this book to describe not only superconducting transmission lines but also the electromagnetics of a superconductor itself. With that in mind, this appendix serves as a quick review of non-superconducting transmission lines and how they are modeled.

A simple transmission line consists of two parallel conductors (Fig. A.1a), and we are concerned with guided electrical waves propagating along the length of the line, which we take as the z direction. The relevant variables are the voltage $v(z)$ between the two conductors and the current $i(z)$, which points in the $+z$ direction in the top conductor and returns along the $-z$ direction in the lower conductor. The two conductors are coupled inductively and capacitively, with a distributed series inductance per unit length given by L (in H/m) and a shunt capacitance per unit length given by C (in F/m). There may also be a series resistance per unit length R (in Ω/m) due to the conductors and a leakage shunt conductance per unit length G (in S/m = Ω^{-1} m^{-1}) across the dielectric between the two conductors. It is conventional to represent the line with L and R in the top conductor and with the lower conductor as an equipotential "ground" but these parameters actually represent the series combination of both conductors (Fig. A.1b).

The problem may be addressed most generally in the frequency domain in terms of complex phasors $V(\omega)$ and $I(\omega)$, where $v(t) = \mathrm{Re}[V(\omega)e^{j\omega t}]$ and $i(t) = \mathrm{Re}[I(\omega)e^{j\omega t}]$. Then, there is a series impedance per unit length $Z = R + j\omega L$ and a shunt admittance per unit length $Y = G + j\omega C$. Applying Kirchhoff's laws to a differential element dz, we have

$$\frac{dI}{dz} = -YV \qquad \frac{dV}{dz} = -ZI \qquad (A.1)$$

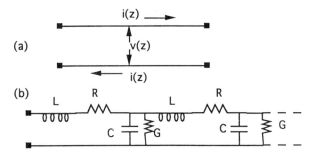

Figure A.1. (*a*) Schematic of transmission line with current *i*(z) and voltage *v*(z) along two conductors. (*b*) Circuit model of transmission line with distributed series inductance *L* and resistance *R* and distributed shunt capacitance *C* and conductance *G* (all per unit length).

If we take the derivative of the second equation and substitute from the first, we have

$$\frac{d^2 V}{dz^2} = -Z\frac{dI}{dz} = +ZYV \tag{A.2}$$

which has the standard solution $V = V_0 \exp(\pm \gamma z)$, where $\gamma = \sqrt{ZY}$ is the propagation constant, which is normally complex. We get exactly the same equation for the current *I*. (This should not be a surprise; the duality transformation on the transmission line yields the same transmission line, but with $Y \leftrightarrow Z$.) The two solutions correspond in general to attenuated waves propagating in the $\pm z$ directions. Since $dV/dz = \pm \gamma V$ but also $= -ZI$ from Eq. (A.1), we get

$$\frac{V}{I} = \pm \frac{Z}{\gamma} = \pm \sqrt{Z/Y} \tag{A.3}$$

where $Z_0 = \sqrt{Z/Y}$ is known as the characteristic impedance of the transmission line.

The simplest case is the one where there is no dissipation, so that both $R = 0$ and $G = 0$. Then, we have $\gamma = \alpha + j\beta = j\omega\sqrt{LC}$, which corresponds in the time domain to wave solutions of the form

$$v(z, t) = V_0 \cos(\omega t \pm \beta z) \tag{A.4}$$

with $\pm$ corresponding to propagation in the minus or plus direction. This wave exhibits no attenuation or dispersion, having a phase velocity $u = \omega/\beta = 1/\sqrt{LC}$ for all frequency components. Also, $Z_0 = \sqrt{L/C}$ for all frequencies, so that *v* and *i* are always in phase for a single direction of wave propagation. In fact, in this case any function $f(z \pm ut)$ is a solution to the wave equation.

More generally, if either R or G is nonzero, then both attenuation and dispersion of the wave are present. An important approximation is the low-loss (high-frequency) limit, where $R \ll \omega L$ and $G \ll \omega C$. In that case, to lowest order Z_0 and β are unchanged, while $\alpha = \frac{1}{2}(R/Z_0 + GZ_0)$. In the other limit, for large losses and low frequencies, one has simply a distributed resistor network with input resistance $Z_0 = \sqrt{R/G}$ and attenuation constant $\alpha = \sqrt{RG}$.

Real Transmission Lines

The most common transmission line is probably the coaxial line (Fig. A.2a), with a center conductor of outer radius a and an outer conductor with inner radius b, separated by a lossless dielectric with permittivity $\varepsilon = \varepsilon_r \varepsilon_0$ and permeability μ_0. If we assume that the fields are confined within the dielectric (a good assumption for electric fields, not quite so good for magnetic fields), we can calculate C and L rather simply. Consider a short length l of transmission line with a dc voltage V applied between the two conductors. If the charge on the center line is Q, then the electric field from Gauss's law is given by $E = Q/2\pi\varepsilon r l$, where r (between a and b) is the radius in cylindrical coordinates. Integrating E gives $V = (Q/2\pi l)\ln(b/a)$, so that $C = Q/lV = \varepsilon[2\pi/\ln(b/a)]$. To determine L, consider the flux for a short-circuited length l of transmission line. From Ampere's law, the circumferential magnetic field is given by $H = I/2\pi r$, so that $\Phi = \int B \cdot dS = \mu_0 l \int H\, dr = (\mu_0 I l/2\pi)\ln(b/a)$ and $L = \Phi/Il = \mu_0[\ln(b/a)/2\pi]$. Then, the propagation constant $\gamma = j\omega\sqrt{\mu_0 \varepsilon}$, corresponding to a wave velocity $u = c/\sqrt{\varepsilon_r}$, where c is the speed of light in a vacuum. The characteristic impedance of the line is $Z_0 = \sqrt{\mu_0/\varepsilon}\ln(b/a)/2\pi = (337\,\Omega/\sqrt{\varepsilon_r})\ln(b/a)/2\pi$, where $377\,\Omega = \sqrt{\mu_0/\varepsilon_0}$ is the "impedance of free space."

Another structure that is commonly analyzed is a parallel-plate transmission line (Fig. A.2b), where the width $w \gg d$, the separation between the lines. Then, neglecting edge effects, the capacitance per unit length $C = \varepsilon w/d$ and the inductance $L = \mu_0 d/w$. The propagation constant $\gamma = j\omega\sqrt{\mu_0\varepsilon}$, the wave

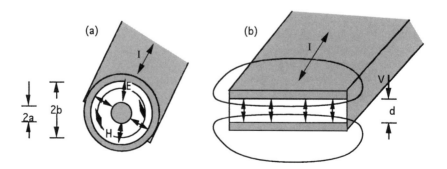

Figure A.2. Common geometries for transmission lines: coaxial (*a*) and parallel plate (*b*).

velocity is $c/\sqrt{\varepsilon_r}$, and $Z_0 = \sqrt{(\mu_0/\varepsilon)}(d/w)$. A structure that is more common in practice is a microstrip transmission line, consisting of a strip above a ground plane with air above the strip. This is similar to the parallel-plate line, except that the fringe fields at the edge may lie partly in the air, changing the properties slightly.

The resistance parameter R of the transmission line is associated with the surface resistance R_s of both conductors; for the parallel-plate line, $R = 2R_s/w$, while for the coaxial line, $R = (R_s/2\pi)(1/a + 1/b)$. This is normally a function of frequency, since $R_s = 1/\sigma\delta$ for a conductor thicker than the ac skin depth $\delta = 1/\sqrt{\pi f \mu_0 \sigma}$. The change in the skin depth also brings about a small change in the inductance as well, leading to some dispersion of the wave.

Transmission Lines and Electromagnetic Waves

It is not an accident that the velocity of the wave on the transmission line is $u = c/\sqrt{\varepsilon_r}$ for both line geometries given, the same as that of an unguided plane electromagnetic wave in the same dielectric. In fact, in the low-loss limit, the wave on the transmission line is precisely a TEM (transverse electromagnetic) wave. For a homogeneous (and isotropic) dielectric, the capacitance and inductance parameters are always given by $C = \varepsilon/F$ and $L = \mu_0 F$, where F is a dimensionless geometrical parameter, typically of order unity. This factor cancels out in the velocity $u = 1/\sqrt{LC}$ but is maintained in the line impedance $Z_0 = F\sqrt{\mu_0/\varepsilon} = \eta F$, where $\eta = 377\,\Omega/\sqrt{\varepsilon_r} = E/H$ is the intrinsic impedance of the medium.

One can turn the problem around and use the equivalent transmission line to describe the medium itself (Fig. A.3). Assume for simplicity a parallel-plate geometry with $w = d = a$ and also assume that the electric field remains vertical, even at the edge. (This is equivalent to a "magnetic wall" boundary condition at the two sides.) Then the factor $F = 1$, so that $L = \mu_0$, $C = \varepsilon$, and $Z_0 = \eta$. In this

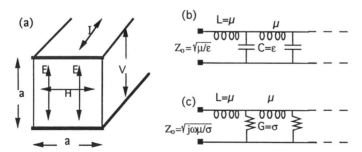

Figure A.3. Transmission line model for plane electromagnetic wave. (a) Equivalent geometry of voltage, current, and fields. (b) Model for wave in dielectric with permittivity ε. (c) Model for wave in conductor with conductivity σ.

way, we have simplified a three-dimensional wave equation to a one-dimensional circuit equation that describes the same physical phenomena. The longitudinal current in the transmission line $I = Ha$ describes the transverse magnetic field in the medium, and the voltage on the line $V = Ea$ describes the electric field. So it should be no surprise that $V/I = E/H = \eta$.

We can go one step further here and use the same construction to model electromagnetic wave propagation in any medium, not simply a good dielectric. A lossy dielectric can be modeled in terms of a conductance parameter $G = \sigma$ in parallel with the capacitance $C = \varepsilon$. Equivalently, one could also use either an effective complex permittivity $\varepsilon_{\text{eff}} = \varepsilon + \sigma/j\omega$ or an effective complex conductivity $\sigma_{\text{eff}} = \sigma + j\omega\varepsilon$. This will, of course, lead to attenuation of the wave. If the medium is a good conductor ($\sigma \gg \omega\varepsilon$), the capacitance will be negligible (Fig. A.3c). In this limit, the propagation constant is

$$\gamma = \sqrt{ZY} = \sqrt{j\omega\mu_0\sigma} = \frac{1+j}{\delta} = \alpha + j\beta \tag{A.5}$$

where $\delta = \sqrt{2/\omega\mu_0\sigma}$ is the ac skin depth. So for a thick conductor, the field does not propagate even a single wavelength ($\lambda = 2\pi/\beta = 2\pi\delta$) into the surface. Furthermore, the characteristic impedance of the line is

$$Z_0 = \sqrt{\frac{Z}{Y}} = \sqrt{\frac{j\omega\mu_0}{\sigma}} = (1+j)\sqrt{\frac{\omega\mu_0}{2\sigma}} = R_s + j\omega L_s, \tag{A.6}$$

where $R_s = 1/\sigma\delta$ is the surface resistance and $L_s = \frac{1}{2}\mu_0\delta$ is the surface inductance associated with penetration of the magnetic field into the conductor. These standard results are more typically derived through the solution of somewhat messy vector differential equations.

Input Impedance and Finite Lengths

Of course, the greatest utility of a transmission line model is in determining the (complex) reflection coefficient of a wave at a termination or interface (Fig. A.4). If a line with characteristic impedance Z_0 is terminated with a lumped impedance Z_1, then a wave V_i moving to the right will cause a proportional reflected wave $V_r = \rho V_i$ to be generated at the termination, according to the following boundary condition:

$$V_i + V_r = (I_i + I_r)Z_1 = \frac{(V_i - V_r)Z_1}{Z_0} \tag{A.7}$$

which yields the well-known result

$$\rho = \frac{Z_1 - Z_0}{Z_1 + Z_0} \tag{A.8}$$

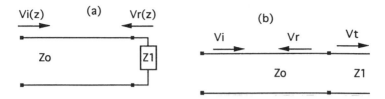

Figure A.4. (a) Schematic of transmission line with wave reflected from termination. (b) Schematic of interface between two transmission lines giving rise to both reflected and transmitted waves.

This has the limits one would expect; no reflection if there is a "matched load" $Z_1 = Z_0$ and complete reflection ($\rho = \pm 1$) for an open circuit ($Z_1 = \infty$) or a short circuit ($Z_1 = 0$). If the termination is really a long length of a second transmission line with characteristic impedance Z_1, then the same formula applies, since the input impedance of a long transmission line is equal to its characteristic impedance. The only difference is that there is also an additional transmitted wave given by the boundary condition that $V_t = V_r + V_i$. The analogy with TEM wave propagation continues to apply as well; this also gives the E-field reflection coefficient for normal incidence of a plane wave on a plane interface between two media.

We can also consider the spatial dependence of V, I, and their ratio $Z = V/I$ at a distance z from the load at the end of a transmission line. These take the standard forms

$$V(z) = V(0)[\cosh(\gamma z) + \frac{Z_0}{Z_1}\sinh(\gamma z)] \quad (A.9)$$

$$I(z) = I(0)[\cosh(\gamma z) + \frac{Z_1}{Z_0}\sinh(\gamma z)] \quad (A.10)$$

$$Z(z) = Z_0 \frac{[Z_1 + Z_0\tanh(\gamma z)]}{Z_0 + Z_1\tanh(\gamma z)} \quad (A.11)$$

where $z = 0$ represents the load so that $Z_1 = V(0)/I(0)$. For the matched load, or for a long line with attenuation, $Z(z) = Z_0$.

For a very short length of transmission line (compared to a wavelength), one can return back to the lumped equivalents that make it up. For example, the input impedance of a short length l is $Z_{in} = Zl + 1/Yl \approx 1/Yl$. One can also apply this to the transmission line model of the conductor; the surface impedance of a thin conducting film of thickness d is simply $Z_{in} = 1/\sigma d$, as expected.

For intermediate lengths, where there may be reflections off both the front and back connections of a transmission line, a two-port lumped-element equivalent must be used (Fig. A.5). This can be given in terms either of the T-network

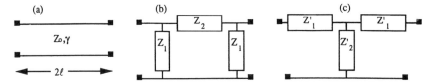

Figure A.5. Finite length of transmission line with equivalent two-port circuits. (*a*) Schematic of line of length 2*l* with characteristic impedance Z_0 and propagation constant γ. (*b*) Π-equivalent, with $Z_1 = Z_0 \coth(\gamma l)$ and $Z_2 = Z_0 \sinh(2\gamma l)$. (*c*) T-equivalent, with $Z'_1 = Z_0 \tanh(\gamma l)$ and $Z'_2 = Z_0/\sinh(2\gamma l)$.

or of the Π-network; the two are related via the Y–Δ transformation. For the Π-equivalent of a length 2*l* of a line with characteristic impedance Z_0 and propagation constant γ, the parameters (see Fig. A.5) are

$$Z_1 = Z_0 \coth(\gamma l) \qquad Z_2 = Z_0 \sinh(2\gamma l) \tag{A.12}$$

For the T-equivalent, the corresponding parameters are

$$Z'_1 = Z_0 \tanh(\gamma l) \qquad Z'_2 = \frac{Z_0}{\sinh(2\gamma l)} \tag{A.13}$$

These parameters permit the calculation of the reflection coefficient for a thin layer between two media of different intrinsic impedances as a (relatively) simple circuit problem.

We can also determine the spatial dependence of V and I along the line. Taking $z = 0$ at the center, $V_1 = V(-l)$, and $V_2 = V(l)$, we have

$$V(z) = \frac{V_1 + V_2}{2} \frac{\cosh(\gamma z)}{\cosh(\gamma l)} + \frac{V_2 - V_1}{2} \frac{\sinh(\gamma z)}{\sinh(\gamma l)} \tag{A.14}$$

$$I(z) = \frac{V_1 + V_2}{2Z_0} \frac{\sinh(\gamma z)}{\sinh(\gamma l)} + \frac{V_2 - V_1}{2Z_0} \frac{\cosh(\gamma z)}{\cosh(\gamma l)} \tag{A.15}$$

where V_1 and V_2 are the voltages at the two ends.

Consider a simple example (Fig. A.6). A planar surface of a conductor with resistivity 10 $\mu\Omega$-cm is subjected to normal incidence (from air) by microwaves with frequency 10 GHz. What fraction of the incident power is reflected and what fraction absorbed? The conductor corresponds to a transmission line with impedance $Z_1 = \sqrt{(j\omega\mu_0/\sigma)} = 63(1 + j)$ mΩ; the air corresponds to $Z_0 = 377\ \Omega$. To lowest order, the wave is completely reflected; the conductor impedance is very small, practically a short. Looking more closely, the voltage reflection coefficient is $\rho = (Z_1 - Z_0)/(Z_1 + Z_0) \approx -1 + 2Z_1/Z_0$. The reflected power goes as $|\rho|^2 = \rho\rho^* \approx 1 - 4\,\text{Re}(Z_1/Z_0) = 1 - 1.7 \times 10^{-4}$. A very small portion of

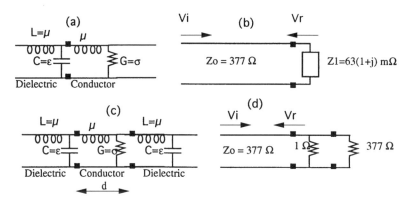

Figure A.6. (a) Transmission line equivalent for TEM wave reflecting from conducting surface. (b) Lumped-element equivalent of (a). (c) Transmission line equivalent for wave normally incident on thin conducting layer. (d) Limped-element circuit for (c) used in calculation.

the incident power, 0.017%, is absorbed in a thickness $\delta = 1/R_s\sigma = 1.6$ μm; the remainder is reflected. In contrast, if we consider the same problem but with a foil $d = 0.1$ μm thick ($\ll \delta$), then the effective impedance $Z_1 = 1/\sigma d = 1\,\Omega$. The reflection coefficient $\rho \approx -1 + 2/377$, so that $1 - 4/377 \approx 99\%$ of the power is reflected. Most of the remainder is absorbed in the foil, but a small portion is transmitted through the foil to the air on the other side. This can be modeled with a transmission line with $Z_0 = 377\,\Omega$ on the other side of the foil, which will draw 1/378 of the "current" at the interface. The portion of the power transmitted is the same as that dissipated in the 377 Ω resistance in Fig. A.6d; $4/377^2 \approx 3 \times 10^{-5}$.

A finite length of transmission line can also be used as a distributed resonator, which exhibits standing waves at resonance. For example, a length l with both ends shorted has resonant modes for voltage oscillations identical to those of a vibrating string, with an integral number of half-wavelengths along the length. Similar, if both ends are open, one has the same relations with the roles of current and voltage reversed. For a low-loss transmission line, the Q-factor of such a resonator takes the particularly simple form $Q = \beta/2\alpha$. This follows from the fact that $2\pi/Q$ is equal to the fraction of the stored energy lost in a period. A traveling wave traverses one wavelength λ in a period and loses a fraction of voltage $\alpha\lambda$ and a fraction of energy (which goes as V^2) $2\alpha\lambda = 4\pi\alpha/\beta$. The standing waves at resonance are simply linear combinations of traveling waves, so the same relation holds true. If the loss is dominated by conductor loss R, then $Q = Q_c = \omega L/R$ (as in the case of the lumped-element series LCR resonator), so that Q is inversely proportional to the surface resistance of the conductor. If the loss is dominated by dielectric (leakage) loss G, then $Q = Q_d = \omega C/G$, as for the parallel LCR resonator. If the region between the two conductors is filled with a dielectric characterized by a complex permittivity $\varepsilon = |\varepsilon|\exp(-j\delta)$, then $G + j\omega C \sim j\omega\varepsilon$, and $Q_d = 1/\tan\delta$, where $\tan\delta$ is the dielectric loss tangent. If both contribute, then $Q^{-1} = Q_c^{-1} + Q_d^{-1}$, so that the total Q is less than that due to either contribution.

These same principles are used in Chapter 2 and elsewhere in the text to describe wave propagation on superconducting transmission lines and current and field distributions in superconductors.

REFERENCES

S. E. Schwarz, *Electromagnetics for Engineers* (Saunders, Philadelphia, 1990).

F. Ulaby, *Fundamentals of Applied Electromagnetics* (Prentice-Hall, Englewood Cliffs, NJ, 1997).

PROBLEMS

A.1. Lossy transmission line. A coaxial transmission line has a characteristic impedance of 50 Ω, an insulator with a dielectric constant $\varepsilon_r = 2$, and a center conductor with diameter 1 mm. A 100-MHz traveling wave on the line is observed to attenuate to half its amplitude in a distance of 50 m.

(a) Determine the wave velocity on this line.

(b) Determine the inductance and capacitance per unit length, L and C, and the diameter of the outer conductor.

(c) Determine the attenuation factor α and the resistance per unit length R.

(d) Determine the resistivity of the conductor and the skin depth at 100 MHz.

A.2. Transmission line model of good conductor. A metallic film 1 μm thick with resistivity 2 μΩ-cm is plated on glass, a lossy dielectric with $\varepsilon_r = 4$. A plane electromagnetic wave at 1 GHz is normally incident from air onto the film.

(a) Determine the equivalent transmission line parameters of each of the three media and the complex characteristic impedance and propagation constant Z_0 and γ of each.

(b) What is the attenuation length (i.e., the skin depth) of the metallic film at this frequency? Is this in the thick limit, the thin limit, or neither?

(c) Using the appropriate approximation, determine the reflection coefficient from the film.

(d) Estimate the fraction of power that is absorbed in the film and the fraction that is transmitted into the glass.

A.3. Transmission line resonator. Consider a length $l = 10$ cm of a transmission line with $Z_0 = 100$ Ω and velocity $u = 10^8$ m/s. The attenuation length decreases as $\sqrt{f}$ and is 1 m at $f = 1$ GHz.

(a) Identify the resonant frequencies of this line, assuming that the ends of the line are open.

(b) Determine the Q-factor of each mode.
(c) Using one of the lumped-element equivalents, determine the effective input impedance Z_{in} at the end of the line. Plot $|Z_{in}(f)|$ over the first few resonances. (A math package that computes hyperbolic functions of complex arguments would be helpful.)
(d) Examine the maximum values Z_{max}/Z_0 and the normalized widths Δf of the resonant peaks and compare to Q.

APPENDIX B

COMPUTER SIMULATIONS OF JOSEPHSON CIRCUITS

In designing complex circuits using conventional semiconductor devices, it is essential to simulate the detailed behavior of the devices in the circuit. This is generally carried out using standard computerized simulation packages, the most common of which is known as SPICE, for Simulation Program with Integrated Circuit Emphasis (see, e.g., Rashid, 1995). Variants of this program are now very widespread and available for virtually all types of computers; a version for desktop computers (both PC and Macintosh) is known as PSpice. SPICE incorporates standard linear devices and sources and all of the common semiconductor devices such as diodes and transistors. It provides a numerical solution to the nodal differential equations associated with a given circuit. It does not, however, include a Josephson junction as a standard element.

As Josephson circuits have become more complex, it has become equally essential to use a sophisticated simulation tool (Gaj et al., 1998). One approach has been to develop a custom system specially for superconducting circuits; an example is PSCAN (Polonsky et al., 1991, 1997) (see Table B.1). However, a more common approach has been to incorporate Josephson junctions into various versions of SPICE (Whitely, 1991, 1993), so that the other aspects of this very familiar program can be directly carried over (JSPICE, JSim). Several versions of these are available; some are available free or are in the public domain. We will describe a straightforward way to implement some simple Josephson circuits using SPICE and briefly mention some of the more sophisticated superconducting circuit design tools that have recently been developed. Finally, for completeness we will briefly refer to the earlier techniques of analog simulation of Josephson junctions and circuits.

COMPUTER SIMULATIONS OF JOSEPHSON CIRCUITS   353

Table B.1. Selected Josephson Circuit Simulation Packages

Name	Available From	Computer Platform
JSim	University of California at Berlekey	UNIX
JSpice3	Whitely Research, Inc.	UNIX
PSCAN	State University of New York at Stony Brook	DOS, UNIX
SPICE[a]	UC Berkeley, also commercial	UNIX
PSpice[a]	MicroSim, Inc., now part of OrCAD	DOS, Windows, Mac.

[a] For standard versions of SPICE, the Josephson junction must be added as a subcircuit.

SPICE and Josephson Junctions

A unique aspect of the Josephson junction as a circuit element is its nonlinear dependence on the phase difference ϕ, or equivalently on the flux Φ, through the relation

$$I_s = I_c \sin(\phi) = I_c \sin\left(\frac{2\pi\Phi}{\Phi_0}\right) \tag{B.1}$$

where I_s is the supercurrent, I_c is the critical current of the junction, and $\Phi_0 = h/2e$ is the flux quantum. From Faraday's law, the flux is the time integral of the voltage across the junction:

$$\Phi(t) = \int V \, dt = \frac{\Phi_0 \phi}{2\pi} \tag{B.2}$$

So to simulate a Josephson junction, one needs simply to integrate the voltage to obtain the flux and to invoke a current source whose value depends on the sine of the flux. This can be achieved in SPICE by defining a subcircuit and using the appropriate behavioral modeling command to express the function in Eq. (B.1). One way to model the integration is with a simple series LR network, with $L_{int} = 1$ H and $R_{int} = 1\,\Omega$, as shown in Fig. B.1 (alternatively, a series RC network could be used). For transient effects on time scales much less than $\sim L/R = 1$ s, most of the voltage drop is across the inductor, and the voltage across the resistor is

$$V_R = R_{int} I_{int} = \frac{R_{int}}{L_{int}} \int V_L \, dt \approx \int V_J \, dt = \Phi \tag{B.3}$$

Then, we can express the supercurrent through the junction as

$$I_s = I_c \sin\left(\left(\frac{2\pi}{\Phi_0}\right) V_R\right) \tag{B.4}$$

which can be implemented using a generalized nonlinear dependent current source that is available on most versions of SPICE.

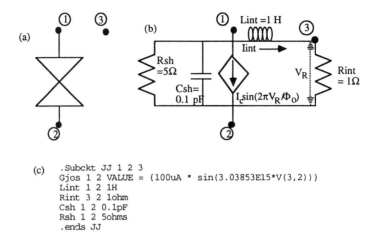

Figure B.1. Simulating a Josephson junction using a standard numerical circuit simulation package such as SPICE. (a) Circuit symbol with nodes labeled. (b) Circuit schematic. For fast voltage transients associated with the Josephson junction, the voltage V_R across the resistor Rint is equal to the integral of the junction voltage, i.e., the flux Φ. The nonlinear flux-dependent current can be implemented using a voltage-dependent current source. A shunt resistor and capacitor are also included, consistent with CRSJ model. (c) Subcircuit listing with syntax for current source Gjos appropriate for PSpice.

The resulting Josephson junction model can be expressed as a subcircuit named JJ, as shown in the listing in Fig. B.1c. Nodes 1 and 2 are the junction nodes; node 3 provides a way to access the flux or the phase across the junction. The syntax of the dependent current source Gjos is for PSpice (the free evaluation version was used) and may be somewhat different on other versions. This is for a critical current $I_c = 100\ \mu A$ with a shunt resistance $R_{sh} = 5\ \Omega$ and a shunt capacitance $C_{sh} = 0.1$ pF added in accordance with the CRSJ shunted junction model. A nonlinear shunt resistance can also be included, although that is somewhat more involved to specify.

It is important to note that the parameters L_{int} and R_{int} are not part of the physical Josephson junction; they are just used with SPICE to calculate the integral of the voltage across the junction. In particular, L_{int} is *not* the same as the Josephson inductance $L_j = \Phi_0/2\pi I_c = 3$ pH. In fact, for this simulated circuit to work properly, one must have $L_{int} \gg L_J$, so that the LR network does not shunt away a significant portion of the current. This also follows from the characteristic times: $L_{int}/R_{int} = 1$ s, while the characteristic time for junction dynamics is $L_J/R_{sh} \sim 1$ ps.

The subcircuit can be invoked by the main SPICE program as an element starting with the letter X using the syntax X1 2 0 1 jj, where X1 refers to a particular Josephson junction in the circuit, connecting nodes 2 and 0, with auxilliary node 1. This can be used multiple times within a circuit as long as each junction has its own distinct variable name (X2, Xleft, etc.).

The standard versions of SPICE have three major modes of operation: dc, ac, and transient. A Josephson junction model can only operate in the transient mode. The ac mode is out because the device is nonlinear and the dc mode because of the Josephson oscillation. (However, see below for a way to obtain the *average* dc voltage.) For the transient mode to work properly here, the independent current or voltage source must start out at zero but can quickly increase. Two simple examples are shown: the oscillating Josephson current for a voltage-biased junction in Fig. B.2 and the SFQ voltage pulse for a current-biased junction in Fig. B.3, both with the same junction parameters as in Fig. B.1.

Of course, the real utility of Josephson circuit simulators is for circuits containing many junctions. Indeed, this is essential to verify the proper operation of complex RSFQ circuits, with picosecond pulses moving between tens and hundreds of Josephson junctions. These simulations take many times longer to run than those for a single junction but are well within the realm of modern high-performance workstations. Furthermore, most modern SPICE packages include a "schematic editor" that can generate the "netlist" (i.e., the SPICE circuit specification) from a schematic drawn using a graphical interface, and this, too, is available for Josephson circuits.

In many cases, we are interested not only in the transient behavior but also in the dc average under steady-state operation of a Josephson circuit. For example, we may want the dc $V(I)$ relation for a current-biased junction or the $V(\Phi)$ characteristics for a SQUID. This requires averaging over multiple Josephson periods, typically over tens or hundreds of picoseconds. There are a number of ways to accomplish this, but one natural approach within SPICE is to use an appropriate low-pass output filter, as suggested in Fig. B.4. Then, if the bias current is ramped slowly on this time scale, the resulting output voltage should reflect the dc average voltage for the given current. In this way, one can use SPICE to generate dc characteristics. The run times tend to be very long,

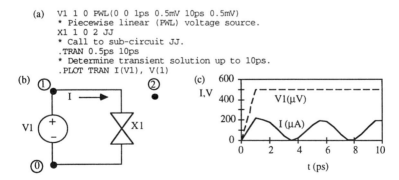

Figure B.2. Voltage-biased Josephson junction in SPICE using subcircuit of Fig. B.1. (*a*) PSpice circuit listing. (*b*) Circuit schematic. (*c*) Plot of bias voltage and current oscillation at Josephson frequency.

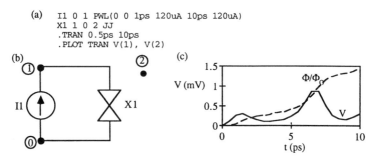

Figure B.3. Current-biased Josephson junction in SPICE using subcircuit of Fig. B.1. (a) PSpice circuit listing. (b) Circuit schematic. (c) Plot of voltage and flux for $I = 1.2 I_c$ showing SFQ pulse.

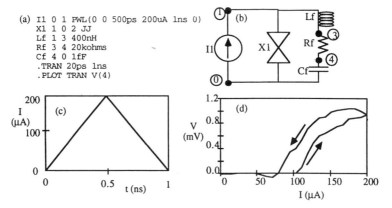

Figure B.4. Use of a low-pass filter to measure dc average voltage in Josephson circuit simulations. (a) SPICE listing for same junction subcircuit as in Fig. B.1. (b) Circuit schematic. (c) Slow current ramp up and down on 1-ns time scale, as compared to 20 ps filter time and several picoseconds junction dynamical time. (d) DC $V(I)$ curve measured across capacitor of output filter. Note the apparent hysteresis between ramp up and ramp down due to the ramp rate being too fast.

however, since the complete sweep may take > 1 ns (of simulation time) while the appropriate numerical time step is still subpicosecond.

The dc average curves obtained in this way are shown in Fig. B.4 for the same junction parameters as in Fig. B.1. Here, we have used a second-order passive LCR filter with $L/R = RC = 20$ ps, much longer than the dynamical time of several picoseconds. The total time for the current ramp is 1 ns, much longer than the filter time. Fig. B.4d shows the voltage across the capacitor of the output filter and represents the dc $V(I)$ relation for a single junction. This is similar to the expected analytic dependence $V = R\sqrt{I^2 - I_c^2}$ for a resistively shunted junction but exhibits a significant amount of apparent hysteresis. This hysteresis is an artifact of sweeping the current too fast; 1 ns is still too fast compared to the filter time of 20 ps. The total time for the simulation would have to be at least a factor of ten longer to eliminate this artifact, making this a very long run indeed. The internal time step in the calculation is ~ 0.2 ps. Still, these remain quite practical with modern computer workstations.

Other Simulations Tools

In addition to Josephson circuit simulation packages, a number of related software packages have been developed. Many of these have also been based on or modeled after similar packages for semiconductor integrated circuits. The design process typically starts with a schematic circuit, which is then used to generate a "netlist" for simulation, such as the listing for SPICE. The circuit parameters are then optimized for proper performance with acceptable margins, that is, percent range of acceptable parameter variations. This is particularly important for superconducting materials, where junction fabrication margins may not always be as tight as in more mature technologies. The schematic is then used to generate the circuit layout on the chip. Even before circuit fabrication, this layout can be used to extract the junction and circuit parameters more accurately, permitting more accurate simulation and refinement of the layout. For ultrafast RSFQ circuits, the presence of parasitic inductances in the circuit are particularly important and must be taken into account. Computer-aided design tools have been developed for each of these aspects in the design of superconducting digital circuits (Gaj et al., 1998).

As superconducting circuits have been developed over several decades, there have also been developments in a completely different class of simulators, those based on analog rather than on digital computation. Given the increasing power, speed, and interactive graphics of modern digital computer, these analog circuit simulators are perhaps mostly obsolete. However, they will be mentioned here for completeness and because for certain types of problems these analogs may still be more convenient to use. An analog Josephson junction simulator is a real physical circuit that displays the same kind of general behavior as a Josephson junction, but scaled to kilohertz frequencies and V amplitudes rather than gigahertz frequencies and sub-millivolt amplitudes as for real junctions (Henry, Prober, Davidson, 1981). Furthermore, other junction and circuit parameters (R, L, C) can also be scaled to convenient component values. The circuits for these analogs can be made using standard semiconductor integrated circuits; there was even a model available commercially back around 1980.

The primary advantage of such a Josephson junction analog is that one can make direct use of standard laboratory instruments, including oscilloscopes, function generators, spectrum analyzers, and dc x–y recorders. For example, Shapiro steps in the dc I–V characteristics are particularly easy to generate and observe, and even the effects of noise on a Josephson junction can be simulated with a standard laboratory noise generator. On the other hand, it is probably impractical to model large multijunction circuits in this way.

REFERENCES

K. Gaj, Q. P. Herr, V. Adler, A. Kraniewski, E. G. Friedman, and M. J. Feldman, "Tools for the Computer-Aided Design of Multi-Gigahertz Superconducting Digital Circuits," submitted to *IEEE Trans. Appl. Supercond.* (1998). See also "Survey of Superconducting Digital Electronics Design Tools" on University of Rochester Web Site: http://www.ee.rochester.edu/users/sde/cad/CLS.html.

R. W. Henry, D. E. Prober, and A. Davidson, "Simple Electronic Analog of a Josephson Junction," *Am. J. Phy.* **49**, 1035 (1981); "Electronic Analogs of Double-Junction and Single-Junction SQUIDs," *Rev. Sci. Instrum.* **52**, 902 (1981).

S. Polonsky, V. Semenov, and P. Shevchenko, "PSCAN'96: New Software for Simulation and Optimization of Complex RSFQ Circuits," *IEEE Trans. Appl. Supercond.* **8**, 2685 (1997). See also SUNY RSFQ Laboratory Web Page: http://pavel.physics.sunysb.edu/RSFQ/.

S. Polonsky, P. Shevchenko, A. Kirichenko, D. Zinoviev, and A. Rylyakov, "PSCAN: Personal Superconductor Circuit Analyzer," *Supercond. Sci. Technol.* **4**, 667 (1991).

M. H. Rashid, *SPICE for Circuits and Electronics Using PSpice*, 2nd ed. (Prentice-Hall, Englewood Cliffs, NJ, 1995).

S. R. Whitely, "Josephson Junctions in SPICE3," *IEEE Trans. Magnet.* **27**, 2902 (1991).

S. R. Whitely, *JSPICE3 User's Manual*, Version 2.2 (Whitely Research, Sunnyvale, CA, 1993). See also Whitely Research, Inc. Web site at http://www.srware.com.

APPENDIX C

CRYOGENIC TECHNOLOGY

This appendix is intended as a brief introduction to the technology of achieving and maintaining the cryogenic temperatures needed for the operation of superconducting devices. This is of course an entire field in itself, so that we will only present the absolute minimum of details here. In particular, we will describe the properties of the most important cryogenic fluids, the capabilities of some cryocoolers, and some considerations in the design of cryogenic test probes.

Cryogenic Fluids

The easiest way to cool a superconducting device, at least in the laboratory, is to immerse it in a bath of boiling cryogenic liquid. The properties of several such liquids are given in Table C.1. The most important of these are nitrogen and helium, which boil at 77 K and 4.2 K, respectively. Figure C.1 shows an approximate phase diagram for both materials. Nitrogen is the main constituent of air, so it is very cheap to produce. Note that liquid nitrogen (LN_2) has a large "latent heat of vaporization" (i.e., the heat needed to boil a given quantity), making it a very effective refrigerant. Given its large latent heat, LN_2 can be collected in a styrofoam cup exposed to the atmosphere at room temperature and transferred with a simple hose or tube. A bath of LN_2 can actually be cooled well below 77 K by pumping on the vapor, thus moving down the liquid–vapor coexistence curve. However, the bath will freeze at 63 K (known as the "triple point"), making further cooling difficult.

In contrast to nitrogen, helium is a relatively rare gas, mostly formed deep in the earth from radioactive decay of uranium (alpha particles are He-4 nuclei). However, there are a few natural gas wells that are relatively rich in helium, and these provide the major sources of helium on the world market. Liquid helium

Table C.1. Properties of Cryogenic Fluids

	Boiling Temperature (K)	Latent Heat (kJ/liter)	Density (kg/liter)	Cost[a] ($/liter)
Nitrogen	77	161	0.81	0.4
Helium	4.2	2.6	0.125	4
Oxygen	90	243	1.14	0.8
Hydrogen	20	31	0.07	1.3
Neon	27	103	1.2	100
Argon	87	220	1.4	2
Helium-3	3.2	0.5	0.06	10,000

[a] Rough estimate of cost in the United States for comparative purposes only.

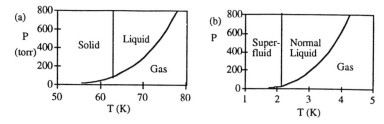

Figure C.1. Approximate phase diagrams for nitrogen (a) and helium (b). The phase boundaries in the pressure–temperature plane of the solid, liquid, and gas phases are noted and, for helium, the superfluid phase. Pressure is in units of Torr, where 760 Torr = 1 atm.

(LHe) boils at 4.2 K and can be pumped down to achieve temperatures as low as about 1 K. Note that the latent heat of boiling is two orders of magnitude smaller than that of nitrogen. In fact, most of the "cooling power" of LHe is associated with the cold boiloff gas, which can be exploited in a properly designed cryogenic system. Given the small latent heat of LHe, greater care must be taken in both its storage and transport. Storage vessels are always vacuum insulated dewars, and vacuum insulated "transfer tubes" must be used to transfer LHe from one vessel to another.

If one follows the boiling curve of LHe down below 2.2 K, one gets not a solid, but rather a superfluid, sometimes called He-II. In the superfluid phase (which is essentially a condensed quantum state similar to a superconductor), the liquid is essentially a perfect thermal conductor, so that it does not exhibit the bubbling that is otherwise characteristic of a boiling liquid. Furthermore, cooling across surfaces is even more efficient than with regular (non-superfluid) liquid helium. In certain applications that may dissipate heat, cooling by immersion in superfluid LHe may yield substantial advantages.

In passing, let us briefly mention another (stable) isotope of helium, He-3. This is extremely expensive, being available mostly as the decay by-product of tritium (H-3) used for thermonuclear weapons. He-3 boils at 3.2 K, and a bath of He-3 (surrounded by a bath of conventional He-4) can be pumped down to temperatures as low as 0.3 K. (Incidentally, He-3 does not become a superfluid until much lower temperatures of ~ 3 mK.) Furthermore, mixtures of He-3 and He-4 are available that form the basis for a "helium dilution refrigerator," which can get down to even lower temperatures of a few millikelvin. These are not needed for standard superconducting devices but are widely used in ultra-low-temperature research.

Other liquid cryogens are used somewhat less. Oxygen is cheap but is somewhat more difficult to handle than nitrogen—pure liquid oxygen promotes combustion, despite its low temperatures. Neither oxygen nor argon offer key advantages over LN_2. Hydrogen is sometimes used but can be highly explosive; liquid hydrogen and liquid oxygen together are standard rocket fuels. The properties of liquid neon would seem to be ideal for medium-temperature cryogenic operation, but neon is rare and is rather expensive.

Entropy and Closed-Cycle Refrigerators

In a laboratory where liquid cryogens are readily available, their use is fairly routine. However, for many other applications, it would be preferable to have a cryocooler that could generate the required temperatures given simply a supply of ac power. A number of systems that accomplish this are available, and their performance and reliability continue to improve. Before addressing them, let us briefly discuss the thermodynamic limitations of such systems.

Heat flows spontaneously from a hot body to a cold body. If one wants to make a refrigerator, one must do work to make heat flow in the reverse direction. The simplest conceptual cryocooler transfers heat from a cold thermal bath at temperature T_c to a hot thermal bath at temperature T_h, typically room temperature (Fig. C.2). This can be addressed using the concept of entropy, usually represented by S. The entropy change ΔS in any process can be quantified in terms of the heat flow through the equation

$$\Delta S_c = \frac{\Delta Q}{T} \qquad (C.1)$$

where ΔQ is the heat transferred in the process and T is the absolute temperature (in kelvin). For example, if one extracts $\Delta Q_c = 1$ J at a temperature of 4 K, this corresponds to an entropy change $\Delta S_c = -0.25$ J/K. The second law of thermodynamics states that the entropy of any closed system must increase with time. If the entropy of part of a system is decreasing, then the entropy must be increasing elsewhere by at least this amount.

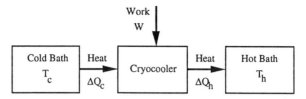

Figure C.2. Schematic energy flow in a simplified cryocooler. In order to extract heat ΔQ_c from the cold bath at T_c, work must be applied to the cryocooler, and the larger amount of heat $\Delta Q_h = \Delta Q_c + W$ will then be delivered to the hot bath at T_h. The efficiency of this process $\varepsilon = \Delta Q_c/W$ is limited by the second law of thermodynamics.

For the cryocooler system, the total entropy change is

$$\Delta S_{tot} = \Delta S_c + \Delta S_h \geq 0 \tag{C.2}$$

Therefore, the entropy change at the hot end must be

$$\Delta S_h \geq |\Delta S_c| = 0.25 \text{ J/K} = \frac{\Delta Q_h}{T_h} \tag{C.3}$$

The minimum heat that must be deposited at room temperature (300 K) to give the required entropy change is $\Delta Q_h = (300 \text{ K}) (0.25 \text{ J/K}) = 75$ J. By conservation of energy, we have

$$\Delta Q_h = \Delta Q_c + W \tag{C.4}$$

where W is the net mechanical (or electrical) work put into the system during this transfer of heat, so for this example, $W = 75 - 1 = 74$ J. The efficiency of this refrigeration process is given in terms of these energies or the equivalent powers by

$$\varepsilon = \frac{\Delta Q_c}{W} = \frac{P_c}{P_i} = \frac{T_c}{T_h - T_c} \tag{C.5}$$

where P_c is the cooling power and P_i is the input power. For the present example $\varepsilon = 1/74 = 1.3\%$. This sounds very inefficient; one must supply 74 W of electrical power at 300 K, for example, to obtain 1 W of cooling at 4 K.

In fact, this calculation is for a highly idealized refrigerator; the real situation is much worse. To assist in heat transfer at the two ends, T_c is normally somewhat below the desired cooling temperature and T_h is normally above room temperature, reducing the efficiency at both ends. The calculation also assumes that the total entropy of the system does not increase, which can only occur (asymptotically) for very slow, "reversible" changes. Any real system that is designed to operate at a reasonable rate has irreversible losses that lower the efficiency further. Typical efficiencies for real closed-cycle coolers are more typically $\varepsilon \sim 0.1\%$ or less; more than 1 kW of power is generally needed to obtain 1 W of cryogenic cooling at 4 K.

Most cryocoolers are based on a cycle of mechanical compression and expansion of gases, much like a conventional refrigerator. A key difference, of course, is that the working gas must remain a fluid at the low-temperature end, which limits the choice to helium; everything else has already frozen solid. A helium liquifier works on very similar principles; it is just a refrigerator that condenses helium gas on the cold end. For these and other cryocoolers, it is common to have multistage systems where the cold end of the first stage is the hot end for the second, and so forth. These multistage systems can obtain lower temperatures with increased efficiency, at the expense of greatly increased complexity and size.

Ironically, the most common application at present for commercial cryocoolers is not even strictly a cryogenic application. This is the cryopump, where gases in a vacuum system condense onto the "cold head" maintained at ~ 10 K via a closed-cycle helium refrigerator. This is a very clean way to obtain a good vacuum without the risk of contamination by pump oil and has become very widespread in semiconductor manufacturing. These cryopumps have also developed a reputation for long-term (years) continuous operation without maintenance, something that will also be necessary for practical commercial cryocoolers.

Some novel designs for closed-cycle cryocoolers have been explored recently. One approach uses a "pulse tube" and has no moving parts at the low-temperature end, which should improve reliability. This design is based in part on the well-known observation of thermoacoustic oscillations that appear when gas in a narrow tube is placed in a large thermal gradient. (This effect is applied in a common probe to measure the liquid level in a helium dewar.) This is in effect a heat engine, in which heat transferred down the gradient promotes the oscillation in the gas. Properly designed driven acoustic oscillations can cause this to operate in reverse as a refrigerator. While not yet available commercially, pulse-tube cryocoolers appear to be promising for cooling to the 30–60 K range.

We have been focusing on closed-cycle refrigerators, but open-cycle systems may also have some applications. One very simple cryocooler requires just a cylinder of high-pressure gas and no electrical power or moving parts at all. This is based on Joule–Thomson expansion of the gas through a constriction, which causes the gas to cool down. One recent implementation of this uses micromachining techniques to make a cold-end heat exchanger on the scale of a microchip. Cooling of small cryogenic circuits to liquid nitrogen temperatures and below can be achieved with these "microminiature refrigerators."

Other types of refrigerators have also been proposed and may be practical in certain regimes. Thermoelectric refrigerators are particularly simple and reliable (no moving parts at all), but seem to be limited to the high-temperature end of the cryogenic regime. Multistage thermoelectric elements can achieve temperatures down to 180 K, unfortunately not low enough for even the highest HTS devices.

Another novel approach uses paramagnetic materials in a varying magnetic field to achieve cooling by a process known as "adiabatic demagnetization." The basic principle is that the magnetic material decreases its entropy when a large magnetic field is applied. This must be accompanied by an equal increase in the entropy of its environment, that is, heating, where the amount of heat is given by $\Delta Q = T \Delta S$. If the magnetic material is subsequently moved to a low-field region (or the field turned off), the internal entropy then increases again, which requires it to absorb the same ΔQ from its environment. This can be carried out repeatedly to form a closed-cycle cooler. Such magnetic refrigeration has been used with superconducting magnets to obtain ultralow temperatures (millikelvin and below) for research, starting from LHe temperatures. More recently, using different types of magnetic materials, a similar process has been used to demonstrate magnetic cooling in an intermediate temperature range. This may show some promise for practical cryocoolers for HTS devices.

Cryogenic Materials and Cryostat Design

In designing systems to operate at cryogenic temperatures, a number of special considerations must be taken into account. First, it is generally important that the superconducting device be maintained at a constant cryogenic temperature with only minimal "heat leakage" from room temperature. In some cases, constant temperature may be obtained by direct immersion in an appropriate cryogenic fluid, either liquid or cold gas. In other cases, the device is mounted in a cooled evacuated container with active thermostatic control of temperature, typically using a thermometer and a small electrical heater (see Fig. C.3).

A number of different devices can sometimes be used to measure cryogenic temperature. Probably the most common is a simple Si pn junction diode, which may be calibrated to cover the range from 1.5 K up to room temperature and above, with an accuracy of 0.1 K or less. Diodes specially prepared for this purpose are available commercially from several suppliers. In operation, such a diode is typically biased in the forward direction with 10 μA of current, and the voltage (~ 1 V, rising at low T) is measured. Other common cryogenic thermometers are based on the resistance of doped Ge (for lower T) or the resistance of Pt wire (for higher T).

It is also important to consider the thermal conductivity $\kappa(T)$ of the various materials that constitute a cryogenic probe. In particular, κ is very different for metallic and nonmetallic (insulating) materials and also depends strongly on the purity and crystallinity of the material. As indicated in Table C.2, there can be a difference of up to six orders of magnitude between the most disordered insulator and the purest metal at 4 K. In insulators, heat is carried by phonons, whose density drops drastically at low T. Amorphous materials (with low phonon mobilities) such as plastics or glasses tend to have the lowest value of κ.

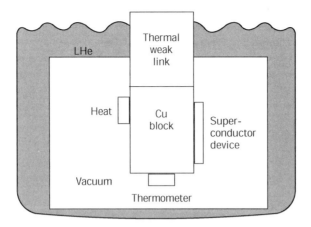

Figure C.3. Schematic design of generic temperature-controlled cryoprobe with sample and thermometer mounted on isothermal copper block inside vacuum can be immersed in LHe. Inclusion of a small electric heater on the block permits control of temperature above that of the LHe bath.

Table C.2. Ranges of Typical Thermal Conductivities for Materials at 4 K

Material Class	Thermal Conductivity (W/m-K)
Plastics and glasses	0.01–0.1
Crystalline insulators	0.1–100
Metallic alloys	0.1–10
Metallic elements	10–10,000
Interface conductance	$\sim 10^4$ W/m^2-K

Crystalline insulators have higher κ, with polycrystalline ceramics at the lower end and single crystals (such as Si, sapphire, or crystalline quartz) at the upper end. Thermal conduction in metals at 4 K is dominated by the electrons, where it is mostly the collisions with impurities that limit the electronic mobility. But the thermal conduction between a metal and an insulator is again dominated by phonons; a typical value at 4 K for many pairs of materials is $\sim 10^4$ W/m^2-K, which can be a limiting factor in some situations.

In general, the electronic thermal conductivity κ_e of a metal is directly proportional to the electrical conductivity σ, with the proportionality given by the Wiedemann–Franz law

$$\kappa_e = LT\sigma = \frac{\pi^2 k_B^2}{3e^2} T\sigma \qquad (C.6)$$

where $L = 2.45 \times 10^{-8}$ W-Ω/K^2 is known as the Lorenz constant. So a disordered alloy, with σ weakly dependent on T, will exhibit a value of κ that decreases linearly with T. Values of thermal conductivity at 4 K can vary from 0.3 W/m-K for stainless steel, a disordered alloy, to more than 1000 W/m-K for ultrapure Cu. A superconductor for $T \ll T_c$ is a notable exception to Eq. (C.6); since Cooper pairs cannot transport heat, κ_e goes to zero exponentially at low temperatures and κ exhibits a much smaller lattice thermal conductivity similar to that of insulators. This has been used in the design of a thermal switch, where a magnetic field above the critical field of the superconductor permits a sharp and reversible switching between a low-κ superconductor and a high-κ normal metal.

In a cryogenic probe, a copper block is typically used as an isothermal mount for a superconducting device and its thermometer (Fig. C.3); if a metal is not acceptable, a single-crystal insulator may be used. In contrast, the link connecting this block to the rest of the system should be a poor thermal conductor, either stainless steel, or possibly plastic. Even the wires linking the device to the outside must be considered. For temperatures of 4 K and below, Cu wires may conduct too much heat; higher resistance Cu-alloy wires are often used instead.

Two other important mechanisms of heat transfer are radiation and gaseous conduction (and convection). Direct radiational heating from room temperature to cryogenic surfaces can be very significant, so that one or more stages at intermediate values of T are typically placed in a dewar to block or reflect the warmer radiation. In fact, one standard construction material is a thin layer of aluminized mylar, sometimes referred to as "superinsulation." Optimum thermal isolation can be achieved by placing multiple layers of this superinsulation between the warm and cold surfaces.

It is equally important that a good vacuum be maintained in order to preserve thermal isolation; even a low gas pressure can be quite effective at conducting heat across a gap to cold surfaces. Of course, the cold surface itself forms an effective cryopump, so if there are no leaks, a good vacuum can be maintained indefinitely. But it is not only leaks from room temperature that are important. If one is working in a bath of superfluid helium, which exhibits zero viscosity, then even an extremely small leak to the superfluid can prove to be critical.

Finally, mechanical properties of materials can also be quite different than at room temperature. For example, elastomers (rubbers and plastics) become rigid and even brittle and are not generally used in the cryogenic regime. Glues and other sealants are often ineffective, since differential thermal contraction on thermal cycling can cause them to crack. An alternative method must be found to provide vacuum seals; metallic indium "o-rings" are sometimes used. But many other standard materials behave quite well as structural materials at cryogenic temperatures, including steel, aluminum, and fiberglass, for example.

REFERENCES

F. Pobell, *Matter and Methods at Low Temperatures*, 2nd ed. (Springer-Verlag, Berlin, 1996).

R. C. Richardson and E. N. Smith, Eds., *Experimental Techniques in Condensed Matter Physics at Low Temperatures* (Addison-Wesley, Reading, MA, 1988).

T. P. Sheahen, *Introduction to High-Temperature Superconductivity*, Chaps. 3 and 24 on Refrigeration (Plenum, New York, 1994).

K. White, *Experimental Techniques in Low-Temperature Physics*, 3rd ed. (Oxford University, Oxford, 1979).

APPENDIX D
ELECTROMAGNETIC UNITS AND FUNDAMENTAL CONSTANTS

The standard system of units and formulas for electromagnetics is known as SI (Système International). Alternatively, this is sometimes known as the MKSA system, for meter, kilogram, second, and ampere, the basic units of the system. The SI units include all of the usual amperes, ohms, volts, etc., as well as the standard metric mechanical units for energy and power (joules and watts). These are the units (with a few exceptions) that we consistently use throughout this book.

For historical reasons, there is another completely distinct system, which is still very common in the physics literature related to superconductivity, particularly on the theoretical end. This is typically known as the "Gaussian cgs system," or sometimes simply Gaussian units, which are based on the centimeter and gram instead of the meter and kilogram. But the Gaussian electromagnetic units are quite different from those of SI, and not simply by a power of ten. Some Gaussian magnetic units remain in common use, most prominently the gauss, which is the unit of magnetic flux density B in the Gaussian system (the proper SI unit is the tesla).

The comparison between these two systems of units is complicated by the fact that it is not merely the units that are different. The basic formulas generally differ by various constant factors, even though the physical phenomena they are describing are of course the same. The SI formulas make use of $\mu_0 = 4\pi \times 10^{-7}$ and $\varepsilon_0 = 1/\mu_0 c^2$, whereas the Gaussian formulas do not have μ_0 and ε_0 but rather use various factors of c and 4π. Rather than try to give a set of general rules to convert from one system to the other, Table D.1 lists some common electromagnetic and superconducting formulas in both systems. A few of these are the same in both systems, but most others differ by key factors. These may be useful when comparing to some of the references that make use of the older Gaussian system.

ELECTROMAGNETIC UNITS AND FUNDAMENTAL CONSTANTS 369

Table D.1. Common Electromagnetic Formulas in SI and Gaussian Systems

Name of Formula	SI	Gaussian
Coulomb's law	$F = q_1 q_2 / 4\pi\varepsilon_0 r^2$	$F = q_1 q_2 / r^2$
Faraday's law	$V = -d\Phi/dt$	$V = -(1/c)\, d\Phi/dt$
Ampere's law	$\oint \mathbf{B} \cdot d\mathbf{l} = \mu_0 I$	$\oint \mathbf{B} \cdot d\mathbf{l} = 4\pi I / c$
Lorentz force	$\mathbf{F} = q(\mathbf{E} + \mathbf{v} \times \mathbf{B})$	$\mathbf{F} = q[\mathbf{E} + (\mathbf{v}/c) \times \mathbf{B}]$
Vector potential	$\mathbf{B} = \nabla \times \mathbf{A}$	$\mathbf{B} = \nabla \times \mathbf{A}$
Electromagnetic potentials	$\mathbf{E} = -\nabla V - \partial \mathbf{A}/\partial t$	$\mathbf{E} = -\nabla V - (1/c)\, \partial \mathbf{A}/\partial t$
Magnetic field relations	$\mathbf{B} = \mu_0(\mathbf{H} + \mathbf{M}) = \mu \mathbf{H}$	$\mathbf{B} = \mathbf{H} + 4\pi \mathbf{M} = \mu \mathbf{H}$
Maxwell's equations	$\nabla \cdot \mathbf{D} = \rho$	$\nabla \cdot \mathbf{D} = 4\pi\rho$
	$\nabla \cdot \mathbf{B} = 0$	$\nabla \cdot \mathbf{B} = 0$
	$\nabla \times \mathbf{E} = -\partial \mathbf{B}/\partial t$	$\nabla \times \mathbf{E} = -(1/c)\, \partial \mathbf{B}/\partial t$
	$\nabla \times \mathbf{H} = \mathbf{J} + \partial \mathbf{D}/\partial t$	$\nabla \times \mathbf{H} = 4\pi \mathbf{J}/c + (1/c)\, \partial \mathbf{D}/\partial t$
Magnetic energy density	$U = B^2 / 2\mu_0$	$U = B^2 / 8\pi$
Flux quantum	$\Phi_0 = h/2e$	$\Phi_0 = hc/2e$
Josephson frequency	$f_J = 2eV/h$	$f_J = 2eV/h$
Ohm's law	$\mathbf{J} = \sigma \mathbf{E}$	$\mathbf{J} = \sigma \mathbf{E}$
AC skin depth	$\delta = \sqrt{2/\omega\sigma\mu_0}$	$\delta = \sqrt{c^2/2\pi\omega\sigma}$
Magnetic penetration depth	$\lambda_L = \sqrt{m/ne^2\mu_0}$	$\lambda_L = \sqrt{mc^2/4\pi n e^2}$
Josephson penetration depth	$\lambda_J = \sqrt{\Phi_0 / 2\pi J_c(2\lambda + d)}$	$\lambda_J = \sqrt{c\Phi_0 / 8\pi^2 J_c(2\lambda + d)}$
Vortex pinning force	$F_p = \Phi_0 J$	$F_p = \Phi_0 J / c$

Table D.2. Magnetic Units and Conversion Factors in SI and Gaussian Systems

Name of Unit	Symbol	SI	Gaussian	Conversion
Magnetic flux density	B	Tesla (T)	Gauss (G)	10^4 G = 1 T
Magnetic field[a]	H	A/m	Oersted (Oe)	1 A/m = 4π mOe
Magnetic moment	μ	A-m^2	emu = erg/G	1 emu = 1 mA-m^2
Magnetization	M	A/m	emu/cm^3[b]	1 emu/cm^3 = 1 kA/m
Magnetic flux	Φ	Weber (Wb)	G-cm^2	1 Wb = 10^8 G-cm^2

[a] In a nonmagnetic material, B- and H-fields are equivalent. In this case, 1 G = 1 Oe, and these units are often used interchangeably. Similarly, in SI units H is often measured in teslas rather than amperes per meter, where 1 T = $1/\mu_0$ = $10^7/4\pi$ A/m ≈ 800 kA/m.

[b] In Gaussian units, the magnetization $4\pi M$ is often expressed in gauss, where 1 emu/cm^3 = 4π G.

In addition, we also summarize in Table D.2 the common magnetic units in both systems and the conversion factors between them. Magnetic sources and measurement instruments are still most often calibrated in Gaussian units. Finally, Table D.3 gives standard values of most of the important physical

Table D.3. Key Fundamental and Derived Physical Constants

Name of Constant	Symbol	Value
Speed of light	c	2.99792458×10^8 m/s
Permeability of vacuum	$\mu_0 = 4\pi \times 10^{-7}$	1.2566371 μH/m
Permittivity of vacuum	$\varepsilon_0 = 1/(\mu_0 c^2)$	8.8541888 pF/m
Planck's constant	h	$6.6260755 \times 10^{-34}$ J-s
H-bar	$\hbar = h/2\pi$	$1.05457266 \times 10^{-34}$ J-s
Electron charge	e	$1.60217733 \times 10^{-19}$ C
Electron-volt	$eV = e \times 1\text{ V}$	$1.60217733 \times 10^{-19}$ J
Flux quantum	$\Phi_0 = h/2e$	$2.06783461 \times 10^{-15}$ V-s
Josephson f/V ratio	$2e/h = 1/\Phi_0$	483.59767 GHz/mV
Electron mass	m_e	$9.1093897 \times 10^{-31}$ kg
Avagadro's number	N_A	6.0221367×10^{23}/mol
Atomic mass unit	$\text{amu} = 1/(10^3 N_A)$	$1.6605402 \times 10^{-27}$ kg
Boltzmann constant	k_B	1.380658×10^{-23} J/K
Impedance of free space	$\sqrt{\mu_0/\varepsilon_0} = \mu_0 c$	376.7303 Ω
Quantum resistance	h/e^2	4108.236 Ω

constants, to many more significant figures than are likely to be needed for problems and other calculations.

REFERENCES

J. D. Jackson, "Appendix on Units and Dimensions," in *Classical Electrodynamics*, 2nd ed. (Wiley, New York, 1975).

D. R. Lide, Ed., *Handbook of Chemistry and Physics*, 79th ed. (CRC Press, Boca Raton, FL, 1998).

APPENDIX E

SYMBOLS AND ACRONYMS

ROMAN LETTERS

a	Length or radius (m)
A	Area (m^2)
A	Magnetic vector potential (T-m)
B	Magnetic flux density (T); bandwidth (Hz)
c	Speed of light (m/s); heat capacity per volume (J/m^3-K)
C	Heat capacity (J/K); capacitance (F); distributed capacitance (F/m)
d	Thickness (m)
D	Diffusion constant (m^2/s)
$D(\varepsilon)$	Electronic density of states (m^{-3} J^{-1})
D^*	Detectivity (cm $\sqrt{Hz}$/W)
e	Electron charge (C)
E	Electric field (V/m)
f	Frequency (Hz); force density (N/m^3)
F	Force (N)
g	Acceleration due to gravity
G	Conductance ($S = \Omega^{-1}$); distributed conductance (S/m)
$h, \hbar$	Planck's constant (J-s)
h	Height (m)
H	Magnetic field (A/m)
I, i	Current (A)
j, i	$\sqrt{-1}$ (dimensionless)
J	Current density (A/m^2); Bessel function (dimensionless)

k_B	Boltzmann's constant (J/K)
k	Wave vector (m^{-1})
k	Magnetic coupling constant (dimensionless)
l	Electronic scattering length (m); length (m)
L	Inductance (H); distributed inductance (H/m)
m	Mass (kg)
M	Magnetization (A/m); mutual inductance (H)
n	Number density (m^{-3} or m^{-2})
N, n	Number (dimensionless)
N	Demagnetizing coefficient (dimensionless)
p	Momentum (kg-m/s)
P	Power (W); power density (W/m^3)
q	Charge (C)
Q	Q-factor (dimensionless); heat (J)
Q_m	Magnetic charge (A-m^2)
r, R	Radius (m)
R	Resistance (Ω); distributed resistance (Ω/m); reluctance (H^{-1}); responsivity (A/W or V/W)
R_t	Thermal resistance (K/W)
S	S-parameter (dimensionless); entropy (J/K)
t	Time (s)
T	Temperature (K); transmission coefficient (dimensionless); torque (N-m)
u	Velocity (m/s)
U	Energy (J); energy density (J/m^3)
v	Velocity (m/s)
V	Voltage (V)
w	Width (m)
W	Work (J)
x	Distance (m); variable number (dimensionless)
Y	Admittance (Ω^{-1}); distributed admittance (Ω^{-1} m^{-1})
z	Distance (m)
Z	Impedance (Ω)

GREEK LETTERS

α	Attenuation coefficient (m^{-1}); temperature coefficient (K^{-1}); electron–phonon coupling constant (dimensionless)
β	Phase factor (m^{-1}); parameter ratios (dimensionless)
γ	Propagation constant (m^{-1}); relativistic correction factor (dimensionless)
δ	Skin depth (m)
$\delta(t)$	Delta (unit impulse) function (dimensionless)
Δ	Superconducting energy gap (eV)

$\mathscr{E}$	Energy (J or eV)
ε	Permittivity (F/m)
ε_r	Dielectric constant (dimensionless)
η	Viscocity (kg/s)
θ	Phase of wave function (dimensionless)
κ	Ginzburg–Landau parameter (λ/ξ, dimensionless); thermal conductivity (W/m-K); decay constant (m^{-1})
λ	Penetration depth (m); wavelength (m)
Λ	Kinetic inductivity (H-m)
μ	Magnetic moment (A-m^2)
μ_0	Permeability (H/m)
μ	Vortex mobility (m^3/A-s)
ξ	Coherence length (m)
π	Pi (dimensionless)
ρ	Resistivity (Ω-m); reflection coefficent (dimensionless)
σ	Conductivity (Ω^{-1} m^{-1})
τ	Characteristic time (s)
ϕ	Phase difference (dimensionless)
Φ	Magnetic flux (T-m^2 = V-s)
χ	Magnetic susceptibility (dimensionless)
Ψ	Quantum wave function (m$^{-1.5}$)
ω	Angular frequency (rad/s)
Ω	Ohms

ACRONYMS

A/D	Analog-to-digital
BSCCO	Bi–Sr–Ca–Cu–O
CMOS	Complementary metal–oxide–semiconductor
CPU	Central processing unit
CRSJ	Capacitive resistively shunted junction
D/A	Digital-to-analog
DRO	Destructive readout (memory)
DSP	Digital signal processing
EDX	Energy-dispersive x-ray (analysis)
FET	Field-effect transistor
FFT	Fast Fourier transform
FIFO	First-in, first-out (memory)
HTS	High-temperature superconductor
IBAD	Ion-beam-assisted deposition
IBM	International Business Machines, Inc.
JTL	Josephson transmission line
LHe	Liquid helium
LN$_2$	Liquid nitrogen

LTS	Low-temperature superconductor
MEG	Magnetoencephalogram
MITI	Ministry of International Trade and Industry
MVTL	Modified variable threshold logic
NEP	Noise-equivalent power
NDRO	Nondestructive readout (memory)
NDT	Nondestructive testing
PIT	Powder in tube
PSCAN	Personal superconducting analyzer
QFP	Quantum flux parametron
RABiTS	Rolling assisted biaxial textured substrate
RAM	Random-access memory
ROM	Read-only memory
RSFQ	Rapid (or Russian?) single flux quantum
RSJ	Resistively shunted junction
RTS	Room temperature superconductor
SFFT	Superconducting flux flow transistor
SFQ	Single flux quantum
SIS	Superconducting–insulating–superconducting (junction)
SNS	Superconducting–normal–superconducting (junction)
SPICE	Simulation Program with Integrated Circuit Emphasis
SQUID	Superconducting quantum interference device
STJ	Superconducting tunnel junction (detector)
VLSI	Very large scale integration
WIMP	Weakly interacting massive particle
YBCO	Y–Ba–Cu–O

MATERIALS INDEX

A-15 compounds, 143–144. *See also* Niobium compounds
Aluminum (Al)
  metal, 4, 149, 326
  oxide (Al_2O_3), 40, 149–150, 162, 164, 253, 326
Argon (Ar), 148, 165, 360

Bismuthates (bismuth oxides), 170–171
  barium lead-bismuth oxide (BPBO), 171, 329
  barium-potassium bismuth oxide (BKBO), 170
BSCCO, *see* Cuprates

Chevrel phase, 170
Copper (Cu)
  alloys, 146, 366
  oxides, *see* Cuprates
  pure metal, 4, 37, 64, 83, 117–119, 123–124, 134–136, 145, 366
Cuprates (copper oxides), 151–155, 173. *See also* High-temperature superconductors
  bismuth strontium calcium copper oxide (BSCCO), 152, 155–161, 166, 170, 176–177
  lanthanum barium copper oxide (LBCO), 151
  lanthanum strontium copper oxide (LSCO), 152, 155
  mercury barium calcium copper oxide (HBCCO), 4, 152, 155, 161, 166
  neodymium cerium copper oxide (NCCO), 152, 155
  praseodymium barium copper oxide (PBCO), 168
  thallium barium calcium copper oxide (TBCCO), 52, 155–156, 158, 161, 166
  thin films, 161–169
  yttrium barium copper oxide (YBCO), 4, 66, 151–169, 177, 255, 294, 334

Fullerenes (C_{60}), 172
  cesium rubidium fullerene (Cs_2RbC_{60}), 172

Gallium arsenide (GaAs), 285, 294
Germanium (Ge), 364
Gold (Au), 4, 142, 150, 165

Helium
  liquid (LHe), 4, 16, 293, 363–365
  superfluid (He-II), 52, 360
Helium-3 (He^3), 52, 360, 361
High-temperature superconductors (HTS), 4, 135, 151–169, 255, 312. *See also* Bismuthates; Cuprates; Fullerenes
Hydrogen (H_2), 124–125, 130, 360–361

Indium (In), 142, 330, 366
Indium arsenide (InAs), 174
Insulators, *see* Aluminum oxide; Lead oxide; Magnesium oxide; Niobium oxide; Silicon dioxide
Iron (Fe), 120–122. *See also* Steel

Lanthanum aluminate ($LaAlO_3$), 162, 164–167, 250, 330
Lead (Pb), 4, 66, 142, 147, 155, 171
  lead molybdenum sulfide ($PbMo_6S_8$), *see* Chevrel phase
  lead oxide (PbO_2), 147
Liquid cryogens, *see* Helium; Nitrogen; Neon
Low-temperature superconductors (LTS), 4, 124, 127, 129, 135, 141–151. *See also* A-15 compounds; Chevrel phase; Lead; Niobium

Magnesium oxide (MgO), 150, 162, 297
Manganates (manganese oxides), 173
Mercury (Hg), 4, 142
Molybdenum (Mo), 142, 150

375

MATERIALS INDEX

Neodymium gallate (NdGaO$_3$), 162
Neon (Ne), 360
Niobium (Nb),
   alloys, 142–143
   compounds, 143–147, 149
   niobium-germanium (Nb$_3$Ge), 143–144, 150
   niobium nitride (NbN), 143, 147, 149–150, 177, 294–297, 322, 329, 335
   niobium oxide (Nb$_2$O$_5$), 144, 149
   niobium-tin (Nb$_3$Sn), 124, 130, 143–146
   niobium-titanium alloy (Nb–Ti), 93, 143–146, 176, 337
   pure metal, 4, 34, 37, 80, 93, 142–150, 174, 182–184, 195, 250, 253–254, 259–260, 285–287, 305, 308, 326, 334, 336–337
   thin films, 147–150
Nitrogen, liquid (LN$_2$), 4, 135, 359–360

Organic conductors, 171, 172–177. *See also* Fullerenes
bis(ethylenedithio)tetrathiofulvalene (BEDT-TTF), 172, 177
tetramethyl-tetraselenafulvalene (TMTSF), 172
tetrathiofulvalene (TTF), 172
Oxygen (O$_2$), 148–149, 151, 155, 161, 360–361

Perovskites, 154, 163, 171, 173. *See also* Bismuthates; Cuprates; LaAlO$_3$; NdGaO$_3$; SrTiO$_3$
Platinum (Pt), 364

Quartz, *see* Silicon dioxide

Sapphire, *see* Aluminum oxide
Semiconductors, 174, 319, *see* GaAs; Ge; InAs; Si
Silicon (Si), 46, 150, 162, 259, 285–287, 364–365
Silicon dioxide (quartz, SiO$_2$), 40, 150
Silver (Ag), 4, 83, 117, 136, 142, 160, 165
Steel, 225, 366
Strontium titanate (SrTiO$_3$), 162, 165–167, 171
Substrates, 162, 167–168. *See also* MgO; LaAlO$_3$; NdGaO$_3$; SrTiO$_3$; YSZ

Tantalum (Ta), 142, 250
TBCCO, *see* Cuprates
Technetium (Tc), 142
Thallium (Tl), 142, 166
Thin films, 147–151, 161–169, 330
Tin (Sn), 142, 146, 250
Titanium (Ti), 142–143

Yttria-stabilized zirconia (YSZ), 162, 164–165
YBCO, *see* Cuprates

Zirconium (Zr), 142

NAME INDEX

Abelson, L., 150, 175
Anacker, W., 260, 296

Bardeen, J., 8, 35, 53, 100
Bean, C. P., 112
Beasley, M. R., 95, 137, 161, 176
Bednorz, J. G., 151, 176
Benz, S. P., 245
Bocko, M. F., 339
Booth, N. E., 330, 332
Bumble, B., 322, 323, 332
Bunyk, P., 286, 296
Burns, G., 340
Burroughs, C. J., 245

Cabrera, B., 330–332
Carr, P. H., 40, 64
Chen, X. D., 151, 176
Chevrel, R., 170
Clarke, J., 222, 223, 245
Cooper, L. N., 8, 53
Cyrot, M., 15, 176, 340

Davidson, A., 95, 137, 358
Delin, K. A., 64, 137, 245, 340
DiVincenzo, D. P., 339

Feldman, M. J., 296, 308, 333, 339, 358
Fuller, B., 172

Gaj, K., 281, 296, 352, 357, 358
Gallagher, W. J., 173, 176
Gallop, J. C., 223, 245, 340
Gao, G., 339
Geballe, T. H., 142, 176
Ginzburg, V. L., 61

Goldie, D. J., 330, 332
Goldman, A. M., 321, 332
Goto, E., 283, 296
Grant, P. M., 120, 137
Gross, R., 166, 176

Hamilton, C. A., 190, 245, 288, 296
Harada, Y., 283, 296
Hasuo, S., 260, 264, 285, 296
Hayakawa, H., 260, 296
Henry, R. W., 358
Herr, A. M., 322, 332, 339
Hioe, W., 283, 296
Hu, Q., 308, 333

Imamura, T., 260, 296

Jackson, J. D., 370
Johnson, M. W., 64, 322, 327, 332
Josephson, B., 10, 178

Kadin, A. M., 64, 95, 105, 137, 321, 322, 327, 328, 332
Kamerlingh Onnes, H., 4
Kidiyarova-Shevchenko, A. Y., 296
Kirichenko, A. F., 282, 296
Kittel, C., 46, 64, 333
Kleinsasser, A. W., 173, 176

Landau, L., 61
Larbalestier, D. C., 159, 165, 176
Le Duc, H. G., 322, 323, 332
Lee, G. S., 290, 296
Lide, D. R., 370
Likharev, K. K., 192, 238, 245, 270, 278, 281, 290, 296, 339, 340

NAME INDEX

Lindgren, M., 323, 333
Litskevitch, P., 296
Lukens, J. E., 239, 245

Magin, R. L., 126, 137
Matthias, B. T., 142
McAvoy, B. R., 64
McDonald, D. G., 320, 333
Messina, P. C., 339
Montgomery, D. B., 120, 137
Mooij, J. E., 137
Mukhanov, O. A., 294, 296
Müller, K. A., 151, 176

Nahum, M., 315, 333
Nisenoff, M., 340
Nordman, J. E., 101, 137, 238, 245

Orlando, T. P., 32, 64, 80, 90, 108, 110, 118, 133, 134, 137, 233, 245, 340

Pavuna, D., 15, 176, 340
Peck, T. L., 137
Pedersen, N. F., 236, 239, 245
Petersen, D. A., 290, 296
Phillips, J. M., 162, 176
Pobell, F., 367
Polonsky, S., 352, 358
Prober, D. E., 358
Przybysz, J. X., 290, 292, 296

Ralston, R. W., 46, 64, 340
Rashid, M. H., 358
Richards, P. L., 308, 319, 333
Richardson, R. C., 367
Roberts, B. W., 142, 176
Ruggiero, S. T., 245, 340
Rudman, D. A., 340
Rylov, S. V., 293, 296
Ryzhikh, A. I., 296

Sauvageau, J. E., 320, 333
Schrieffer, R., 8, 53

Schwarz, S. E., 350
Semenov, A. D., 323, 333
Semenov, V. K., 270, 278, 296, 358
Shapiro, S. V., 186
Sheahen, T. P., 120, 137, 367
Shimakage, H., 333
Silver, A. H., 292, 294, 296
Simon, R. W., 15
Smith, A. D., 15, 333
Smith, E. N., 367
Sterling, T. L., 339

Taur, Y., 259, 296
Taylor, C. E., 340
Thomasson, S., 150, 176
Tinkham, M., 54, 61, 64, 86, 137, 156, 158, 176, 202, 321, 333, 340
Tucker, J. R., 308, 333
Turner, C. W., 64, 245, 340

Ulaby, F., 350
Uher, C., 315, 333

Van Duzer, T., 29, 32, 35, 54, 61, 64, 199, 245, 287, 291, 294, 296, 340
Verhoeve, P., 327, 333

Wada, Y., 266, 296
Webb, A. G., 137
Weinstock, H., 296, 340
Weisskopf, V. F., 53, 64
Wengler, M. J., 309, 333
White, G. K., 367
White, R. M., 142, 176
Whitely, S. R., 352, 353, 358
Wikswo, J. P., 224, 245
Williams, J. M., 171, 176
Withers, R. S., 46, 64, 176

Yuh, P. F., 293, 296

Zhang, D., 164, 176

SUBJECT INDEX

Analog-to-digital converters, 290–292, 299. *See also* Digital circuits
Anisotropy
 and cuprate superconductors, 152–161
 and organic superconductors, 171, 177
Astronomical applications of superconductors, *see* Radiation detectors

BCS theory, 8, 53–61
Bean model, *see* Critical state model
Bessel functions, 90, 188–190, 307, 334
Bolometers, 316–323. *See also* Radiation detectors
Brillouin zone, 47–50

Capacitance
 in interdigital fingers, 39
 in Josephson junction, 201, 206, 252–254
 in transmission line, 27, 36, 344
Characteristic impedance of transmission line, 21–23, 27, 33, 36, 343–346, 350
Clock schemes for digital superconducting systems
 RSFQ, 280–283
 voltage-state, 264–266, 283–285
Coherence length, 61–63, 66, 85, 90, 93, 143, 157, 339
Computers, superconducting, 250, 260, 284–287, 338
Condensation energy, 55, 62
Cooper pairs, 8, 52, 61–62, 66
Critical parameters, superconducting
 critical current density J_c, 80, 83, 110, 145–147, 164
 critical current I_c of Josephson junction, 10, 150, 166, 179–183, 212, 229–232, 253
 critical frequency f_c, 5, 33, 58

 critical magnetic field for Type I, 81–82
 critical magnetic fields for Type II, 93, 100, 143, 153, 170, 174, 177
 critical temperature T_c, 4, 54, 142–143, 152, 155, 166, 170–173
Critical state model (Bean model), 111–115
Cryocoolers, 360–364
Cryogenics, 359–367
Cryotron, 250
Crystal structures, 143–144, 152–155, 162, 170–172
Current distribution
 in Type I, 24–29
 in Type II, 111–115
 in vortex, 89–92
Current-voltage characteristics ($I–V$)
 of Josephson junctions, 12, 184, 188–191, 198–205, 236, 256, 262–263, 311, 356
 and self-heating, 83
 of SQUIDs, 12, 211, 212, 216–217
 of tunnel junctions, 59–61, 183, 306–307
 in Type II, 103, 105, 109

de Broglie relations, 8–9, 87, 179
Debye model, 54, 314
Density of states, electronic, 47–51, 55–57, 59–61
Demagnetizing factor, 77
Detectivity D^*, 319. *See also* Radiation detectors
Diamagnetism in superconductors, 72–74, 94, 131, 138, 173
Diffusion constants
 for electrons, 327
 for heat, 118, 139, 313
 for magnetic flux, 118, 139

379

SUBJECT INDEX

Digital circuits, superconducting, 14, 249–299.
  See also Analog-to-digital converters
  digital systems, 284–294
  digital-to-analog (D/A) converters, 292
  single-flux-quantum logic, 270–283
  switching dynamics, 251–259
  voltage-state logic, 259–270
Duality transformations, 94–107
  between charge and flux, 6–7, 98–107, 325
  in circuits, 5–6, 95–98
D-wave pairing, 156

Electric power applications, 128–130
Electron–phonon interaction, 53, 55, 320–327
Electron scattering, 19, 32, 48–49, 62–63, 93, 313, 320, 366
Energy gap
  in semiconductor, 49, 174, 318–319, 324–326
  in superconductor, 8, 46, 50–61, 143, 156, 326
Entropy, 361–362

Fermions and bosons, 52
Fermi surface parameters, 47–51
Flux, magnetic. *See also* Vortices
  conservation, 67–72
  expulsion, *see* Meissner effect
  flux flow, 98–107, 232
  pinning, 7, 107–119
  quantization, 9, 88–89
Flux flow transistors, *see* Transistors, superconducting
Fluxonic photodetection, 328–329, 335
Fluxons. 7, 192. *See also* Single-flux quantum; Solitons; Josephson junction; Vortex
Fourier transforms, 36–37, 229–231, 246, 292

Ginzburg–Landau (GL) theory, 61–63, 85–86, 93
Grain boundaries
  granular superconductor, 144, 231, 246, 329–330
  and HTS wire fabrication, 158, 160–161
  as Josephson junctions, 166–169

Heat capacity, 117–118, 312–317, 327
Heating effects
  hotspots and photon absorption, 327–330
  nonequilibrium ("hot electrons"), 320–321
  self-heating and hysteresis, 83
  stability in wires, 116–119
High-temperature superconductors (HTS), *see* Materials, HTS
Holes (electron vacancies), 49–50, 56, 153–155
Hysteresis, *see also* Heating effects
  in critical state, 113, 115
  in Josephson junction, 201–205
  in SQUID characteristics, 216–217

Inductance
  Josephson, 183–185, 226, 242, 252
  kinetic, 18, 30–32, 39, 64
  magnetic, 1, 39, 65, 123, 218–219
  surface, 21–22, 24, 27, 65
Integrated circuit fabrication, superconducting, 149–151, 165–169, 285–286, 294
Interface energy of S/N interface, 85–86
Instrumentation, superconducting
  samping oscilloscope, 288–289
  SQUID magnetometer systems, 219–225
Intermediate state of superconductor, 84
Irreversibility line and HTS in large fields, 158–159

Josephson effects, 10, 178–183
  resonant peak at gap voltage, 183
Josephson junction, 10–12, 178–248
  arrays of JJs, 190, 239–243
  critical current, 181–183
  effect of magnetic field, 227–231, 246
  as fast switch, 252–258
  HTS junctions, 166–169, 177, 255
  inductance, *see* Inductance, Josephson
  long junction, 225–239, 247
  LTS fabrication, 149–151
  pendulum model for, 185–186, 195–196
  shunted junction models (RSJ and CRSJ), 197–204, 246
  as voltage-controlled oscillator, 12, 179–180, 204–206, 238–242

Kinetic inductance, *see* Inductance, kinetic
Kosterlitz–Thouless transition, 104

London equations, 73, 87
Lorentz force on vortex, 98–99. *See also* Flux
Low-temperature superconductors (LTS), *see* Materials, LTS

Magnetic
  flux, *see* Flux, magnetic
  levitation, 1–2, 16, 132–134, 140
  materials, 95–97, 120–122, 131, 364
  monopoles, 330–331
  penetration depth, *see* Penetration depth, magnetic
  pressure, 81, 123–124, 128, 138
  resonance imaging (MRI), 124–126
  separation, 130–131
  shielding, 77–79, 218–219, 252, 278

susceptibility, 74–75, 131, 138, 220, 222
vector potential, 73, 87–90, 369
Magnetization of superconductors, 75, 84, 113, 115
Magnetometers, SQUID, 218–225
  applications of, 223–225
  flux modulation and feedback, 220–221
  digital SQUID, 293
Magnet technology, 119–136
Materials, superconducting, 3–5, 141–177. *See also* Materials Index
  HTS, 151–169
  LTS, 141–151
  organic and other superconductors, 169–173
Medical applications of superconductors
  magnetic resonance imaging (MRI), 124–126
  magnetic source imaging (SQUID arrays), 223–224
Meissner effect, 23, 72–74, 139. *See also* Diamagnetism
  transmission line model for, 74
Memory cells, superconducting
  arrays for computer systems, 285–286
  SFQ logic, 275, 279
  voltage-state logic, 266–269
Microwave applications of superconductors
  amplifiers, 102, 238–239
  detectors and mixers, 300–311, 323
  filters and resonators, 33–46, 164
  oscillators, 238–242
Mixed state, *see* Type II superconductors
Models of superconductor
  flux insulator, 5–7
  I-picture (currents), 74
  lossless inductor, 1–3
  macroscopic atom, 8–11, 89
  magnetic bubble model, 72, 81
  m-picture (magnetization), 74
  transmission line analog, 19–22
  two-fluid model, 30–35

Noise in superconducting detectors, 215, 222, 240, 253, 312, 319, 326
Nonequilibrium effects in superconductors, 320–323
Normal electrons, *see* Quasiparticles
Nuclear and particle physics applications
  accelerator magnets, 126–128
  digital timer and counter, 294
  fusion magnets, 130
  particle detectors, 330

Optical properties of superconductors, 57–58, 318

Penetration depth
  Josephson, 227, 232, 242, 244, 246
  magnetic (London), 3, 22, 28–29, 32–33, 85–86, 104, 143, 157–158
Persistent currents, 1, 69–72, 137, 250
Phase diagrams
  and carrier doping of HTS, 155
  magnetic, of superconductor, 82, 84, 94, 159
  thermal, of cryogenic liquids, 360
Phase of superconducting wave function, 8–10, 87–90, 178–181, 191
Point contact, 183–184, 216, 247
Proximity effect, superconducting, 62–63, 174
PSCAN, *see* Simulation tools
Pulse generation and propagation, 7, 9, 36–38, 97, 198–200, 236, 243, 257–258, 270–276, 280–281, 287–289, 312, 327, 329
Punchthrough in Josephson digital circuits, 266–270

Quality factor Q
  of Josephson junction, 194–196, 201–203, 234, 252–258
  of resonator, 38–42, 216, 309, 349
Quantum computing, superconducting, 338
Quantum flux parametron (QFP), 283
Quantum interference, *see* SQUID
Quantum wave function, 8–9, 86–89, 178–181. *See also* Phase of superconducting wave function
Quasiparticles, 55–57, 326–328

Radiation detectors, 300–336
  infrared, 301, 317–323, 327–329
  modulation and mixing, 301–312, 323
  single-photon, 324–329
  thermal and quasi-thermal, 312–323
Refrigerators, cryogenic, *see* Cryocoolers
Resistance in superconductor. *See also* Bolometer
  and flux-flow, 100–101, 109–111
  in tunnel junction, 59, 194, 203, 254, 326
  and two-fluid model, 30–32. *See also* Surface resistance
Resonator, superconducting, 38–46
Room-temperature superconductor, 16, 172, 339
RSFQ logic, 270–283
  clock generation, 280–282
  gates, 277–280
  memory cells, 275, 279
  pulse propagation, 270–274

SUBJECT INDEX

Semiconductors, 8, 46–50, 98–105, 174, 284–287, 294, 301–305, 319, 324–326, 364
Shapiro steps in Josephson junction, 187–191, 200–201
Simulation tools for superconducting circuits, 206–207, 352–358
  analogs, 357
  PSCAN, 352–353
  SPICE, 298, 353–357
Single electron transistor (SET), 192
Single flux quantum (SFQ), 6–10, 90, 98, 190–192, 257, 270, 276, 329. *See also* Fluxon; RSFQ logic
SIS mixer, 305–310. *See also* Tunnel junctions
Skin depth in normal metal, 2, 21, 346
Solenoid, superconducting, 1, 15, 71–72, 122–124, 138
Soliton, Josephson, 235–237. *See also* Vortex, Josephson
S-parameters (S_{11} and S_{21}), 42–46
Speed-power diagram for digital circuits, 258–259
SPICE, *see* Simulation tools
SQUIDs, 207–225. *See also* Magnetometers
  coupling circuits, 218–220
  dc SQUIDs, 12, 17, 207–216
  HTS SQUIDs, 222
  rf SQUIDs, 208, 216–218
Stability, electrothermal of superconducting composites, 83, 116–119
Substrates, 150, 162–168. *See also* Thin films
Superconducting magnetic energy storage (SMES), 128
Surface resistance, rf, 3, 33–35, 41, 164, 335
Switches, superconducting, 249–259
  comparison to semiconductors, 258–259
  Josephson junctions, 251–258
  networks, 287–288

Thermal conductivity, 313–315, 364–365
Thermodynamics, second law of, 361
Thermometers, 317–318, 364
Thin films, superconducting. *See also* Substrates
  HTS, 161–169
  LTS, 147–150
  properties, 24–26, 29, 79–82, 321, 328
Time constants
  aperture time in A/D conversion, 290
  electron collision time, 19, 48

  inelastic electron–phonon scattering, 320–322, 328
  for Josephson junctions, 252–259
  L/R time, 2, 69, 134, 250
  thermal time constants, 315–316
Transistors, superconducting, *see also* SQUID
  electric field-effect transistors, 174
  magnetic transistors, 100–106, 140, 238–239
Transmission lines
  conventional, 342–350
  Josephson junction, 235–237, 242–243, 270–272
  lumped element equivalents, 24, 41, 45, 348
  model for electromagnetic propagation, 20–22, 31, 74, 345–346
  resonator, 6, 66, 164, 349
  superconducting, 27–28, 35–38, 65, 177
Two-fluid model, *see* Models of superconductor
Tunnel junctions, superconducting, 58–61, 66, 149–150
  as mm-wave mixers, 305–312
  as photon detectors, 326–327
Type I superconductors, 78–86
Type II superconductors, 92–94, 107–119, 139

Units, electromagnetic, 368–370

Voltage standard, 14, 190–191, 292
Voltage-state logic, 259–270
  dual-rail logic, 263–264
  gates, 261–263, 269–270
  memory cells, 266–268, 286–287
  microprocessor, 284–286
  power supply and clock, 264–266
  threshold characteristics, 260–261, 268
Vortex, superconducting. *See also* Flux, magnetic; Single flux quantum
  in Josephson junction, 232–237
  lattice, 92, 110, 139, 158
  pinning, 7, 107–111
  structure of, 89–93
  vortex-antivortex pairs, 104, 158, 329

Weak links, 6–7, 183–184. *See also* Point contact
Wire, superconducting
  HTS, 159–161
  LTS, 145–147
  multifilamentary conductors, 116–119
Wiedemann–Franz law, 314, 365